The History of Physics
A Biographical Approach

H. Thomas Milhorn, MD, PhD

Mr. Rose

Room 308

"The History of Physics: A Biographical Approach," by H. Thomas Milhorn, MD, Ph.D. ISBN 978-1-60264-202-7.

Published 2008 by Virtualbookworm.com Publishing Inc., P.O. Box 9949, College Station, TX 77842, US. ©2008, H. Thomas Milhorn. All rights reserved. No part of this publication may be reproduced, stored in a retrieval system, or transmitted in any form or by any means, electronic, mechanical, recording or otherwise, without the prior written permission of H. Thomas Milhorn.

Manufactured in the United States of America.

Preface

The history of physics ranges from antiquity to modern string theory. Since early times, human beings have sought to understand the workings of nature—why unsupported objects drop to the ground, why different materials have different properties, and so forth.

Initially, the behavior and nature of the world and celestial phenomena were explained by invoking the actions of the various gods. Then, one god became the explanation for both. And then explanations for earthly events began to be accepted based on speculation as to their cause and nature, while celestial events were still felt to be of a divine nature, requiring no explanation. Eventually, human beings began to look for scientific explanations for both.

The emergence of physics as a science, distinct from natural philosophy, began with the scientific revolution of the 16th and 17th centuries when the scientific method came into vogue. Speculation was no longer acceptable; research was required.

The beginning of the 20th century marks the start of a more modern physics. Physicists began to study the atom, with its electrons and nucleus. The nucleus was found to be composed of neutrons and protons. Then came nuclear physics, where physicists began to look at the forces that hold the nucleus together and the particles that account for the four natural forces—strong nuclear force, weak nuclear force, electromagnetic force, and gravitational force.

Physicist began to look for a unified theory that accounted for all four natural forces, leading to chromodynamics, the electroweak theory, the Standard Model, and string theory.

Although the great body of knowledge we now call physics has come into being because of the work of physicists, many individuals from other disciplines have contributed, including those from mathematics, engineering, chemistry, and medicine. And at least one botanist made a significant contribution.

To those who made significant contributions to physics, and I inadvertently left them out of this book, I humbly apologize.

Those individuals whose contributions have been mainly astronomy or astrophysics are notably absent. An additional volume, *The History of*

Astronomy and Astrophysics, is planned.

I have chosen to approach the history of physics from a biographical point of view, feeling that people are more interesting than things, and the combination of the two are more interesting than the sum of the individual parts. After a brief overview of classical and modern physics, 336 one-page biographies of individuals who have made significant contribution to the field of physics are presented.

H. Thomas Milhorn, M.D., Ph.D.

Contents

Introduction

Physics is the branch of science traditionally defined as the study of matter, energy, and the relation between the two. It was called natural philosophy until the late 19th century. Today, physics is subdivided into classical physics and modern physics.

Classical Physics

Classical physics includes the traditional branches that were recognized, and fairly well developed, before the beginning of the 20th century—mechanics, sound, optics, heat, and electricity and magnetism.

Mechanics

Mechanics is concerned with bodies acted on by forces and bodies in motion. It may be divided into statics and dynamics. Statics is the study of the forces on bodies at rest. Dynamics is the study of bodies in motion.

Dynamics is subdivided into kinematics and kinetics. Kinematics describes the motion of objects without consideration of the masses or forces that brought about the motion. In contrast, kinetics is concerned with the forces and interactions that produce or affect the motion.

Mechanics may also be divided into solid mechanics and fluid mechanics. The latter includes the branches of hydrostatics, hydrodynamics, aerodynamics, and pneumatics.

Sound

Sound is due to longitudinal pressure vibrations traveling through air, water, or some other medium. It can be explained in terms of the laws of mechanics. Ultrasonics is the study of sound waves of very high frequency, beyond the range of human hearing.

Optics

Optics deals with light and vision, chiefly the generation, propagation, and detection of electromagnetic radiation having wavelengths longer than X-rays and shorter than microwaves. Light outside the visible range exhibits all of the phenomena of visible light except visibility.

Optics is divided into geometric optics and physical optics. Geometric optics, or ray optics, describes light propagation in terms of rays. Physical optics, or wave optics, involves interference, diffraction, polarization, and other phenomena for which the ray approximation of geometric optics is not valid.

Heat

Heat is a form of energy, the internal energy possessed by the particles of which a substance is composed. Thermodynamics is the study of the relationships between heat, work, and energy.

Electricity and Magnetism

Electricity and magnetism have been studied as a single branch of physics since the connection between them was discovered in the early 19th century. An electric current gives rise to a magnetic field and a changing magnetic field induces an electric current. Electrostatics deals with electric charges at rest, electrodynamics with moving charges, and magnetostatics with magnetic poles at rest.

Modern Physics

Modern physics is concerned with the behavior of matter and energy under extreme conditions or on the very large or very small scale. The categories of modern physics include of atomic, nuclear, and molecular physics; elementary particle physics; cryogenics; solid-state physics; plasma physics; quantum theory; and relativity.

Atomic, Nuclear, and Molecular Physics

Atomic physics is the field of physics that studies atoms as an isolated system of electrons and an atomic nucleus.

Nuclear physics is concerned with the nucleus of the atom and the nucleus's fundamental particles—protons and neutrons.

Molecular physics is the study of the physical properties of molecules and of the chemical bonds between atoms that bind them.

Nanotechnology is an area of atomic and molecular physics that refers broadly to a field of applied science and technology whose unifying theme is the control of matter on an extremely small scale, normally 1 to 100 nanometers, and the fabrication of devices with critical dimensions that lie within this size range.

Elementary Particle Physics

The physics of elementary particles is on an even smaller scale than atomic and nuclear physics, being concerned with the most basic units of matter. This branch of physics is also known as high-energy physics because of the extremely high energies necessary to produce many types of particles in large particle accelerators.

The two approaches to a uniform quantum field theory in theoretical physics are the Standard Model and String Theory.

Standard Model. The Standard Model is the name given to the theory of fundamental particles and how they interact. This theory includes strong interactions (quantum chromodynamics) and a combined theory of weak and electromagnetic interaction (electroweak theory).

Quantum chromodynamics (QCD) is a theory of the strong interaction (color force), which is a fundamental force describing the interactions of the quarks and gluons found in hadrons (such as protons and neutrons). QCD is a quantum field theory of a special kind called a non-abelian gauge theory. Gauge theories are a class of physical theories based on the idea that symmetry transformations can be performed locally as well as globally. It is called non-Abelian if the law of commutativity does not always hold.

The electroweak theories introduce W and Z bosons as the carrier particles of weak processes and photons as mediators to electromagnetic interactions.

The particles of the Standard Model are grouped into two classes— bosons (particles that transmit forces) and fermions (particles that make up matter). The bosons have particle spin that is either 0 or 1. The fermions have spin 1/2. The particles of the Standard Model are summarized in the following tables:

Name	Spin	Charge	Mass
Photon	1	0	0
Gluon	1	0	0
W^+	1	+1	80 GeV
W^-	1	-1	80 GeV
Z^0	1	0	91 GeV
Higgs	0	0	> 78 GeV

Particles that transmit force (bosons)
(Based on The Official String Theory Web Site, www.superstringtheory.com)

3

Name	Spin	Charge	Mass
Electron	1/2	-1	.0005 GeV
Muon	1/2	-1	0.10 Gev
Tau	1/2	-1	1.8 Gev
Electron neutrino	1/2	0	0?
Muon neutrino	1/2	0	<.00017 GeV
Tau neutrino	1/2	0	<.017 GeV
Up quark	1/2	2/3	0.005 GeV
Charm quark	1/2	2/3	1.4 GeV
Top quark	1/2	2/3	174 GeV
Down quark	1/2	-1/3	0.009 GeV
Strange quark	1/2	-1/3	0.17 GeV
Bottom quark	1/2	-1/3	4.4 GeV

Particles that make up matter (fermions)
(Based on The Official String Theory Web Site, www.superstringtheory.com)

The first six fermion particles (electron, muon, tau, electron neutrino, muon neutrino, tau neutrino) are known as leptons.

The Higgs boson is the only particle predicted by the Standard Model that has not yet been detected.

The Standard Model accounts for three of the four fundamental forces (electromagnetic, strong nuclear, weak nuclear), but fails to account for the fourth (gravity).

String Theory. Theories in particle physics that treat subatomic particle as infinitesimal, one-dimensional string-like objects rather than dimensionless points in space-time are known as string theories. Different vibrations of the strings correspond to different particles.

One of the predictions of string theory is that at higher energy scales we should start to see evidence of a symmetry that gives every particle that transmits a force (a boson) a partner particle that makes up matter (a fermion), and vice versa. This symmetry between forces and matter is called supersymmetry.

Cryogenics

Cryogenics is the science and technology of phenomena and processes at very low temperatures, defined arbitrarily as below 150 K (−190°F). At these temperatures, quantum effects, such as superconductivity and superfluidity, occur.

Superconductivity is a phenomenon occurring in certain materials at extremely low temperatures, characterized by the absence of electrical resistance and the exclusion of an interior magnetic field.

Superfluidity is a state of matter characterized by the complete absence of viscosity.

Solid-state physics

Solid-state physics is the study of rigid matter, or solids. It is the largest branch of condensed matter physics, which is the field of physics that deals with the macroscopic physical properties of matter.

The bulk of solid-state physics theory and research is focused on crystals, largely because the periodicity of atoms in a crystal facilitates mathematical modeling, and also because crystalline materials often have electrical, magnetic, optical, or mechanical properties that can be exploited for engineering purposes.

Plasma physics

Plasma physics is the study of the properties of plasma, which is is an electrically neutral, highly ionized gas composed of ions, electrons, and neutral particles. Because of its unique properties, it is considered to be a distinct state of matter (in addition to gas, liquid, and solid). The free electric charges make the plasma electrically conductive so that it responds strongly to electromagnetic fields.

Quantum Theory

Quantum theory is concerned with the discrete, rather than the continuous, nature of many phenomena at the atomic and subatomic level, and with the complementary aspects of particles and waves in the description of such phenomena. It is concerned with phenomena on such a small-scale that they cannot be described in classical terms, and it is formulated entirely in terms of statistical probabilities.

Uncertainty Principle. The Heisenberg uncertainty principle states that locating a particle in a small region of space makes the momentum of the particle uncertain; and conversely, that measuring the momentum of a particle precisely makes the position uncertain.

Since the position of an atom is measured with a photon, the reflected photon changes the momentum of the atom, and hence the actual momentum is uncertain. The amount of uncertainty can never be reduced below the limit set by the uncertainty principle, regardless of the experimental setup.

In quantum mechanics, the position and momentum of particles do not have precise values, but have probability distributions. There are no states in which a particle has both a definite position and a definite momentum. The narrower the probability distribution is in position, the wider it is in momentum.

Quantum electrodynamics (QED). QED is a relativistic quantum field theory of electrodynamics. QED mathematically describes all phenomena involving electrically charged particles interacting by means of exchange of photons.

Relativity

Relativity, as developed by Albert Einstein, is comprised of two parts—the special theory of relativity and the general theory of relativity.

Special Theory of Relativity. The special theory of relativity postulates that the speed of light is the same for all observers, regardless of their motion relative to the source of the light, and that all observers moving at constant speed should observe the same physical laws. The only way both of these can happen is if time intervals and/or lengths change according to the speed of the system relative to the observer's frame of reference. For example, an atomic clock traveling at high speed in a jet plane ticks more slowly than its stationary counterpart.

This theory is called special because it is limited to bodies moving in the absence of a gravitational field.

The discovery of the relativity of space and time led to an equally revolutionary insight—matter and energy are interrelated as given by $E = mc^2$, where m = mass and c = the speed of light.

General Theory of Relativity. The general theory of relativity unifies special relativity, Newton's law of universal gravitation, and the insight that gravitational acceleration can be described by the curvature of space and time.

1. Cryogenics, Microsoft Encarta Encyclopedia Online, http://encarta.msn.com/ encyclopedia_761563758/Cryogenics, 2007.
2. Physics, The Columbia Encyclopedia, Sixth Edition, www.bartleby.com/65/ph/physics-ent, 2001-2007.
3. Plasma (Physics), Microsoft Encarta Encyclopedia Online, http://ca.encarta.msn.com/ encyclopedia_761568690/Plasma_(physics), 2007.
4. Quantum Chromodynamics, Encyclopedia.com, www.encyclopedia.com/doc/1E1-quantumch.
5. Quantum Electrodynamics, Encyclopedia.com, www.encyclopedia.com/doc/1E1-quantumel.
6. Quantum Theory, Microsoft Encarta Encyclopedia Online, http://encarta.msn.com/ encyclopedia_761559884/Quantum_Theory, 2007.
7. Relativity, Microsoft Encarta Encyclopedia Online, http://encarta.msn.com/ encyclopedia_761558302_1____4/Relativity, 2007.
8. Solid-state Physics, Microsoft Encarta Encyclopedia Online, http://encarta.msn.com/ encyclopedia_761577376/Condensed-Matter_Physics, 2997.
9. Special Relativity, Wikipedia, http://en.wikipedia.org/wiki/Special_relativity.
10. String Theory, Wikipedia, http://en.wikipedia.org/wiki/String_theory.
11. The Standard Model, The Official String Theory Web Site, www.superstringtheory.com/ experm/exper2.
12. Uncertainty Principle, Wikipedia, http://en.wikipedia.org/wiki/Uncertainty_principle.

Early Physics
624 BC to 1599

Thales (c. 624-c. 546 BC)

Attributed with the birth of scientific thought

Thales was born around 624 BC in the city of Miletus, an ancient Ionian seaport on the western coast of Asia Minor. He visited Egypt and probably Babylon, bringing back knowledge of astronomy and geometry.

Thales had a profound influence on other Greek thinkers and therefore on Western history. He investigated almost all areas of knowledge—philosophy, history, science, mathematics, engineering, geography, and politics.

Thales proposed theories to explain many of the events of nature. His questioning approach to the understanding of heavenly phenomena was the beginning of Greek astronomy.

Thales attempted to find naturalistic explanations of the world, without reference to the supernatural or mythological. He explained earthquakes by hypothesizing that the Earth floats on water, and that earthquakes occur when the Earth is rocked by waves.

Thales was known for his innovative use of geometry, both theoretical and practical. He understood similar triangles and right triangles, and he was familiar with ratio of the run to the rise of a slope. It was said that he measured the height of the pyramids by their shadows at the moment his own shadow was equal to his height. A right triangle with two equal legs is a 45-degree right triangle, all of which are similar. The length of the pyramid's shadow, measured from the center of the pyramid at that moment, would have been equal to its height.

Attributed to Thales is the discovery that a circle is bisected by its diameter, that the base angles of an isosceles triangle are equal, and that when two pairs of angles are formed by two intersecting lines the vertical angles are equal to each other. He is credited with inventing deductive mathematics.

Thales' theorem states that if A, B and C are points on a circle, where the line AC is a diameter of the circle, then the angle ABC is a right angle.

Many philosophers followed Thales' lead in searching for explanations in nature (logos) rather than in the supernatural (mythos). He is considered by many to be the father of science.

Thales is said to have died of dehydration around 547 BC while watching a gymnastic contest.

1.   Thales of Miletus (62?-546), The Internet Encyclopedia of Philosophy, www.iep.utm.edu/t
     /thales.
2.   Thales of Miletus (634-546 BC), Eric Weisstein's World of Biography, Wolfram Research,
     http://scienceworld.wolfram.com/ biography/Thales.
3.   Thales of Miletus, Wikipedia, http://en.wikipedia.org/wiki/Thales.

Anaxagoras (c. 500- 428 BC)

Proposed that materials are made of seeds

Anaxagoras was born around 500 BC in Clazomenae in Asia Minor. In early manhood he went to Athens, which was rapidly becoming the center of Greek culture. There he is said to have remained for 30 years.

Anaxagoras correctly explained the phases of the Moon and eclipses of the Moon and the Sun in terms of their movements. He believed that heaven and Earth were brought into existence by the same processes, and were composed of the same materials. The heavenly bodies, he asserted, were masses of stone torn from the Earth and ignited by rapid rotation.

Because his theories brought him into conflict with the popular faith of the day, Anaxagoras's views on heavenly bodies were considered dangerous by the Church. He was arrested and charged with contravening the established religion On release from custody, he was forced to move from Athens to Lampsacus in Ionia.

Anaxagoras believed that all things have existed from the beginning, but originally they existed in infinitesimally small fragments of themselves, endless in number and inextricably combined throughout the universe. All things existed in this mass, but in a confused and indistinguishable form. There were the seeds (miniatures of corn, flesh, gold, and so forth) in the primitive mixture, but the parts of like nature with their wholes had to be eliminated from the complex mass before they could receive a definite name and character. The existing species of things having thus been transferred were multiplied endlessly in number by reducing their size through continued subdivision. At the same time, each thing, as Anaxagoras saw it, was so indissolubly connected with every other that the keenest analysis can never completely sever them.

In trying to explain the processes of nutrition and growth, Anaxagoras theorized that for the food an animal eats to turn into bone, hair, flesh, and so forth it must already contain all of those constituents within it.

Anaxagoras marked a turning-point in the history of philosophy. With him, scientific speculation passed from the colonies of Greece to settle at Athens. By the theory of minute constituents of things, and his emphasis on mechanical processes in the formation of order, he helped pave the way for atomic theory. Anaxagoras died in Lamsacus in Ionia in 428 BC.

1. Anaxagoras of Clazomenae (ca. 500-ca. 428 BC, Eric Weisstein's World of Biography, Wolfram Research, chttp://scienceworld.wolfram.com/biography/Anaxagoras.
2. Anaxagoras, NNDB, www.nndb.com/people/884/000087623.
3. Anaxagoras, Stanford Internet Encyclopedia of Philosophy, http://plato.stanford.edu/entries/Anaxagoras.
4. Anaxagoras, Wikipedia, http://en.wikipedia.org/wiki/Anaxagoras#Biography.

Democritus (460-370 BC)

Believed that matter is made up of atoma

Democritus was born at Abdera in Thrace, a northern territory of Greece, about 460 BC. He was a student of Leucippus (c. 500-450 BC) and co-originator with him of the belief that all matter is made up of various imperishable, indivisible elements, which he called atoma or "indivisible units," from which we get the English word atom.

Democritus and Leucippus contended that the human soul is composed of exceedingly fine and spherical atoma. They held that these atoma move because it is their nature never to be still, and that as they move they draw the whole body along with them and set it in motion. He viewed soul atoma as being similar to fire atoma—capable of penetrating solid bodies.

Democritus explained the three senses in terms of atoma as well. He hypothesized that different tastes were a result of differently shaped atoma in contact with the tongue. Smells and sounds were explained similarly. Vision, according to Democritus, worked by the eye receiving effluences of bodies that emanated from the bodies.

Democritus's theory argued that atoma had several properties—size, shape, and perhaps weight. He considered all other properties, such as color and taste, to be the result of complex interactions between the atoma in our bodies and the atoma of the matter that we are examining. Furthermore, he believed that the properties of atoma determine the perceived properties of matter—something that is solid is made of small, pointy atoma, while something that has water-like properties is made of large, round atoma. He suggested that some types of matter are particularly solid because their atoma have hooks to attach to each other, and some are oily because they are made of very fine, small atoma which can easily slip past each other.

Democritus was also a pioneer of mathematics. He was among the first to observe that a cone or pyramid has one-third the volume of a cylinder or prism, respectively with the same base and height. He realized that the celestial body we perceive as the Milky Way is formed from the light of distant stars. Other philosophers, including Aristotle later argued against this. Democritus was among the first to propose that the universe contains many worlds, some of them inhabited.

1. Berryman, Sylvia, Stanford Encyclopedia of Philosophy, http://plato.stanford.edu/entries/democritus, August 15, 2004.
2. Democritus, Encyclopedia Britannica Online, www.britannica.com/eb/ article-9029904/Democritus.
3. O'Connor, J.J. and E F Robertson, Democritus, MacTutor, www-groups.dcs.st-and.ac.uk/~history/ Mathematicians/Democritus.

Aristotle (384-322 BC)
Formulated a systematic approach to science

Aristotle was born in 384 BC at Stagirus, a Greek colony and seaport on the coast of Thrace. His father, Nichomachus, was court physician to King Amyntas of Macedonia. While he was still a boy, his father died. At age 17 his guardian, Proxenus, sent him to Athens, the intellectual center of the world, to complete his education. From age 17 to 37 he was a pupil of Plato in Athens and a teacher of Alexander the Great.

Aristotle wrote on many subjects, including physics, metaphysics, poetry, theater, music, logic, rhetoric, politics, government, ethics, biology, and zoology.

As a teacher at the Lyceum in Greece, Aristotle held that the universe was divided into two parts—the terrestrial region and the celestial region. In the realm of the Earth Aristotle believed, as did the Greek philosopher Empedocles (c. 490-430 BC), that all bodies were made of combinations of four substances—earth, fire, air, and water. Heavy bodies, like one made of iron, consisted mostly of earth, so they sought to return to the earth when dropped. Less dense objects were thought to contain a larger admixture of the other elements, along with varying amounts of earth. A feather, which contains a large amount of air, would have less attraction for the earth and therefore would fall more slowly. Smoke, containing even more air, would seek its own kind and rise.

In the universe beyond the Earth, Aristotle considered the heavenly bodies to be made of a fifth substance he called quintessence, which was under divine control and required no explanation.

The fundamental assumption in Aristotelian physics was that the natural state of matter is rest. Objects therefore must seek their natural place at rest in the center of the Earth, unless stopped by an impenetrable surface like the ground or a table. An object given an initial push on a table was considered to come to a stop because it desired to be in its natural state of rest.

Aristotle made a distinction between "natural" downward motion, such as a falling rock, and "violent" motion not directed toward the center of the Earth, such as a rock propelled by a catapult.

Aristotle's systematic approach became the method from which western science later arose. He died at Chalcis in Euboea in 322 BC.

1.   Aristotle, Catholic Encyclopedia, www.newadvent.org/cathen/01713a.
2.   Aristotle's Physics, http://aether.lbl.gov/www/classes/p10/aristotle-physics.
3.   Hewitt, Paul G., John Suchocki, and Leslie A. Hewitt, Conceptual Physical Science, Third Edition, Pearson/Addison Wesley, San Francisco, 2003.

Archimedes (287-212 BC)

Discovered the relationship between buoyancy and the weight of displaced fluid

Archimedes was born in the seaport city of Syracuse, Sicily in 287 BC. He was the son of an astronomer, but little else is known about his early life, other than that he studied for a time in Alexandria, Egypt.

Archimedes is best known for his studies of the relationship between buoyancy and the weight of fluid displaced, expressed as **Archimedes principle**:

- When a body is wholly or partially immersed in a fluid, the fluid exerts and upward force on the body equal to the weight of the fluid that is displaced by the body

It is said that the king of Syracuse once asked Archimedes to find a way of determining if one of his crowns was pure gold without destroying the crown in the process. The story goes that as Archimedes lowered himself into a bath he noticed that some of the water was displaced by his body and flowed over the edge of the tub. He realized that dividing the weight of the crown by its volume as determined by emersion would give its density. If the determined density was equal to that of pure gold, he had the proof he needed.

Archimedes contributed to our knowledge of pulley systems and levers. He was asked to reposition a ship on land. He did so easily by designing and using a compound pulley system. For levers, Archimedes demonstrated mathematically that the ratio of the effort applied to the load raised is equal to the inverse ratio of the distances of the effort and load from the pivot or fulcrum of the lever.

King Hieron II commissioned Archimedes to design a huge ship. Since a ship of this size would leak a considerable amount of water through the hull, Archimedes' purportedly designed a machine with a revolving screw-shaped blade inside a cylinder to remove the bilge water. The screw was turned by hand. It also could be used to transfer water from a low-lying body of water into irrigation canals. Versions of the **Archimedes' screw** are still in use today in developing countries.

In 212 BC, Archimedes was killed by a Roman Soldier in the siege of Syracuse.

1. Archimedes, Science Division, http://scidiv.bcc.ctc.edu/Math/Archimedes.
2. Archimedes, Wikipedia, http://en.wikipedia.org/wiki/Archimedes#Discoveries_and_inventions.
3. Sears, Francis W., Mark W. Zemansky, and Hugh D. Young, College Physics, Seventh Edition, Addison Wesley, New York, 1991.

Heron (c.10-70)

Invented the first steam engine, vending machine, wind wheel, and stand-alone fountain

Heron (or Hero) was a Greek philosopher born around 10 AD. He lived in Alexandria, Egypt, where he was an engineer and a mathematician. It is almost certain that Hero taught at the Musaeum, which included the famous Library of Alexandria, because most of his writings appear as lecture notes for courses in mathematics, mechanics, physics, and pneumatics.

Among Heron's most famous inventions were the first documented steam-powered device, a vending machine, a stand-alone fountain, and a wind wheel. His steam-powered device, known as the aerophile or **Heron's engine**, was a rocket-like reaction engine. It was used as a toy and to open temple doors.

With **Heron's vending machine**, when a coin was introduced via a slot on the top of the machine it fell on to a pan attached to a lever. The lever opened up a valve which let a fixed amount of holy water flow out. A counter-weight then snapped the lever back up and turned off the valve.

Heron's Fountain was a standalone fountain that operated under self-contained hydrostatic energy.

Heron's wind wheel was a wind powered device that operated an organ. It was probably the first instance of wind powering a machine in history. Heron also invented many mechanisms for the Greek theater, including an entirely mechanical play that was almost 10 minutes in length. It was powered by a binary-like system of ropes, knots, and simple machines operated by a rotating cylindrical cogwheel.

In optics, Heron formulated the principle of the shortest path of light: If a ray of light propagates from point A to point B within the same medium, the path-length followed is the shortest possible.

In geometry, **Heron's formula** states that the area "A" of a triangle whose sides have lengths "a," "b," and "c" is

$$A = [s(s - a)(s - b)(s - c)]^{1/2}$$

where "s" is the semi-perimeter of the triangle $(a + b + c)/2$.

Also, in geometry a **Heronian triangle** is a triangle whose side lengths and area are all rational numbers.

1. Hero of Alexandria (c. 10–70 AD), The Internet Encyclopedia of Science, www.daviddarling.info/ encyclopedia/H/Hero.
2. Hero of Alexandria, NNDB, www.nndb.com/people/898/000103589.
3. Hero of Alexandria, Wikipedia, http://en.wikipedia.org/wiki/Hero_of_Alexandria.

Ali Al-Hazen (965-1039)

Demonstrated that light travels in a straight line to the eye

Ali Al-Hazen (al-Haitham) was born in 965 in Basrah, Iraq. He received his education in Basrah and Baghdad. He spent most of his life in Spain, where he conducted research in optics, mathematics, physics, medicine, and development of scientific methods.

Al-Hazen studied the refraction of light rays through transparent media, such as air and water, and documented the laws of refraction. He carried out the first experiments on the dispersion of light into colors. He also was the first to describe accurately the various parts of the eye, and he gave a scientific explanation of the process of vision, contradicting the theories of Aristotle (384-322 BC), Euclid (c. 325-270 BC), and Ptolemy (c. 87-150) that the eye sends out visual rays to the object. Al Hazen believed that the rays originate in the object of vision and not in the eye. He also attempted to explain binocular vision.

Focusing on spherical and parabolic mirrors, Al-Hazen observed that the ratio between the angle of incidence and the angle of refraction does not remain constant. He also investigated the magnifying power of a lens. He developed a camera obscura to demonstrate that light and color from different candles pass through a single aperture in straight lines without intermingling at the aperture. A camera obscura consists of small darkened rooms with light admitted through a single tiny hole. The result is an inverted image of the outside scene cast on the opposite wall.

Al-Hazen is regarded as the father of optics for his influential *Book of Optics*.

Al-Hazen gave a correct explanation of the apparent increase in size of the Sun and the Moon when near the horizon. He also studied atmospheric refraction, discovering that the twilight only ceases, or begins, when the Sun is 19 degrees below the horizon. From this, he deduced the height of the homogeneous atmosphere to be 55 miles. The homogeneous atmosphere is a hypothetical atmosphere in which the density is constant with height.

Al-Hazen distinguished astrology from astronomy. He refuted the study of astrology due to the methods used by astrologers, being conjectural rather than empirical.

In mathematics, Al-Hazen developed analytical geometry by establishing linkage between algebra and geometry. He died around 1039.

1. History of Optics, Wikipedia, http://en.wikipedia.org/wiki/History_of_optics.
2. Mirshhahi, Shahrokh, Al-Hazen, www.fravahr.org/spip.php?article80, June 4, 2003.
3. Zahoor, A, ABU ALI HASAN IBN AL-HAITHAM (ALHAZEN) (965 - 1040 AD), www.geog.ucsb.edu/~jeff/115a/history/alhazen..

Shen Kuo (Kua) (1031-1095)

Described the magnetic needle compass

Shen Kuo (or Kua) was a Chinese mathematician, astronomer, meteorologist, geologist, zoologist, botanist, pharmacologist, agronomist, ethnographer, writer, poet, diplomat, hydraulic engineer, inventor, finance minister, and governmental state inspector. He was born in Qiantang (modern-day Hangzhou) China in 1031. His father, Shen Zhou, served in official posts on the provincial level.

In 1054, when Shen himself began serving in minor local governmental posts, his natural abilities to plan, organize, and design became evident. He designed and supervised the hydraulic drainage of an embankment system, which converted 100,000 acres (400 km²) of swampland into prime farmland.

In 1072, Shen was appointed the head official of the Bureau of Astronomy. He proposed many reforms to the Chinese calendar. He accurately mapped the orbital paths of the Moon and the planets in an intensive five-year project that rivaled the later work of the Danish astronomer Tycho Brahe (1546–1601). He proposed that heavenly bodies were spheres, based on his observations of the waxing and waning of the Moon. And he supported the hypothesis that the Moon was reflective rather than producing light itself.

In his *Dream Pool Essays* of 1088, Shen was the first to describe the magnetic needle compass, which was later described in Europe in 1187 by Alexander Neckam (1157-1217). Shen also discovered the concept of true north in terms of magnetic declination towards the North Pole by experimentation with suspended magnetic needles. He improved the Meridian by measuring the distance between the North Star and true north. The Meridian is an imaginary great circle on the earth's surface that passes through the North and South geographic poles. All points on the same meridian have the same longitude. Shen's work made compasses more useful for navigation—a concept unknown in the west for 400 more years.

Shen was the first in China to mention the use of the drydock to repair boats suspended out of water, and also wrote of the effectiveness of the relatively new invention—the canal pound lock.

Chen died in Ching-k'ou, China in 1095. His body was interred at a tomb in Yuhang District of Hangzhou, at the foot of the Taiping Hill.

1. China's Greatest Scientist—Shen Kuo, Historum, www.historum.com/showthread.php?t=2597.
2. Shea, Marilyn, Shen Kuo, http://hua.umf.maine.edu/China/astronomy/tianpage/ 0017Shen_Kuo_9266w, May 2007.
3. Shen Kuo, Wikipedia, http://en.wikipedia.org/wiki/Shen_Kuo.

Abd Al-Khazini (12th Century)

The first to apply experimental scientific methods to the fields of statics and dynamics

Abd al-Rahman al-Khazini was a born sometime in the twelfth century. He was a Muslim physicist, astronomer, chemist, biologist, mathematician, and philosopher of Byzantine Greek descent.

As a slave of the Seljuq Turks, Al-Khazini was taken to Merv, a province in Persia, after the Seljuq victory over the Byzantine Emperor, Romanus IV. His master there, al-Khazin, gave him the best possible education in mathematical and philosophical subjects.

Al-Khazini later became a mathematical practitioner under the patronage of the Seljuk court, under Sultan Ahmed Sanjar. Little else is known about his personal life.

Al-Khazini's *Book of the Balance of Wisdom*, completed in 1121, contained studies of the hydrostatic balance, including its construction and uses and the theories of statics and hydrostatics that lie behind it as developed by his predecessors, his contemporaries, and himself.

Al-Khazini and Abu al-Biruni (973-1048) were the first to apply experimental scientific methods to the fields of statics and dynamics, particularly for determining specific weights, such as those based on the theory of balances and weighing.

Al-Khazini and his Muslim predecessors unified statics and dynamics into the science of mechanics, and they combined the fields of hydrostatics with dynamics to give birth to hydrodynamics. They were the first to generalize the theory of the center of gravity, and also the first to apply it to three-dimensional bodies. They also created the science dealing with gravity, which was later further developed in medieval Europe.

The contributions of al-Khazini and his Muslim predecessors to mechanics laid the foundations for the later development of classical mechanics in Renaissance Europe.

Al-Khazini recorded the specific gravities of fifty substances, including various stones, metals, liquids, salts, amber, and clay. The accuracy of his measures were comparable to modern values. He also discovered that the density of water is greater nearer the earth's center more than a 100 years before Roger Bacon (1220-1294) propounded and proved the same hypothesis.

Al-Khazini appears to have been the first to propose that the gravity of a body varies with its distance from the center of the Earth.

1. Al-Khazini, Wikipedia, http://en.wikipedia.org/wiki/Al-Khazini.
2. Merv's Physicist, Muslim Heritage.comAl-Khazini-http://www.muslimheritage.com/topics/default.cfm?ArticleID=493

Roger Bacon (c. 1214-1294)

A proponent and practitioner of the experimental method

Roger Bacon was born in Ilchester in Somerset, England, possibly in 1214. His parents were well-off land owners, but during the stormy reign of Henry III their property was despoiled and several members of the family were driven into exile.

Bacon's initial studies covered the trivium of grammar, logic, and rhetoric. He then progressed to the quadrivium—geometry, arithmetic, music, and astronomy. He received a Master's Degree from the University of Oxford, and remained there teaching until around 1241.

Wishing to re-introduce the teachings of Aristotle (384-322 BC), which had been banned on the grounds that Aristotle was not a Christian, the University of Paris, sometime between 1237 and 1245, hired Bacon, who at Oxford had become an expert on Aristotle.

About 1256, Bacon became a Friar in the Franciscan Order. Sometime between 1277 and 1279 he was placed under house arrest by Jerome of Ascoli, the Minister-General of the Franciscan Order. His arrest was thought to be due to the Condemnations of 1277, which banned the teaching of certain philosophical doctrines, including deterministic astrology. Sometime after 1278 Bacon returned to the Franciscan House at Oxford, where he continued his studies.

Bacon's *Opus Majus* contains treatments of mathematics, optics, alchemy, the manufacture of gunpowder, and the positions and sizes of celestial bodies. It anticipates later inventions, such as microscopes, telescopes, spectacles, flying machines, hydraulics, and steam ships.

The study of optics in part five of *Opus Majus* includes a discussion of the physiology of eyesight and the anatomy of the eye and brain. It considers light, distance, position, and size, direct vision, reflected vision, refraction, mirrors, and lenses. Bacon first recognized the visible spectrum in a glass of water, centuries before Isaac Newton (1643-1727) discovered that prisms could disassemble and reassemble white light. Bacon is credited with the invention of the magnifying glass.

Bacon placed considerable emphasis on empiricism, and is known as one of the earliest advocates of the modern scientific method of acquiring knowledge about the world. He died in Oxford, England in 1294 without important followers, was quickly forgotten, and remained so for a long time.

1.   Roger Bacon, MacTutor, ://www-groups.dcs.st-and.ac.uk/~history/Biographies/Bacon.
2.   Roger Bacon, Microsoft Encarta Encyclopedia Online, http://encarta.msn.com/
      encyclopedia_761569765/roger_bacon, 2007
3.   Roger Bacon, Wikipedia, http://en.wikipedia.org/wiki/Roger_Bacon.

Jean Buridan (1300-1358)

Developed the concept of impetus, the first step toward the modern concept of inertia

Jean Buridan was born in Béthune, France in 1300. He studied at the University of Paris under William of Ockham (c. 1288-c. 1347). After his studies were completed, he was appointed professor of philosophy, and later rector, at the same university.

Buridan spent his entire career as a teaching master in the arts faculty at the University of Paris, lecturing on logic and the works of Aristotle (384-322 BC). Bacon was a charismatic figure, with reputed numerous amorous affairs and adventures.

The concept of inertia was unknown to the physics of Aristotle. Aristotle and his followers held that a body was only maintained in motion by the action of a continuous external force. Thus, in the Aristotelian view a projectile moving through the air would owe its continuing motion to eddies or vibrations in the surrounding medium applying a sustaining force to the projectile. In the absence of this force, the body would come to rest almost immediately.

Buridan felt that motion was maintained by some property of the body, imparted when the body was set in motion. He named the property impetus. Moreover, he rejected the view that the impetus dissipated spontaneously, asserting that a body came to a stop because of the forces of air resistance and gravity, which might be opposing its impetus. He further held that the impetus of a body increased with the speed with which it was set in motion and with its quantity of matter.

Buridan used the theory of impetus to give an accurate qualitative account of the motion of projectiles. His impetus is closely related to the modern concept of momentum. His work, through his theory of impetus, prepared the way for Galileo Galilei (1564-1642).

Buridan's former mentor, Ockham, disagreed on the subject of impetus, which led to a falling out between the two around 1340.

Buridan also wrote on solutions to paradoxes, such as the liar paradox. The liar paradox encompasses paradoxical statements in which self-reference results in a sentence or phrase being true if and only if it is false. Buridan died in 1358.

Subsequent followers of Ockham succeeded in having Buridan's writings placed on the *Index Librorum Prohibitorum* from 1474 to 1481.

1. Jean Buridan, Stanford Encyclopedia of Philosophy, http://plato.stanford.edu/entries/buridan.
2. Jean Buridan, Wikipedia, http://en.wikipedia.org/wiki/Jean_Buridan.
3. Jean Buridan, Columbia Encyclopediawww.encyclopedia.com/doc/1E1-Buridan, 2007.

William Gilbert (1544-1603)

Studied magnets and hypothesized that the Earth is a giant magnet

William Gilbert was born in Colchester, England on May 24, 1544. He entered St. John's College, Cambridge in 1558 where he obtained a B.A., an M.A., and finally an M.D. in 1569. He then became a senior fellow of the college.

In 1600, Gilbert was appointed physician to Queen Elizabeth I, and a few months before his death from the plague in 1603 he was appointed physician to James I.

Gilbert's *De Magnete* (On the Magnet), published in 1600, was one of the first scientific works based on observation and experiment. It quickly became the standard work throughout Europe on electrical and magnetic phenomena. Although Europeans were making long voyages across oceans guided by compasses, little was known about lodestone or magnetized iron. Gilbert made the first clear distinction between magnetism and static electricity.

Gilbert discovered various methods for producing and strengthening magnets. In addition, he observed that the magnetism of a piece of material was destroyed when the material was heated. He also coined the word "electricity."

From experiments involving a spherical lodestone, the most powerful magnet then available, Gilbert concluded that the Earth was a huge magnet, with a north and a south magnetic pole. He likened the polarity of the magnet to the polarity of the Earth.

Gilbert also studied static electricity using amber, which is called "elektron" in Greek, so Gilbert decided to call its effect the "electric force." He is credited as one of the originators of the term *electricity*, and many regard him as the father of electrical engineering or father of electricity.

Gilbert argued that electricity and magnetism were not the same thing. It took James Clerk Maxwell (1831-1879) to show that both effects were aspects of a single force.

On his death, Gilbert was buried in Holy Trinity, an Anglican church in Colchester. The inscription on his tomb reads "He composed a book, concerning the magnet, celebrated among foreigners and among those engaged in nautical affairs."

1. Encyclopedia of World Biography, William Gilbert, www.bookrags.com/biography/ william-gilbert.
2. Manley, D. Mark, Famous Physicists, http://cnr2.kent.edu/~manley/physicists, August 30, 2006.
3. Van Helden, Al, The Galileo Project, http://galileo.rice.edu/sci/gilbert, 1995.
4. William Gilbert, Incredible People, http://profiles.incredible-people.com/william-gilbert/.

Francis Bacon (1561-1626)

Established and popularized an inductive methodology for scientific inquiry

Francis Bacon (later Lord Verulam and the Viscount St. Albans) was born at York House in Strand, London, England on January 22, 1561. His father was Lord Keeper of the Great Seal under Elizabeth I (1533-1603). Bacon entered Trinity College, Cambridge in 1573 at the age of twelve, living there for three years.

Bacon interrupted his studies to take a position in the diplomatic service in France as an assistant to the ambassador. In 1579, while he was still in France, his father died, leaving him virtually without support. With no position, no land, no income, and no immediate prospects he returned to England and resumed the study of law, becoming a barrister in 1582. Two years later he took a seat in the House of Commons.

His opposition in 1584 to Queen Elizabeth's tax program retarded his political advancement. With the accession of James I (1566-1625), however, a number of honors were bestowed on Bacon. He was knighted in 1603, made Solicitor General in 1604, Attorney General in 1613, and Lord Chancellor in 1618.

In 1621, while serving as Chancellor, Bacon, who was always pressed for money, was indicted on charges of bribery, and forced to leave public office. He then retired to his estate, where he devoted himself full time to his continuing literary, scientific, and philosophical work.

Bacon's studies led him to the conclusion that the methods, and thus the results, of the science of his day were erroneous. Up to and during Bacon's time there existed philosophies rooted not so much in reason but in pure faith—philosophies promoted by the church. Bacon was violently opposed to such speculative philosophies, and argued that the only knowledge of importance to man was empirically rooted in the natural world; and that a clear system of scientific inquiry would assure man's mastery over the world.

Bacon delineated the principles of the inductive thinking method, which constituted a breakthrough in the approach to science. It was this approach that brought about the great discoveries of Copernicus (1473-1543) and Galileo (1564-1642).

Bacon died on April 9, 1626, and was buried at Saint Michael's Church in St. Albans, just north of London in Hertfordshire.

1. Francis Bacon (1561-1626), The Internet Encyclopedia of Philosophy, www.iep.utm.edu/b/bacon.
2. Francis Bacon, Biographies, Blupete, www.blupete.com/Literature/Biographies/Philosophy/Bacon.
3. Francis Bacon, Wikipedia, http://en.wikipedia.org/wiki/Francis_Bacon.

Willebrord Snell (1580-1626)
Discovered the law of refraction

Willebrord Snell was born in Leiden, Netherlands in 1580. Snell's father, Rudolph Snell, was professor of mathematics at the University of Leiden. From about 1600 he traveled to various European countries, mostly discussing astronomy. In 1602 he went to Paris where his studies continued. He received his degree from Leiden in 1607, and in 1613 Willebrord succeeded him as professor of mathematics.

In 1615, Snell planned and carried into practice a new method of finding the dimensions of the Earth by triangulation. This work is the foundation of geodesy, which is the science of measuring the size, shape, and gravity field of the Earth. Geodesy supplies positioning information about locations on the Earth, and this information is used in a variety of applications, including civil engineering, boundary demarcations, navigation, resource management and exploration, and geophysical studies of the dynamics of the Earth.

In 1617, Snell published *Eratosthenes Batavus*, which contains his methods for measuring the Earth.

Snell also improved the classical method of calculating approximate values of π by polygons. Using his method, 96 sided polygons give π correct to seven places, while the classical method yields only two places.

Although Snell discovered the law of refraction, a basis of modern geometric optics in 1621, he did not publish it, and only in 1703 did it become known when Christiaan Huygens (1629-1695) published Snell's result in *Dioptrica*.

Snell is best known for his law of refraction. The law of refraction (**Snell's law**) relates the degree of bending of light to the properties of the refractive materials (for example, from air to water). Snell's law is given by:

$$n_1 \sin\theta_1 = n_2 \sin\theta_2$$

where "n_1" and "n_2" are the indices of refraction found on either side of a surface and "θ_1" and "θ_2" are the angles of incidence and refraction. Snell also discovered the law of sines.

Snell died at the age of 46 on October 30, 1626.

1. Geodesy, Sci-tech Encyclopedia, Info.com, www.answers.com/topic/geodesy?cat=technology&nr=1.
2. Hewitt, Paul G., John Suchocki, and Leslie A. Hewitt, Conceptual Physical Science, Third Edition, Pearson/Addison Wesley, San Francisco, 2003.
3. Willebrord van Royen Snell, MacTutor History of Mathematics, www-history.mcs.st-and.ac.uk/~history/Printonly/Snell.

Galileo Galilei (1564-1642)

Showed that light objects fall as fast as heavier ones and that a force is not necessary to keep an object in motion

Galileo Galilei was born in Pisa, Tuscany on February 15, 1564, the same year Michelangelo died. He initially studied medicine, but then changed to mathematics. He became interested in motion, and soon was in opposition with others who held to the 2000-year-old idea of Aristotle (384-322 BC) that heavy objects fall faster than lighter ones. To prove his point, the story goes, he dropped two objects of different weight from the Leaning Tower of Pisa and showed that they hit the ground at the same time. In other words, he experimented.

The theory of impetus was set forth and named by the Parisian philosopher, Jean Buridan (1300-1358). It was based on observations that had been made in the sixth century by the Greek philosopher, John Philoponus (c. 490-c. 570) about the motion of objects. Galileo, like Buridan and Philoponus, was uncomfortable with Aristotle's explanation that a force is necessary to keep an object in motion once it has been set in motion. He believed that once an initial force was applied to the object it remained in motion without additional force being necessary.

By 1609 Galileo had determined that the distance fallen by a body is proportional to the square of the elapsed time (the law of falling bodies) and that the trajectory of a projectile is a parabola.

Galileo ran afoul of the church when he became an advocate of the theory of the solar system advanced by Nicolas Copernicus (1473-1543); that is, the Earth, rather than being the center of the solar system, actually revolves around the Sun. He was warned by the church not to teach this view. After 15 years of restraint Galileo published his findings. For this, he was arrested and forced to stand trial. He was found guilty and under threat of torture renounced his discoveries that supported Copernicus's views. He was sentenced to house arrest for the remainder of his life at his home at Arcetri. Nevertheless, he completed his studies on motion, and his writings were smuggled out of Italy and published in Holland. He died four years later.

In 1992, Pope John Paul II officially conceded that that Galileo had been correct—the Earth is not stationary. It revolves around the Sun.

1. Galileo (1564-1642), Wolfram Research, http://scienceworld.wolfram.com/biography/Galileo
2. Galileo Galilei, Biographies/Images of Physicists and Astronomers, www.mlahanas.de/Physics/Bios/GalileoGalilei.
3. Galileo Galilei, Starry Messenger, www.hps.cam.ac.uk/starry/galileo.
4. Hewitt, Paul G., John Suchocki, and Leslie A. Hewitt, Conceptual Physical Science, Third Edition, Pearson/Addison Wesley, San Francisco, 2003.

René Descartes (1596-1650)
Discovered the law of reflection

René Descartes was born in La Haye en Touraine, Indre-et-Loire, France on March 31, 1596. His father, Joachim, was a judge in the High Court of Justice. His mother died of tuberculosis when he was a year old.

Descartes studied at the University of Poitiers, where he earned a degree in law in 1616, but he never practice law. Instead, he turned his interests to mathematics and physics.

By applying infinitesimal calculus to the tangent line problem, Descartes' provided the basis for the calculus of Isaac Newton (1642-1727) and Gottfried Leibniz (1646-1716), thus permitting the evolution of that branch of modern mathematics. As the inventor of the Cartesian coordinate system, Descartes founded analytic geometry, the bridge between algebra and geometry which was also crucial to the invention of calculus and analysis.

Descartes also discovered the law of conservation of momentum, and he created exponential notation, indicated by numbers written in what is now referred to as superscript. His rule of signs is also a commonly used method in modern mathematics to determine possible quantities of positive and negative zeros of a function.

Descartes also made contributions to the field of optics. He showed by using geometric construction and the law of refraction (also known as **Descartes's law**) that the angle subtended at the eye by the edge of a rainbow and the ray passing from the Sun through the rainbow's center is 42 degrees. He also independently discovered the law of reflection. His essay on optics was the first published mention of this law.

Although Descartes had at first been inclined to accept the theory of Nicolas Copernicus (1473-1543) that the universe consisted of a system of spinning planets revolving around the Sun, he abandoned this theory when it was pronounced heretical by the Roman Catholic Church.

Descartes has been dubbed the Father of Modern Philosophy and the Father of Modern Mathematics. His most famous statement is "I think, therefore I am."

Descartes died on February 11, 1650 in Stockholm, Sweden, where he had been invited as a teacher for Queen Christina. The cause of death was said to be pneumonia.

1. René Descartes, Microsoft Encarta Online Encyclopedia, http://encarta.msn.com/ encyclopedia_761555262/ Rene_Descartes.html#s3, 2007.
2. René Descartes, Wikipedia, http://en.wikipedia.org/wiki/Ren%C3%A9_Descartes.
3. Smth, Kurt, Descartes' Life and Works, Stanford Encyclopedia of Philosophy, http://plato.stanford.edu/entries/descartes-works, 2007.

Otto von Guericke (1602-1686)

Established the field of vacuums

Otto von Guericke was born to an aristocratic family of Magdeburg, Germany on November 20, 1602. At the age of 15, he entered the Faculty of Arts at the Leipzig University. When he was 18 his father died. In 1621 he went to Jena to study at the university there. He completed his studies in Leiden, Netherlands in 1623.

Guericke was a scientist, inventor, and politician who served as the mayor of Magdeburg from 1646 to 1676.

In 1650, Guericke invented a vacuum pump consisting of a piston and cylinder with one-way flaps designed to pull air out of whatever vessel it was connected to. He used it to investigate the properties of the vacuum in many experiments.

Guericke demonstrated the force of air pressure by joining two 51-centimeter-diameter copper hemispheres and pumping the air out of the enclosure. Then he harnessed a team of eight horses to each hemisphere. The horses were unable to separate the hemispheres. When air was again allowed into the enclosure, the hemispheres were easily separated. He repeated this demonstration in 1663 at the court of Friedrich Wilhelm I of Brandenburg in Berlin using 24 horses.

Guericke proved that substances were not pulled by a vacuum; rather they were pushed by the pressure of the surrounding atmosphere or other fluids. With his experiments, Guericke disproved the hypothesis that nature abhors a vacuum, which for centuries had been a problem for philosophers and scientists.

Guericke applied the barometer to weather predictions, and thus prepared the way for meteorology. He also invented the first electrostatic generator. It was made of a large sulfur ball cast inside a glass globe, mounted on a shaft. The ball was rotated by means of a crank, and a static electric spark was produced when a pad was rubbed against the rotating ball. The globe was then removed and used as source of electricity for experiments.

Guericke died in Hamburg, Germany on May 21, 1686. The Otto von Guericke University of Magdeburg is named after him

1.   Otto von Guericke (1602 - 1686), Corrosion Doctors, www.corrosion-doctors.org/Biographies/ GuerickeBio.
2.   Otto von Guericke (1602 - 1686), www.uni-magdeburg.de/magdeburg/guericke_eng.
3.   Otto von Guericke, Biographies/Images of Physicists and Astronomers, www.mlahanas.de/ Physics/ Bios/OttoVonGuericke.
4.   Otto von Guericke, Wikipedia, http://en.wikipedia.org/wiki/Otto_von_Guericke.

Evangelista Torricelli (1608-1647)

Invited the barometer

Evangelista Torricelli was born in Faenza Papal States (Italy) on October 15, 1608. His father was a textile worker, and the family was fairly poor. Torricelli entered a Jesuit College in 1624 and studied mathematics and philosophy there until 1626.

After Galileo's death in 1642, Grand Duke Ferdinando II de' Medici asked Torricelli to succeed Galileo as the grand-ducal mathematician and professor of mathematics in the University of Pisa.

In 1643, to measure atmospheric pressure, Torricelli created a tube one meter long, which was sealed at the top end, filled with mercury, and set vertically into a basin of mercury. The column of mercury fell to about 76 centimeters, leaving a **Torricellian vacuum** above. This was the first barometer.

Torricelli's equation was created by Torricelli to find the final velocity of an object moving with a constant acceleration without having a known time interval. Mathematically, it is expressed as $v_f^2 = v_i^2 + 2a\Delta d$, where "$v_f$" is the final velocity, "v_i" is the initial velocity, "a" is acceleration and "d" is distance.

Torricelli's law states that the speed of a fluid flowing out of an opening under the force of gravity is proportional to the square root of the product of twice the acceleration of the gravity "g" multiplied by the vertical distance between the level of the surface and the center of the opening "h"

$$v = \sqrt{2gh}$$

This speed coincides with the speed the fluid would have in a free-fall from the height "h."

Gabriel's (Torricelli's) Horn is a figure which has infinite surface area but finite volume. The name refers to the tradition identifying the archangel Gabriel with the angel who blows the horn to announce Judgment Day.

On October 25, 1647, Torricelli died in Florence, Italy a few days after having contracted typhoid fever. He was buried in the Basilica of St. Lawrence. The **torr**, a unit of pressure, was named in his honor.

1. Evangelista Torricelli, Biographies/Images of Physicists and Astronomers, www.mlahanas.de/Physics/Bios/EvangelistaTorricelli.
2. Evangelista Torricelli, Wikipedia, http://en.wikipedia.org/wiki/Evangelista_Torricelli.
3. Torricelli's Equation, Wikipedia, http://en.wikipedia.org/wiki/Torricelli%27s_equation.
4. Torricelli's Law, Chemie.De Information Services, Life Science Network, www.chemie.de/lexikon/e/Torricelli's_Law.

Blaise Pascal (1623-1662)

Formulated the relationship between applied pressure and the pressure within a liquid

Born in France on June 19, 1623, Blaise Pascal was a mathematician, physicist, and religious philosopher. He helped create two major new areas of research: (1) At the age of 16 he wrote a significant treatise on projective geometry, and (2) he corresponded with Pierre de Fermat on probability theory, strongly influencing the development of modern economics and social science.

At age 18, to help his father with this work, Pascal constructed a mechanical calculator called the Pascaline. It was capable of addition and subtraction, and arguably was the first computer.

Pascal in 1653 recognized the fact that if the pressure at the top of a liquid in a container is increased in any way, the pressure at any depth of the liquid must increase by the same amount. This relationship is known as **Pascal's law**:

- Pressure applied to an enclosed fluid is transmitted undiminished to every portion of the fluid and the walls of the containing vessels.

Mathematically, this is express by:

$$p = p_0 + \rho g h$$

where "p" is the pressure within the fluid, "p_0" is the applied pressure, "ρ" is the density of the fluid, "g" is the gravitational constant (9.8m/s^2), and "h" is the depth within the liquid.

Following a mystical experience in 1654, Pascal left mathematics and physics and devoted himself to writing about philosophy and theology. He suffered from ill-health throughout his life, which ended in an early death on August 19, 1662—two months after his 39th birthday. He died in intense pain after a malignant growth in his stomach spread to the brain.

In honor of his scientific contributions, the name Pascal has been given to the SI unit of pressure (1 **Pascal**) = 1 N/m^2) to Pascal's law described above, and to a computer programming language, **Pascal**.

1.   Blaise Pascal, Famous Physicists and Astronomers, www.phy.hr/~dpaar/fizicari/xpascal.
2.   Hewitt, Paul G., John Suchocki, and Leslie A. Hewitt, Conceptual Physical Science, Third Edition, Pearson/Addison Wesley, San Francisco, 2003.
3.   Pascal, Wikipedia, http://en.wikipedia.org/wiki/Blaise_Pascal.
4.   Sears, Francis W., Mark W. Zemansky, and Hugh D. Young, College Physics, Seventh Edition, Addison Wesley, New York, 1991.

Robert Boyle (1627-1691)

Developed the relationship between pressure and volume at a constant temperature

Robert Boyle was born in Lismore castle, in the province of Munster, Ireland on January 25, 1627. He was the 14th child of Richard Boyle, Earl of Cork. After an extended stay abroad he returned to England in 1644 to find his wealthy father had died and left him well off. From that time, unhampered by financial restraints, he devoted his life to scientific research. He was to become a chemist, a physicist, and an inventor.

In 1657, Boyle read about the air pump that Otto von Guerick (1602-1686) had developed and set about, with the assistance of Robert Hooke (1635-1703), to devise improvements in its construction. When finished, he began a series of experiments on the properties of air, and discovered that the volume of a gas varies inversely as the pressure when the temperature is maintained constant. This relationship is now known as **Boyle's law**, and is expressed mathematically as

$$PV = k$$

where "P" is pressure, "V" is volume, and "k" is a proportionality constant.

Boyles's other important work in physics included the discovery of the role of air in the propagation of sound, and studies on the expansive force of freezing water, specific gravities, refractive powers, crystals, electricity, color, and hydrostatics.

Understanding the distinction between mixtures and compounds, Boyle made considerable progress in detecting their ingredients, a process which he designated by the term "analysis." He further supposed that the elements were ultimately composed of particles of various sorts and sizes, and that they could not to be destroyed in any known way. He also studied the chemistry of combustion and of respiration.

In 1669, Boyle's health began to fail, and he gradually withdrew from his public engagements. He died on December 30, 1691, and was buried in Saint Martins-in-the-Fields, London, England. He is considered one of the founders of modern chemistry.

1.   Robert Boyle, Biographies/Images of Physicists and Astronomers, www.mlahanas.de/Physics/ Bios/RobertBoyle.
2.   Robert Boyle, Wikipedia, http://en.wikipedia.org/wiki/Robert_Boyle.
3.   Robert Boyle: Mighty Chemist, The Woodrow Wilson National Fellowship Foundation, www.woodrow.org/teachers/chemistry/institutes/1992/Boyle.Sears.
4.   Wattenberg, Frank, Boyle's Law, www.math.montana.edu/frankw/ccp/before-calculus/function/boyle/body.

Christiaan Huygens (1629-1695)

Formulated a geometrical method for finding from a known shape of a wave front the wave front at a later time

Christiaan Huygens was born in the Hague Netherlands in 1629. He was the leading proponent of the wave theory of light, and in 1678 stated a geometrical method for finding, from the know shape of a wave front at some instant the shape of the wave front at some later time. The principle is now known as **Huygens' principle**:

- Every point of a wave front may be considered the source of secondary wavelets that spread out in all directions with a speed equal to the speed of propagation of the wave.

The new wave front is then found by constructing a surface tangent to the secondary wavelets or, as it is called, the envelope of the wavelets.

Huygens also made important contributions to mechanics, stating that in a collision between bodies, neither body loses nor gains momentum. He stated that the center of gravity moves uniformly in a straight line. He also derived the equation for the expression for centrifugal force:

$$F = mv^2/r$$

where "m" is the mass of the object, "v" is its velocity, and "r" is the radius of the circle about which it moves.

Huygens also worked on more accurate clocks, suitable for naval navigation. In 1658, he published a book on the subject called *Horologium*. He worked with pendulum clocks and invented the clock with spiral spring or cycloid pendulum, which enables a more precise time measurement. In 1675, he patented a pocket watch.

After Blaise Pascal encouraged him to do so, Huygens wrote the first book on probability theory, which he published in 1657.

Huygens' work in astronomy included accurately describing the rings of Saturn and discovering Saturn's moon, Titan. He moved back to The Hague in 1681 after suffering a serious illness. He died there on July 8, 1695.

1. Christiaan Huygens (1629-1695), Eric Weissstein's World of Biography, Wolfram Research, http://scienceworld.wolfram.com/biography/Huygens.
2. Christiaan Huygens, Biographies/Images of Physicists and Astronomers, www.mlahanas.de/Physics/Bios/ChristiaanHuygens.
3. Christiaan Huygens, Biographies/Images of Physicists and Astronomers, www.mlahanas.de/Physics/Bios/ChristiaanHuygens.
4. Christiaan Huygens, www.surveyor.in-berlin.de/himmel/Bios/Huygens-e.
5. Sears, Francis W., Mark W. Zemansky, and Hugh D. Young, College Physics, Seventh Edition, Addison Wesley, New York, 1991.

Robert Hooke (1635-1703)

Discovered the relationship between the force applied to a spring and the resulting displacement

Robert Hooke, the son of a clergyman, was born in Freshwater, Isle of Wight, England on July 18, 1635. He played an important role in the scientific revolution.

At the University of Oxford, Hooke impressed other scientists with his skills at designing experiments and building equipment, and soon became an assistant to Robert Boyle (1627-1691).

In 1662, Hooke was named Curator of Experiments of the newly formed Royal Society of London. In this job, he was responsible for demonstrating new experiments at the Society's weekly meetings. He later became Professor of Geometry at Gresham College, London, where he had a set of rooms in which he lived for the rest of his life.

Hook was the inventor of, among other things, the iris diaphragm in cameras, the universal joint used in motor vehicles, and the balance wheel in a watch.

Hooke devised the compound microscope and illumination system, which was one of the best such microscopes of his time. With it, he observed organisms as diverse as insects, sponges, bryozoans, foraminifera, and bird feathers. He was the originator of the word "cell" in biology.

Hooke's first major published work was *Micrographia* in 1965. It established the foundation of using microscopy to advance biological science.

Hooke is best known for **Hooke's Law**—to keep a spring stretched an amount "x" beyond its unstretched length, we have to apply a force with magnitude "F" at each end. If the elongation is not too great, F is directly proportional to x:

$$F = kx$$

where "k" is a constant known as the spring constant. The fact that the elongation is directly proportional to force, for small elongations, was discovered by Hooke in 1678.

Hooke died in London on March 3, 1703.

1. Robert Hooke, www.microscopy-uk.org.uk/mag/artmar00/hooke1.
2. Robert Hooke, www.roberthooke.org.uk.
3. Sears, Francis W., Mark W. Zemansky, and Hugh D. Young, College Physics, Seventh Edition, Addison Wesley, New York, 1991.

Isaac Newton (1642-1727)

Formulated the three fundamental laws of motion

Isaac Newton was born in Lincolnshire, near Grantham, England on December 25, 1642. He came from a family of modest yeoman farmers. His father died several months before he was born. He was educated at Trinity College, Cambridge. Afterward, he was appointed professor of mathematics at Cambridge. He held the post for 28 years.

Seeing an apple fall from a tree led Newton, at the age of 23, to consider the force of gravity extending to the Moon and beyond. He formulated the law of universal gravitation. The Moon he saw as falling around the Earth an exact amount to match the earth's curvature.

In his law of universal gravitation Newton stated that every mass attracts every other mass with a force "F" that, for any two masses, is directly proportional to the product of the masses and inversely proportional to the square of the distance separating them. Expressed mathematically $F = Gm_1m_2/r^2$, where "G" is the gravitational constant, "m_1" and "m_2" are the two masses, and "r" is the distance between them.

Newton co-invented calculus independently of Gottfried Liebniz (1646-1716) and, extending Galileo's work, formulated the three fundamental laws of motion:

- **Newton's first law of motion:** Every object continues in a state of rest, or in a state of motion in a straight line at a constant speed, unless it is compelled to change that state by forces exerted on it.
- **Newton's second law of motion:** The acceleration produced by a net force on an object is directly proportional to the net force, is in the same direction as the net force, and is inversely proportional to the mass of the object.
- **Newton's third law:** Whenever one object exerts a force on a second object, the second object exerts and equal and opposite force on the first object.

Newton also did work with optics. He formulated a theory of the nature of light, and demonstrated with prisms that white light is composed of all colors. He proposed that light was composed of tiny particles called corpuscles.

In 1705, Newton was knighted by Queen Anne. He died at the age of 85 at Kensington, London on March 20, 1727. The SI unit of force, the **Newton** (N), is named for him.

1. Isaac Newton, Biographies/Images of Physicists and Astronomers, www.mlahanas.de/Physics/Bios/IsaacNewton.
2. Newton, Sir Isaac, Famous Physicists and Astronomers, www.phy.hr/~dpaar/fizicari/xnewton.
3. Wilkins, David R., Isaac Newton, Mathematicians of the Seventeenth and Eighteenth Centuries, www.maths.tcd.ie/pub/HistMath/People/Newton/RouseBall/RB_Newton.

Olaf Rømer (1644-1710)

Discovered that light travels at a definite speed

Olaf Christensen Rømer was born in Århus, Denmark on September 25, 1644. His father was a merchant and skipper. In 1662, Rømer graduated from the University of Copenhagen.

After graduation, Rømer was then employed by the French government. Louis XIV made him teacher for the Dauphin (heir apparent of the throne of France). During this time, he also took part in the construction of the magnificent fountains at Versailles.

In 1681, Rømer returned to Denmark and was appointed professor of astronomy at the University of Copenhagen and royal mathematician. He was also an observer at the University Observatory at Rundetårn, where he used improved instruments of his own construction.

In his position as royal mathematician, Rømer introduced the first national system for weights and measures in Denmark in 1683. He also developed one of the first temperature scales. Daniel Fahrenheit (1686–1736) visited him in 1708 and improved on the Rømer scale, the result being the Fahrenheit temperature scale still in use today in a few countries.

In 1705, Rømer was made the second Chief of the Copenhagen Police, a position he kept until his death in 1710. As one of his first acts, he fired the entire force, convinced that the morale was alarmingly low. He invented the first street lights (oil lamps), and worked hard to control the beggars, poor people, unemployed, and prostitutes of Copenhagen.

Rømer joined the observatory of Uranienborg on the island of Hven, near Copenhagen, in 1671. Over a period of several months, he and Jean Picard (1620-1682) observed about 140 eclipses of Jupiter's moon, Io. By comparing the times of the eclipses, they calculated the difference in longitude of Paris to Uranienborg.

In 1672, Rømer went to Paris and continued observing the satellites of Jupiter. He noted that times between eclipses got shorter as Earth approached Jupiter, and longer as Earth moved farther away. It became obvious to Rømer that the speed of light was finite. However, he did not calculate a value for the speed of light.

Rømer died in Copenhagen, Denmark on September 19, 1710.

In 1809, Jean Baptiste Delambre (1749-1822), using Rømer's approach, reported the time for light to travel from the Sun to the Earth as 8 minutes and 12 seconds, yielding a value for the speed a little more than 300,000 km/second.

1. Ole Rømer, Wikipedia, http://en.wikipedia.org/wiki/Ole_R%C3%B8mer.
2. Rømer, Olaus or Ole, Infoplease.com, www.infoplease.com/ce6/people/A0842330.

Gottfried Leibniz (1646-1716)

Invented calculus independent of Isaac Newton

Gottfried Wilhelm Leibniz was born in Leipzig, Germany on July 1, 1646. When he was six years old, his father, a Professor of Moral Philosophy at the University of Leipzig, died, leaving a personal library to which Leibniz was granted free access from age seven onwards. By age 12, he had taught himself Latin, which he used freely all his life.

Leibniz entered the University of Leipzig at age 14, and completed his undergraduate degree by age 20, specializing in law. He obtained his doctorate in law from the University of Altdorf in 1666.

After obtaining his doctorate, Leibniz served Johann Philipp von Schönborn, archbishop elector of Mainz, in a variety of legal, political, and diplomatic capacities. In 1673, when the elector's reign ended, Leibniz moved to Paris. During his three years there he devoted his time to the study of mathematics, science, and philosophy.

In 1676, Leibniz was appointed librarian and privy councilor at the court of Hannover. For the next 40 years, until his death, he served Ernest Augustus, duke of Brunswick-Lüneburg and later elector of Hannover; George Louis, elector of Hannover; and even later George I, king of Great Britain and Ireland.

Leibiz's work encompassed not only mathematics and philosophy but also theology, law, diplomacy, politics, history, physics, and philology (the scientific study of the relationship of languages to one another and their history).

Leibniz discovered, in 1675, the fundamental principles of infinitesimal calculus. This discovery was arrived at independently of the work of Isaac Newton (1642-1727), whose system of calculus was invented in 1666.

Leibniz developed a rudimentary binary system, transposing numerals into seemingly infinite rows of ones and zeros. The fundamental idea of the binary yes-no/on-off principle is the foundation of the modern computer. In 1672, Leibniz invented a calculating machine (the stepped wheel calculator) capable of multiplying, dividing, and extracting square roots. He is considered a pioneer in the development of mathematical logic. Leibniz died on November 14, 1716.

1. Gottfried Leibniz, Wikipedia, http://en.wikipedia.org/wiki/Gottfried_Leibniz.
2. Gottfried Wilhelm Leibniz (1646 - 1716), Pioneers, KerryR.net, www.kerryr.net/pioneers/leibniz.
3. Gottfried Wilhelm Leibniz, Microsoft Encarta Encyclopedia Online, http://encarta.msn.com/encyclopedia_761576058/Leibniz, 2007.
4. O'Connor J. J. and E F Robertson, Gottfried Wilhelm von Leibniz, MacTutor, www-groups.dcs.st-and.ac.uk/~history/Biographies/Leibniz.

Stephen Gray (1666-1736)

Discovered that electricity can flow

Stephen Gray was born in Canterbury, Kent, England on December 26, 1666. His father, Mathias Gray, was a cloth dyer. Gray had no formal education in a University, but he managed to acquire a working knowledge of Latin. He became a dyer like his father, but cultivated science as a hobby.

Gray eventually turned his interests to electricity, and in a letter to Hans Sloane (1660-1753) in 1708 he described the use of down feathers to detect electricity. He was fascinated by lights produced by rubbing a glass tube to charge it, and realized electricity and the lights were related.

Gray's glass tubes were 3.5 feet long and 1.2 inches in diameter. When rubbed with a dry hand or dry paper, the glass would obtain an electric charge. He noticed that the cork in the end of one of his tubes generated an attractive force on small pieces of paper when he rubbed the tube. When he extended the cork with a small fir stick plugged into the middle, the charge was evident at the end of the stick. So he tried longer sticks, and finally he added a length of thread, connected at the other end to an ivory ball. In the process, he had discovered that electricity, which he called electric virtue, would carry over distance, and that the ivory ball would attract light objects as if it were the electrified glass tube.

Gray subsequently extended the reach of his thread-wire to a distance of 800 feet. In the process, he discovered the importance of insulating the thread wire from earth contact. From these experiments came an understanding of the role played by conductors and insulators.

Gray went on to do more electrical experiments, including inducing electrical polarity in suspended objects. He realized that his electric virtue was the same as lightning, many years before Franklin formulated his flying-kite theory.

Eventually falling on hard times, the Prince of Wales nominated Gray in 1719 to become a pensioner of the Charterhouse.

In 1732, Gray was admitted as a member of the Royal Society, but he died destitute a few years later, on February 6, 1736. He is believed to be buried in a common grave in an old London cemetery in an area reserved for pauper pensioners of Charterhouse.

1. Stephen Gray (Scientist), Wikipedia, http://en.wikipedia.org/wiki/Stephen_Gray_(scientist).
2. Stephen Gray, Conduction, Insulation and Electric Current – 1729, Spark Musem, www.sparkmuseum.com/BOOK_GRAY.HTM.
3. Stephen Gray, Institute of Chemistry, http://chem.ch.huji.ac.il/history/gray.html, April 12, 2003.

H. Thomas Milhorn, MD, PhD

Gabriel Fahrenheit (1686-1736)

Invented the Fahrenheit thermometer

Daniel Gabriel Fahrenheit was born in Danzig in the Polish-Lithuanian Commonwealth on May 24, 1686. He was the son of a well-to-do merchant. After losing both parents to mushroom poisoning in 1701, he apprenticed to a shopkeeper in Amsterdam. After completing a term of four years, he turned to physics and became a glassblower and manufacturer of precision instruments, including barometers, altimeters and thermometers.

From 1718 onward, Fahrenheit gave lectures in chemistry in Amsterdam, and traveled widely, spending a considerable amount of time in England, where he became a member of the Royal Society.

Fahrenheit is best known for developing precision thermometers. He filled his first thermometers with alcohol, beginning in 1709. Then, in 1814, he became the fist to use mercury. Fahrenheit set the low end of his scale at the temperature of a water and ice mixture (32°F) and high end at the boiling point of water (212°F). Normal human body temperature therefore became the familiar 98.6°F. Fahrenheit's temperature scale was divided into 180 equal increments (degrees) between the freezing and boiling points of water.

Until the 1970s, the Fahrenheit scale was in general common use in the English-speaking countries of the world; the Celsius scale was employed in most other countries, and for scientific purposes worldwide. Since that time, English-speaking countries have officially adopted the Celsius scale for scientific work. However, the Fahrenheit scale is still used for everyday temperature measurements by the general population in the United States. The conversion formula for a temperature that is expressed on the Celsius scale is $F = (9/5)C + 32$.

Fahrenheit discovered that the boiling points of liquids vary with atmospheric pressure. He also discovered the phenomenon of super cooling of water; that is, cooling water to below its normal freezing point without it being converted to ice.

Fahrenheit died, unmarried, on September 16, 1736 in The Hague, Netherlands.

1. Daniel Gabriel Fahrenheit, Chem209, http://chem.oswego.edu/chem209/Misc/fahrenheit, *October 21, 2004.*
2. Encyclopedia of World Biography on Gabriel Daniel Fahrenheit, Bookrags, www.bookrags.com/biography/gabriel-daniel-fahrenheit.
3. Gabriel Fahrenheit, Wikipedia, http://en.wikipedia.org/wiki/Gabriel_Fahrenheit.
4. Sears, Francis W., Mark W. Zemansky, and Hugh D. Young, College Physics, Seventh Edition, Addison Wesley, New York, 1991.

Pieter van Musschenbroek (1692-1761)

Invented the leyden jar (capacitor)

Pieter van Musschenbroek was born in Leiden (Leyden), Netherlands on March 14, 1692. His father, Johann van Musschenbroek was a maker of physical apparatus. At the time of Pieter's birth, the family was turning to the making of scientific instruments, such as air pumps, microscopes, and telescopes. This may have stimulated Pieter's interest in science.

Musschenbroek studied at the University of Leiden, where he received his degree in medicine in 1715 and later his doctor of philosophy degree in natural philosophy. He then visited England in 1717 and met Isaac Newton.

From 1719 to 1723 Musschenbroek was professor of Mathematics and Philosophy in Duisberg, Germany, where he worked with Gabriel Fahrenheit (1686-1736). In 1721, Musschenbroek was additionally appointed professor of Medicine. He was promoted to the chair of astronomy at Utrecht in 1732.

In 1729, Musschenbroek used the word "physics," which had never been used before.

Musschenbroek provided the first approach to scientific study of electrical charge and its properties. Transient electrical energy could be generated by friction machines, but there was no way to store it. Musschenbroek discovered that energy could be stored in a glass jar filled with water into which a brass rod had been placed; and that the energy could be released only by completing an external circuit between the brass rod and another conductor in contact with the outside of the jar.

A German scientist, Ewald von Kleist (1715-1759), had independently constructed a similar device in late 1745, shortly before Musschenbroek, but failed to publicize his invention in time.

The Leyden jar was an immediate sensation to scientists and nonscientists throughout the world. Today, this device is recognized as the first capacitor, and Musschenbroek is given credit for its invention.

The invention of the Leyden jar to many at the time was proof of the fluid theory of electricity.

Musschenbroek's *Elementa Physica*, published in 1726, played an important part in the transmission of the ideas of Isaac Newton (1642-1727) to Europe. Musschenbroek died in Leyden on September 19, 1761.

1. Currier, Dean P., Petrus Musschenbroek, Adventures in Cybersound, www.acmi.net.au/ AIC/VAN_MUSSCHENBROEK_BIO.
2. Pieter (Petrus) van Musschenbroek, http://chem.ch.huji.ac.il/history/musschenbroek.
3. Pieter van Musschenbroek, Wikipedia, http://en.wikipedia.org/wiki/Pieter_van_Musschenbroek.

Charles du Fay (1698–1739)

Discovered the existence of two types of electricity

Charles François du Fay was born in Paris, France on September 14, 1968. Virtually nothing is known about him until he began his experiments with electricity in the 1730s. He discovered the existence of two types of electricity, one produced by glass (vitreous) and the other by resin (resinous). These terms were used for 15 years until they were replaced with "positive" and "negative"—coined independently by William Watson (1715-1787) and Benjamin Franklin (1706-1790). Du Fay considered electricity to come in two varieties, which cancelled each other. He expressed this in terms of a two-fluid theory.

Du Fay also noted the difference between conductors and insulators, calling them "electrics" and "non-electrics" for their ability to produce contact electrification. He also discovered that like-charged objects repelled each other and unlike-charged objects attracted each other. He also disproved certain misconceptions regarding electric charge, such as that electric properties of a body depend on its color.

Using a Leyden jar, which could store electricity, Du Fay was able to perform some interesting demonstrations. In one, he passed an electrical discharge through 180 soldiers who had joined hands in a circle.

Du Fay was one of the first to perform experiments with an electroscope, which makes use of the fact that like-charges repel each other and that the strength of the repulsive force is proportional to the electrical charge. Therefore, two small metal leaves will be held at an angle against the pull of gravity if they both have the same electrical charge, and the angle separating them will be greater if the charge is greater.

Du Fay's other work included showing that all bodies can be electrically charged by heating and rubbing, except metals and soft or liquid bodies; all bodies, including metal and liquid, can be charged by influence (induction); the electrical properties of an object unique to color are affected by the dye, not the color itself; glass is as satisfactory as silk as an insulator; and thread conducts better wet than dry.

Du Fay's work was influential in the later work of Benjamin Franklin (1706-1790) with electricity. Du Fay died in 1739 after a brief illness.

1. C. G. du Fay, Wikipedia, http://en.wikipedia.org/wiki/C._F._du_Fay.
2. Charles François de Cisternay DuFay (1698 - 1739), Two Kinds of Electrical Fluid: Vitreous and Resinous – 1733, Spark Museum, www.sparkmuseum.com/BOOK_DUFAY.
3. Science and Its Times on Charles-François De Cisternay Du Fay, BookRags, www.bookrags.com/ C.F._du_Fay.

18th Century Physics
(1700-1799)

Daniel Bernoulli (1700-1782)

Developed the fundamental relationship of pressure, flow velocity, and height of an ideal fluid

Daniel Bernoulli was born on January 29, 1700 in Gröningen, Netherlands. He was the second son of Jean Bernoulli, a noted mathematician who began the use of the symbol "g" for the acceleration of gravity. Over the years, his father became jealous of Daniel's success, which led to conflict between the two.

Daniel was sent to Basel University at the age of 13 to study philosophy and logic. He obtained his baccalaureate degree in 1715 and his master's degree in 1716. His father then sent him back to Basel University to study medicine..

Bernoulli is best known for his studies on the fluid flow in pipes. His discovery, called **Bernoulli's principle**, for a level pipe with laminar flow, can be stated as:

• Where the speed of a fluid increases, internal pressure in the fluid decreases

Bernoulli's principle is a consequence of the conservation of energy. More speed and kinetic energy mean less pressure, and more pressure means less speed and kinetic energy.

One form of **Bernoulli's equation**, when the pipe is not level, states that the work per unit volume of fluid is equal to the sum of the changes in kinetic and potential energies per unit volume that occur during the flow:

$$p_1 - p_2 = (1/2)\rho(v_2^2 - v_1^2) + \rho g(y_2 - y_1)$$

where "$p_1 - p_2$" is the pressure difference at two points in the pipe, "ρ" is the density of the fluid, "v_1" and "v_2" are the fluid velocities at the two pints, "g" is the gravitational constant, and "$y_2 - y_1$" is the height difference between the two points. Bernoulli published this work in *Hydrodynamica* in 1738.

Bernoulli also worked on a number of other subjects, including vibrating strings, ocean tides, and the kinetic theory of gases. He also designed an hour glass to be used at sea so that the trickle of sand was constant even when the ship was rolling in heavy seas.

Bernoulli died in Basel, Switzerland on March 17, 1782.

1. Daniel Bernoulli, BookRags, www.bookrags.com/Daniel_Bernoulli.
2. Hewitt, Paul G., John Suchocki, and Leslie A. Hewitt, Conceptual Physical Science, Third Edition, Pearson/Addison Wesley, San Francisco, 2003.
3. Sears, Francis W., Mark W. Zemansky, and Hugh D. Young, College Physics, Seventh Edition, Addison Wesley, New York, 1991.

Anders Celsius (1701-1744)

Invented the Celsius temperature scale

Anders Celsius was born in Uppsala, Sweden on November 27, 1701. His father, Nils Celsius, was professor of astronomy at Uppsala University. Early on, Anders became engaged in the general problem of weights and measures, including temperature measurements. As a student, he assisted astronomy professor, Erik Burman, in meteorology observations. At that time there were a variety of thermometers, each based on a different scale.

Celsius became professor of astronomy at Uppsala University in 1730. He remained there until 1744. From 1732 to 1735 he traveled, visiting notable observatories in Germany, Italy, and France. Soon after his return to Uppsala, he participated in the famous expedition to Torneå in the most northern part of Sweden. The expedition was headed by French astronomer, Pierre-Louis Maupertuis (1698-1759). Its aim was to measure the length of a degree along a meridian close to the pole and compare the result with a similar expedition to Peru, near the equator. The two expeditions confirmed the belief of Isaac Newton (1643-1727) that the shape of the Earth is an ellipsoid, flattened at the poles.

In 1742, in a paper to the Royal Swedish Academy of Sciences, Celsius proposed the Celsius temperature scale. His thermometer had 100 divisions between the freezing point of water ($100^{\circ}C$) and the boiling point of water ($0^{\circ}C$). The scale was reversed by Carolus Linnaeus in 1745 to how it is today. Celsius temperature can be converted from Fahrenheit by $C = (5/9)(F - 32)$.

Celsius was the first to perform and publish careful experiments aimed at the definition of an international temperature scale on scientific grounds. In his paper *Observations of two Persistent Degrees on a Thermometer* he reported on experiments to check that the freezing point is independent of latitude and pressure. He determined the dependence of the boiling point of water on atmospheric pressure. He further determined a rule for determining the boiling point if the barometric pressure deviates from a given standard pressure.

Celsius died of tuberculosis in Uppsala on April 5, 1774. He was 43 years old. His grave is next to that of his grandfather, Magnus Celsius, in the Old Uppsala Church.

1. Anders Celsius, Biographies/Images of Physicists and Astronomers, www.mlahanas.de/Physics/ Bios/AndersCelsius.
2. Anders Celsius, History, Uppsala Astronomical Observatory, www.astro.uu.se/history/ Celsius_eng.
3. Anders Celsius, Wikipedia, http://en.wikipedia.org/wiki/Anders_Celsius.
4. Beckman, Olaf, History of the Celsius Temperature Scale, www.astro.uu.se/history/celsius_scale.

Benjamin Franklin (1706-1790)

Verified that lightening is an electrical discharge

Benjamin Franklin was born in Boston, Massachusetts on January 17, 1706. He was the tenth son of soap maker, Josiah Franklin. Benjamin was to become one of the most important and influental Founding Fathers of the United States and a leading author, political theorist, politician, printer, scientist, inventor, civic activist, and diplomat.

In 1729, Franklin bought a newspaper, the *Pennsylvania Gazette*. He not only printed the paper, but often contributed pieces to the paper under aliases. His newspaper soon became the most successful in the colonies.

In 1733, Franklin started publishing *Poor Richard's Almanac* under the guise of a man named Richard Saunders, a poor man who needed money to take care of his nagging wife. Many of Franklins famous sayings, such as "A penny saved is a penny earned" were published in his almanac.

After his success in the printing business, in 1749 Franklin retired from business and started concentrating on science. He had already invented a heat-efficient stove, called the **Franklin stove**, to help warm houses efficiently. Because he invented the stove to help improve society, Franklin refused to take out a patent. Among Franklin's other inventions were swim fins, the glass armonica (a musical instrument), odometer, and bifocals.

In the early 1750s Franklin turned to the study of electricity. His observations, including his kite experiment, demonstrated that lightning is an electrical discharge. He discovered that there were two types of electric charges, which he and William Watson (1715-1787) named independently "positive" and "negative."

Franklin discovered experimentally the law of conservation of charge. He also discovered that electricity prefers to move through pointed objects rather than rounded, and thus was led to invent the lightning rod. Franklin's studies brought him international fame as the leading authority on electricity. He, like others before him, considered electicity to consist of a fluid which flowed in conductors.

Franklin died on April 17, 1790, at the age of 84. He was interred in Christ Church Burial Ground in Philadelphia, Pennsylvania.

1.   A quick biography of Benjamin Franklin, The Electric Ben Franklin, www.ushistory.org/ franklin/info.
2.   Benjamin Franklin, Wikipedia, http://en.wikipedia.org/wiki/Benjamin_Franklin
3.   Hewitt, Paul G., John Suchocki, and Leslie A. Hewitt, Conceptual Physical Science, Third Edition, Pearson/Addison Wesley, San Francisco, 2003.

Leonhard Euler (1707-1783)

Made fundamental contributions to fluid dynamics and mechanics

Leonhard Euler was born on April 15, 1707 in Basel, Switzerland. He made contributions to a wide range of mathematics and physics areas, including analytic geometry, trigonometry, geometry, calculus, and number theory. When considering vibrating membranes, Euler was led to the Bessel equation, which he solved by introducing Bessel functions. He made substantial contributions to differential geometry, investigating the theory of surfaces and curvature of surfaces. Other geometric investigations led him to fundamental ideas in topology, such as the **Euler characteristic** of a polyhedron.

In 1736, Euler published *Mechanica*, which provided a major advance in mechanics. He was the first to appreciate the importance of introducing uniform analytic methods into mechanics, thus enabling its problems to be solved in a clear and direct way.

Euler considered the motion of a point mass, both in a vacuum and in a resisting medium. He analyzed the motion of a point mass under a central force, and also considered the motion of a point mass on a surface.

Mechanica was followed by a two-volume work on naval science. Euler applied variational principles to determine the optimal ship design, and first established the principles of hydrostatics.

In 1765, Euler published another major work on mechanics *Theoria motus corporum solidorum* in which he decomposed the motion of a solid into a rectilinear motion and a rotational motion.

Euler published a number of major pieces of work on fluid dynamics, setting up the main formulas for the topic—the continuity equation, the Laplace velocity potential equation, and the Euler equations for the motion of an inviscid incompressible fluid.

Euler also did important work in astronomy, including determining the orbits of comets and planets. His partial solution for a more perfect theory of lunar motion, published in 1753, assisted the British Admiralty in calculating lunar tables, which were of importance in attempting to determine longitude at sea.

After becoming blind in his later years Euler was forced to perform elaborate calculations in his head. He died of a stroke on September 18, 1783 in St. Petersburg, Russia.

1. Leonhard Euler, Encyclopedia Britannica Online, www.britannica.com/eb/article-9033216/ Leonhard-Euler.
2. Manley, D. Mark, Famous Physicists, http://cnr2.kent.edu/~manley/physicists, August 30, 2006.
3. O'Conner, J. J. and E. F. Robertson, Leonhard Euler, www-groups.dcs.st-and.ac.uk/~history/ Biographies/Euler.

Mikhail Lomonosov (1711-1765)

Stated the idea of conservation of matter

Mikhail Vasilyevich Lomonosov was born in the village of Denisovka in the Arkhangelsk Governorate on an island not far from Kholmogory in the Far North of Russia on November 19, 1711. His father was a peasant fisherman.

In 1736, Lomonosov attended the University of the Imperial Academy of Sciences. There, he studied physical science, philological science, and French and German. He ultimately received a two year grant and studied at the universities of Marburg and Freiberg in Germany.

On his return to Russia in 1745, Lononosov was appointed professor of chemistry at the Slavic Greek Latin Academy in Moscow. He then bcame a chemistry professor at Marburg University in Hesse, Germany. He later joined the faculty at St. Petersburg University, where he ultimately became rector.

Lomonosov made important contributions to literature, education, and science. He regarded heat as a form of motion, suggested the wave theory of light, contributed to the formulation of the kinetic theory of gases, and stated the idea of conservation of matter—"All changes in nature are such that inasmuch is taken from one object insomuch is added to another. So, if the amount of matter decreases in one place, it increases elsewhere. This universal law of nature embraces laws of motion as well, for an object moving others by its own force in fact imparts to another object the force it loses."

In 1745, Lomonosov published a catalogue of over 3,000 minerals. In 1748, he created a mechanical explanation of gravity. He was the first person to record the freezing of mercury and to hypothesize the existence of an atmosphere on Venus. A poet himself, Lomonosov compiled a Russian grammar, which did much to improve the rhythm of Russian verse.

Eager to improve Russian education, Lomonosov and Ivan Shuvalov founded Moscow State University in 1755. In 1760, Lomonosov explained the formation of icebergs, and published the first history of Russia.

In 1764, Lomonosov was appointed Secretary of State. Often called the father of Russian science, Lomonosov died in St Petersburg on April 15, 1765.

1. Mikhail Lomonosov, Microsoft Encarta Encyclopedia Online, http://encarta.msn.com/encyclopedia_761567307/lomonosov, 2007.
2. Mikhail Lomonosov, NNDB, www.nndb.com/people/437/000042311.
3. Mikhail Lomonosov, Wikipedia, http://en.wikipedia.org/wiki/Mikhail_Lomonosov.

Roger Boscovich (1711-1787)

Formulated atomic theory using principles of Newtonian mechanics

Roger Joseph Boscovich was a physicist, astronomer, mathematician, philosopher, diplomat, poet, and Jesuit. He was born in the Republic of Ragusa (now Dubrovnik, in Croatia) on February 13, 1711, and lived in the Italian Peninsula, France, and England. His father, Nikola Bošković, a merchant, died when Roger was 10 years of age.

At age 15 Roger entered the Society of Jesus. On completing his noviciate, which was spent at Rome, he studied mathematics and physics at the Collegium Romanum, and in 1740 was appointed professor of mathematics in the college.

Boscovich is best known for his formulation of atomic theory using principles of Newtonian mechanics. In Vienna in 1758 he published his famous work *Theoria Philosophiae Naturalis Redacta ad Unicam Legem Virium in Natura Existentium* (Theory of Natural philosophy derived to the single Law of forces, which exist in Nature) containing his atomic theory and his theory of forces. In this work, Boscovich stated that the ultimate elements of matter are indivisible points (atoms), which are centers of force, and this force varies in proportion to distance.

Boscovich's work inspired Michael Faraday (1791-1867) to develop field theory for electromagnetic interaction, and was also a basis for the attempt by Albert Einstein (1879-1955) to develop a unified field theory. A unified field theory is a type of field theory that allows all of the fundamental forces between elementary particles (strong nuclear force, weak nuclear force, electromagnetic force, gravitational force) to be written in terms of a single field. So far, no field theory has been successful in doing this.

In 1764, Boscovich was called to serve as the chair of mathematics at the University of Pavia, and he held this post along with the directorship of the observatory of Brera in Milan for six years.

Boscovich also gave many important contributions to astronomy, including the first geometric procedure for determining the equator of a rotating planet from three observations of a surface feature and for computing the orbit of a planet from three observations of its position. He is also credited for discovering the absence of atmosphere on the Moon.

Boscovich died on February 13, 1787.

1. Roger Joseph Boscovich, Answer.com, www.answers.com/topic/roger-joseph-boscovich.
2. Roger Joseph Boscovich, Classic Encyclopedia, www.1911encyclopedia.org/ Roger_Joseph_Boscovich.
3. Roger Joseph Boscovich, S. J., www.faculty.fairfield.edu/jmac/sj/scientists/boscovich.
4. Roger Joseph Boscovich, Wikipedia, http://en.wikipedia.org/wiki/Rudjer_Boscovich.

William Watson (1715-1787)

Showed that the capacity of the Leyden jar could be increased by coating it inside and out with lead foil

William Watson was born on April 3, 1715 in London, England. The son of a London tradesman, he was apprenticed to an apothecary from 1731 to 1738. In 1762, he was appointed physician to the Foundling Hospital.

Watson's early work was in botany. The **Watsonia** (Bugle Lily), a genus of plants in the iris family native to South Africa, is named after him.

Watson is best know for his work in electricity, and was first to investigate the passage of electricity through a rarefied gas. This was based on a series of experiments with the Leyden jar, discovered by Pieter van Musschenbroek (1692-1761) in 1746. Watson not only improved the device by coating the inside of it with metal foil, but also realized that the pattern of discharge of the jar suggested that electricity was simply a single fluid or, as he termed it, an "electrical ether." Watson felt that normally bodies have an equal density of this fluid, so that when two such bodies meet there is no electrical activity. If, however, their densities are unequal, the fluid will flow, and there will be an electric discharge. That is, electricity can only be transferred from one body to another, but it cannot be created or destroyed (conservation of charge). This theory was also developed with greater depth at about the same time by Benjamin Franklin (1706-1790).

Watson coined the term "circuit," was the first to observe the flash of light from the discharge of a Leyden jar, and provided the first demonstration of the passage of electricity through a vacuum.

In 1747, Watson set out to determine the speed of electricity. The general belief at that time was that electricity was faster than sound, but no accurate test had as yet been devised to measure it. Watson, in fields north of London, laid out a line of wire, supported by dry sticks and silk, running 6.4 km. Even at this length, the speed of electricity appeared to be instantaneous.

Watson proposed a theory that electrical ether was not created or destroyed, only transferred. This idea, charge conservation, was announced by Benjamin Franklin (1706-1790) a few months later.

Watson eventually decided not to pursue his electrical experiments, concentrating instead on his medical career. He died on May 10, 1787.

1. Sir William Watson, www.geocities.com/neveyaakov/electro_science/watson.html?20083, October 19, 2002.
2. Watson, Spark Museum, www.sparkmuseum.com/BOOK_WATSON.
3. William Watson, Wikipedia, http://en.wikipedia.org/wiki/William_Watson_(scientist).

Jean d'Alembert (1717-1783)

Developed a method for solving a special case of the wave equation

Jean le Rond d'Alembert was a mathematician, mechanician, physicist, and philosopher. He was born in Paris, France on November 16, 1717, the illegitimate child of the writer Claudine Guérin de Tencin (1682-1749) and the chevalier Louis-Camus Destouches (1668-1726), who was an artillery officer.

A couple of days after birth, his mother left him on the steps of the Saint-Jean-le-Rond de Paris church. According to custom, he was named after the patron saint of the church. D'Alembert was placed in an orphanage, but was soon adopted. Destouches secretly paid for the child's education, but did not want his parentage officially recognized.

In 1739, d'Alembert pointed out the errors he had detected in Charles René Reynaud's *L'analyse démontrée*, published 1708. It was a standard work, one which d'Alembert himself had used to study the foundations of mathematics.

In 1740, d'Alembert submitted *Mémoire sur la réfraction des corps solides*. In this work he theoretically explained refraction. He also wrote about what is now called **d'Alembert's paradox**—that the drag on a body immersed in an inviscid, incompressible fluid is zero.

D'Alembert's principle is a statement of the fundamental classical laws of motion—a powerful new interpretation of Newton's third law. The principle states that the sum of the differences between the forces acting on a system and the time derivatives of the momenta of the system itself along a virtual displacement consistent with the constraints of the system, is zero.

The wave equation is an important second-order linear partial differential equation that describes the propagation of a variety of waves, such as sound waves, light waves, and water waves. A solution to the wave equation that is valid when the string is so long that it may be approximated by one of infinite length was obtained by d'Alembert, and is known as the **d'Alembert solution**.

D'Alembert suffered bad health for many years, and died on October 29, 1783, the result of a bladder illness. As a known unbeliever, he was buried in a common unmarked grave.

1. d'Alembert, Jean-le-Rond (1717-1783), Eric Weisstein's World of Biography, Wolfram Research, http://scienceworld.wolfram.com/biography/dAlembert.
2. Jean le Rond d'Alembert Wikipedia, http://en.wikipedia.org/wiki/Jean_le_Rond_d'Alembert.
3. O'Connor, J. J. and E F Robertson, MacTutor, Jean Le Rond d'Alembert, www-groups.dcs.st-and.ac.uk/~history/Printonly/D'Alembert.

John Michell (1724-1793)

Studied magnetism and built an apparatus for determining the value of Newton's gravitational constant

John Michell was born on December 25, 1724. He was educated at Queens' College, Cambridge, obtaining his M.A. in 1752 and Bachelor of Divinity degree in 1761.

In 1762, Michell was appointed Woodwardian Professor of Geology at Cambridge, and in 1767 he became rector of Thornhill, West Yorkshire. Michell's work spanned a wide range of subjects from astronomy to optics, to gravitation, to geology.

Mitchell's work spanned a wide range of subjects, from astronomy to geology, optics, and gravitation.

In 1750, Michell published an article entitled *A Treatise of Artificial Magnets* in which he showed an easy and expeditious method of making magnets which were superior to the best natural ones (lodestones). In addition to a description of the method of magnetization, the work contained a variety of accurate magnetic observations, and explained the nature of magnetic induction.

In 1760, Michell constructed a theory of earthquakes as wave motions in the interior of the Earth, and suspected a connection between earthquakes and volcanism. He is known as the father of seismology for his studies on earthquakes and vibrations within the Earth.

In a letter to Henry Cavendish (1731-1810), published in 1784, Michell discussed the effect of gravity on light. He is now credited with being the first to study the case of a heavenly object massive enough to prevent light from escaping. Such an object would not be directly visible, but could be identified by the motions of a companion star if it were part of a binary system. This concept was the predecessor of the modern idea of a black hole.

Michell also suggested using a prism to measure the gravitational weakening of starlight due to the surface gravity of the source, a phenomenon now known as gravitational shift.

Michell also invented and built a torsion balance for an experiment to measure Newton's gravitaional constant "G," but didn't live to put it to use, dying on April 29, 1793. His apparatus passed to Cavendish, who performed the experiment in 1798.

1. John Michell (1724 - April 21, 1793), SEDS, www.seds.org/messier/xtra/Bios/michell.
2. John Michell, Microsoft Encarta Encyclopedia Online, http://encarta.msn.com/encyclopedia_761594885/michell_john., 2007.
3. John Michell, Wikipedia, http://en.wikipedia.org/wiki/John_Michell.

Joseph Black (1728-1799)

A founder of thermochemistry

Joseph Black was born in Bordeaux, France on April 16, 1728. His father, who was from Belfast, Ireland but of Scottish descent, was engaged in the wine trade, as was his mother's family. He was one of 13 children. At age 18, he entered the University of Glasgow, and four years later he went to Edinburgh to further his medical studies.

A founder of thermochemistry, Black developed many pre-thermodynamics concepts, such as heat capacity, specific heat, and latent heat. He also laid the foundations of chemistry as an exact science in his investigations on magnesium carbonate, during which he discovered carbon dioxide, which he called "fixed air."

In about 1750, Black developed the analytical balance, based on a light-weight beam balanced on a wedge-shaped fulcrum. Each arm carried a pan on which either a sample or standard weights was placed. It far exceeded the accuracy of any other balance of the time and became an important scientific instrument in most chemistry laboratories.

In 1757, Black was appointed Regius Professor of the Practice of Medicine at the University of Glasgow. And in 1761, he discovered that when ice melts it absorbs heat without changing temperature. From this, he concluded that the heat must have combined with the ice particles and become latent. This hypothesis he verified quantitatively by experiments performed at the end of 1761. In 1764, he measured the latent heat of steam, although it is said not very accurately.

Black also noticed that different bodies in equal masses require different amounts of heat to raise them to the same temperature, and so founded the doctrine of specific heats. He also showed that equal additions or subtractions of heat produced equal variations of bulk in the liquid of his thermometers.

Black was one of the teachers of James Watt (1736-1819), who made important improvements to the steam engine, and kept up a constant correspondence with him.

Black Black never married. He died in Edinburgh on Dcember 6, 1799 at the age of 71, and is buried in Greyfriars Kirkyard. The chemistry buildings at both the University of Edinburgh and the University of Glasgow are named after him.

1. Cooper, Alan, Joseph Black, M.D., Chemistry History, www.chem.gla.ac.uk/dept/black, February/.March, 1999.
2. Joseph Black, Classic Encyclopedia, www.1911encyclopedia.org/Joseph_Black.
3. Joseph Black, Wikipedia, http://en.wikipedia.org/wiki/Joseph_Black.

Henry Cavendish (1731-1810)

First to determine the value of Newton's gravitational constant

Henry Cavendish was born into wealthy British royalty on October 10, 1731 in Nice, France, where his family was living at the time. He was morbidly shy of women and strangers and avoided speaking to them. His appearance was odd as well. He was a tall man with a thin, squeaky voice, and spoke with hesitation and difficulty, especially when embarrassed. He wore a coat of faded velvet and a three-cornered cocked hat from the previous century. At home, his servant was instructed by written notes what to prepare for dinner.

Cavendish is generally credited with the discovery of "inflammable air." The gas had been collected and studied by others, starting with Robert Boyle (1627-1691), for over a century; however, Cavendish's careful studies, involving specific gravity, established the gas as an individual substance— hydrogen.

From hundreds of analyses, Cavendish established an accurate composition of the atmosphere—79.167 percent phlogisticated air (nitrogen plus argon) and 20.833 percent dephlogisticated air (oxygen).

Cavendish is most famous for being the first person to determine the value of the gravitational constant "G" of Isaac Newton (1643-1727). He did so in 1793 by measuring the tiny force between small lead masses with an extremely sensitive torsion balance. After calibration, as the torsion balance twisted due to the attraction between the two masses, a light beam was reflected off an attached mirror and shined on a scale so that the distance the weights moved could be measured. Cavendish determined the value of G to be 6.673×10^{-11}. The equipment he used was designed by John Michell (1724-1793), who died before he could put it to use.

A simpler method for determining the value of G was later developed by Philipp von Jolly (1809-1884), who attached a spherical flask of mercury to one arm of a sensitive balance. After the balance was put into equilibrium by adding weights to the other end, a 6-ton lead sphere was rolled beneath the mercury flask. The gravitational force between the two was determined by the additional weight that had to be added to the balance to again achieve equilibrium.

Cavendish is also known for his measurement of the earth's density and early research into electricity. He died on March 10, 1810.

1. Henry Cavendish, 1731-1820, http://mattson.creighton.edu/History_Gas_Chemistry/ Cavendish, September 25, 2001.
2. Hewitt, Paul G., John Suchocki, and Leslie A. Hewitt, Conceptual Physical Science, Third Edition, Pearson/Addison Wesley, San Francisco, 2003.
3. Sears, Francis W., Mark W. Zemansky, and Hugh D. Young, College Physics, Seventh Edition, Addison Wesley, New York, 1991.

Charles Coulomb (1736-1806)

Established experimentally the nature of the force between two charges

Charles Augustin de Coulomb was born on June 14, 1736 in Angoulême, France. His father, Henri Coulomb, was inspector of the Royal Fields in Montpellier. His mother, Catherine Bajet, came from a wealthy family in the wool trade. When Coulomb was a boy the family moved to Paris.

Coulomb was educated at the École du Génie in Mézieres, graduating in 1761 as a military engineer with a rank of First Lieutenant. He served in the West Indies for nine years, where he supervised the building of fortifications in Martinique. He spent eight years directing the work, during which time he contracted tropical fever.

In 1774, Coulomb shared the Paris Academy of Science's first prize for his paper on magnetic compasses, and also received first prize for his classic work on friction. During the next 25 years, he presented 25 papers to the Academy on electricity, magnetism, torsion, and applications to the torsion balance. He also investigated the strengths of materials and determined the forces that affect objects on beams, thereby contributing to the field of structural mechanics. In addition, he contributed to the field of ergonomics. His research provided a fundamental understanding of the ways in which people and animals can most efficiently do work.

Coulomb is best known for his work with electrical charges. In 1784, he used a torsion balance similar to the one Cavendish had used 13 years earlier. He found that, for two charged objects that are much smaller than the distance between them, the force between them varies directly as the product of the charges and inversely as the square of the separation distance. **Coulomb's law** is expressed mathematically as:

$$F = \frac{kq_1q_2}{r^2}$$

where "k" is a proportionality constant (9×10^9 N.m^2/C^2), "q_1" and "q_2" are the magnitudes of the two charges, and "r" is the distance between the two charges.

The **coulomb** "C" is a unit of charge named after Charles Coulomb. It turns out that one coulomb is the charge associated with 6.25×10^{18} electrons. Coulomb died on August 23, 1806 in Paris, France.

1. Charles Coulomb (1736-1806), Biographies, www.brookscole.com/physics_d/templates/ student_resources/003026961X_serway/essays/coulomb.
2. Hewitt, Paul G., John Suchocki, and Leslie A. Hewitt, Conceptual Physical Science, Third Edition, Pearson/Addison Wesley, San Francisco, 2003.
3. Sears, Francis W., Mark W. Zemansky, and Hugh D. Young, College Physics, Seventh Edition, Addison Wesley, New York, 1991.

James Watt (1736-1819)

Made improvements to the steam engine

James Watt, the son of a merchant, was on January 19, 1736 in Greenock, a seaport on the Firth of Clyde, Scotland. His father was a shipwright, ship owner and contractor.

Watt had little formal education, mostly schooled at home by his mother. Early on, he developed an interest in trying to make things work better. In 1763, while working for Glasgow University, which gave him accommodation and a workshop, he was asked to repair an early steam engine known as a Newcomen engine, which was very inefficient. Watt conceived some changes to the engine that made it faster, safer, and more fuel-efficient.

Watt went into business with Matthew Boulton, a Birmingham engineer, producing engines based on this new approach. Watt's engines were initially used for pumping water from Cornish tin and copper mines.

Watt continued to experiment, and in 1781 produced a rotary-motion steam engine. Whereas his earlier machine, with its up-and-down pumping action, was ideal for draining mines, this new steam engine could be used to drive many different types of machinery. Later, cotton mills switched to steam. Engineers from all the industrialized countries flocked to see the steam engine factory.

The increased power-to-weight ratio of the new engines also permitted their use for marine propulsion. In 1788, a steam-powered catamaran was taken across Dalswinton loch by William Symington.

Watt was also a renowned civil engineer, making several surveys of canal routes. In 1976, he invented an attachment that adapted telescopes for use in measurement of distances on Earth. He also coined the term "horsepower."

Watt was a fellow of the Royal Society of London, the Royal Society of Edinburgh, the Batavian Society, and one of only eight Foreign Associates of the French Academy of Sciences.

When Watt died on August 19, 1819 he was a very rich man. In Watt's honor, the unit of power (1 **Watt** = 1 joule/s) is named after him. His improvements to the steam engine were fundamental to the changes brought by the Industrial Revolution.

1. Hewitt, Paul G., John Suchocki, and Leslie A. Hewitt, Conceptual Physical Science, Third Edition, Pearson/Addison Wesley, San Francisco, 2003.
2. James Watt, Science in the streets, http://level2.phys.strath.ac.uk/ScienceOnStreets/jameswatt.
3. James Watt, Spartacus Educational, www.spartacus.schoolnet.co.uk/SCwatt, December, 2007.
4. James Watt, The Great Idea Finder, www.ideafinder.com/history/inventors/watt.
5. Sears, Francis W., Mark W. Zemansky, and Hugh D. Young, College Physics, Seventh Edition, Addison Wesley, New York, 1991.

Joseph Lagrange (1736-1813)

Developed new methods of analytical mechanics

Joseph Louis Lagrange, although of French extraction, was born in Turin, Sardinia-Piedmont, Italy on January 25, 1736. His father, who had charge of the Kingdom of Sardinia's military chest, was of good social position and wealthy, but before Joseph grew up he had lost most of his property in speculations, and young Lagrange had to rely on his own abilities. As a result, Joseph was largely self-taught, and did not have the benefit of studying with leading mathematicians.

In 1754, at the age of 19, Lagrange was appointed professor of geometry in the Royal School of Artillery. The following year, he sent Leonhard Euler (1707-1783) a better solution than Euler's own for deriving the central equation in the calculus of variations. These solutions, and Lagrange's applications of them to celestial mechanics, were so monumental that by age 25 Lagrange was regarded by many of his contemporaries as the greatest living mathematician. In spite of his fame, Lagrange was always a shy and modest man.

With the aid of the Marquis de Saluces and the anatomist G. F. Cigna, Lagrange founded in 1758 a society which became the Turin Academy of Sciences.

In 1776, on the recommendation of Euler, Lagrange was chosen to succeed Euler as the director of the Berlin Academy. During his stay in Berlin, Euler's work covered many topics, including astronomy, the stability of the solar system, mechanics, dynamics, fluid mechanics, probability, the theory of numbers, and the foundations of the calculus.

After 20 years in Berlin, Lagrange moved to Paris. Napoleon was a great admirer of Lagrange, and showered him with honors—count, senator, and Legion of Honor.

One of Lagrange's most famous works is a memoir, *Mecanique Analytique*, in which he reduced the theory of mechanics to a few general formulas from which all other necessary equations could be derived.

Lagrange also established the theory of differential equations, and invented the method of solving differential equations known as variation of parameters. He died on April 10, 1813 in Paris France.

1. Green, Nick, Lagrange, Joseph (1736 - 1813), About.com, http://n479ad.doubleclick.net/adi/abt.education/education_space;svc=;site=space;t=8;bt=0;bts=1; pc=4;
2. Joseph-Luis Lagrange, Biographies/Images of Physicists and Astronomers, www.mlahanas.de/Physics/Bios/JosephLouisLagrange.
3. Joseph-Luis Lagrange, www.stetson.edu/~efriedma/periodictable/html/Lr.
4. Seikala, Nahla, Joseph-Louis Lagrange, Mathematicians, http://math.berkeley.edu/ ~robin/Lagrange.

Luigi Galvani (1737-1798)

Discovered that muscle and nerve cells produce electricity

Luigi Galvani was an Italian physician and physicist. He was born in Bologna on September 9, 1737. He attended Bologna's medicine school and became a physician like his father. In 1772, Galvani became president of the University of Bologna.

In about 1766, by using electric current delivered by a Leyden jar or a rotating static electricity generator, Galvani began investigating the action of electricity on the muscles of frogs. From observing the twitching in the muscles of frog legs suspended by copper hooks on an iron rail, he invented of the metallic arc. The arc was made of two different metals, such that when one metal was placed in contact with a frog's nerve and the other in contact with a muscle, a contraction occurred. The observation made Galvani the first investigator to appreciate the relationship between electricity and life. This finding provided the basis our current understanding that electrical energy carried by ions is the impetus behind muscle movement.

Galvani is typically credited with the discovery of bioelectricity. Along with contemporaries, he regarded the activation of the frog legs as being generated by an electrical fluid that was carried to the muscles by the nerves. He did not perceive electricity as separable from biology, and believed that animal electricity came from the muscle.

Galvani's associate, Alessandro Volta (1745-1827), reasoned that the animal electricity was a physical phenomenon that was the same as occurs in metals. In Galvani's honor, Volta coined the term **galvanism** for a direct current of electricity produced by chemical action. Volta went on to build the first battery, which became known as the voltaic pile. The **galvanometer** was also named in Galvani's honor.

On Galvani's refusal on religious grounds to take the oath of allegiance to the Cisalpine Republic in 1797, he was removed from his professorship. Deprived of the means of livelihood, he retired to the house of his brother Giacomo, where he soon fell into a feverish decline and died at Bologna on December 4, 1798.

Galvanization, a metallurgical process that is used to coat steel or iron with zinc, is named after Galvani. His report of his investigations of animal electricity were mentioned by Mary Shelley in her novel *Frankenstein*.

1.   Luigi Galvani (1737-1798), Corrosion Doctors, www.corrosion-doctors.org/Biographies/GalvaniBio.
2.   Luigi Galvani, Catholic Encyclopedia, www.newadvent.org/cathen/06371c.
3.   Luigi Galvani, Wikipedia, http://en.wikipedia.org/wiki/Luigi_Galvani.

René Haüy (1743-1822)

The father of modern crystallography

René Just Haüy was born at Saint-Just-en-Chaussée, France on February 28, 1743. His father was a poor weaver, and René owed his early education to the monks of the Premonstratensian Abbey of saint-Just. Their prior sent Haüy to Paris, where he was subsequently admitted to the College of Navarre. Completing his studies there, he was made one of the teaching staff. A few years later he was ordained priest and became Professor at the college of Cardinal Lemoine.

While examining the crystal collection belonging to a friend, M. De France du Croisset, Haüy accidentally dropped a specimen of calc-spar, which broke into pieces. He examined the fragments and was struck by the forms which they assumed. This accident proved the beginning of the exhaustive studies which made him the father of modern crystallography.

Haüy went on to study many specimens, and found that crystals of the same composition possessed the same internal nucleus, even though their external forms differed. He established the law of symmetry, and was able to show that the forms of crystal are perfectly definite and based on fixed laws. He published the mathematical theory of crystals *Traité de minéralogie*, the importance of which was immediately recognized.

Besides his researches in crystallography, Haüy was also one of the pioneers in the development of pyroelectricity, which is the property of certain crystals to produce a state of electric polarity by a change of temperature.

During the Revolution, Haüy refused to take the oath demanded of him. As a result, his papers were seized, his collection of crystals scattered, and he himself imprisoned at the Seminaire de Saint-Firmin. It was only with difficulty that his colleague and former pupil, Geoffroy Saint-Hilaire, was able to induce him to accept the release he had procured for him.

In 1794, Haüy was appointed curator of the Cabinet des Mines, and in the same year he became professor of physics at the Ecole Normale, followed by and appointment to the chair of mineralogy at the Museum of Natural History in Paris.

After the Restoration, Haüy was deprived of his professorship and spent his last days in poverty. He died in Paris on June 3, 1822. His brother, Valentin Haüy, was the founder of the first school for the blind.

1.   Brock, Henry M., Catholic Encyclopedia, www.newadvent.org/cathen/07152a.
2.   René Just Haüy, Wikipedia, http://en.wikipedia.org/wiki/Ren%C3%A9_Just_Ha%C3%BCy.
3.   René-Just Haüy, NNDB, www.nndb.com/people/149/000101843/

Alessandro Volta (1745-1827)

Developed the forerunner of the modern battery

Alessandro Giuseppe Antonio Anastasio Volta was born in Como in northern Italy on February 18, 1745. By the age of 14, he had made up his mind to be a physicist.

In 1774, Volta became professor of physics at the Royal School (high school) in Como, and in the following year he devised the electrophorus, an instrument that produced charges of static electricity.

From 1776 to 1977, Volta discovered methane, and devised experiments, such as the ignition of gases by an electric spark in a closed vessel. He also studied what we now call capacitance, developing separate means to study both electrical potential V and charge Q, and discovered that for a given capacitor V and Q are proportional to each other. This is known as **Volta's Law of Capacitance**

In 1779, Volta became professor of experimental physics at the University of Pavia, a chair he occupied for almost 40 years.

Around 1791, Volta began studying the "animal electricity" noted by Luigi Galvani (1737-1798) when he connected two different metals in series with a frog's leg and to one another. He realized that the frog's leg served as both a conductor of electricity and as a detector of electricity. He replaced the frog's leg by brine-soaked paper, and detected the flow of electricity by means familiar to him from his previous studies of electricity.

By 1800, Volta had developed the voltaic pile, a forerunner of the modern electric battery. It was made of alternating disks of zinc and copper, with each pair separated by brine soaked cloth. Attaching a wire to either end produced a continuous current of low intensity.

Volta's invention of the voltaic pile gave rise to electrochemistry, electromagnetism, and the modern applications of electricity. In honor of his work in the field of electricity, Napoleon made Volta a count in 1810.

Volta died on March 5, 1927, and is buried at Tempio Voltiano. In his honor, potential energy per charge (1 joule/coulomb) is named **volt** (V) after him.

1.   Alessandro Volta, Biographies of Pioneers of Computing, www.thocp.net/ biographies/ volta_alessandro, November 20, 2006.
2.   Alessandro Volta, Biographies/Images of Physicists and Astronomers, www.mlahanas.de/ Physics/Bios/ AlessandroVolta.
3.   Alessandro Volta, History, The Great Idea Finder, www.ideafinder.com/history /inventors/volta.
4.   Alessandro Volta, Wikipedia, http://en.wikipedia.org/wiki/Alessandro_Volta.
5.   Sears, Francis W., Mark W. Zemansky, and Hugh D. Young, College Physics, Seventh Edition, Addison Wesley, New York, 1991.

Jacques Charles (1746-1823)

Discovered that under constant pressure the volume of an ideal gas
increases linearly with the absolute temperature of the gas

Jacques Alexandre César Charles was born in Beaugency Loiret, France
on November 12, 1746. Beginning as a clerk in the finance ministry, he
turned to science and developed several inventions, including a
hydrometer and reflecting goniometer, and improved the Gravesand
heliostat and the Fahrenheit aerometer.

In 1783, realizing that hydrogen was lighter than air, Charles made
the first balloon using hydrogen gas. It was constructed of silk with a
cover of rubber-solution varnish to keep the hydrogen inside. On August
27, 1783, he ascended to a height of nearly 914 meters. One of the
spectators was the American Ambassador to France, Benjamin Franklin
(1706-1790). Upon landing outside of Paris, the balloon was destroyed by
terrified peasants. On December 1, 1783, Charles and Ainé Roberts
ascended to a height of 549 meters.

Around 1787, Charles discovered that under constant pressure, the
volume "V" of an ideal gas increases linearly with the absolute
temperature "T" of the gas:

$$V = kT$$

where "k" is a proportionality constant. This relationship has become
known as **Charles's law**.

When a gas changes volume because of a change in temperature,
Charles's law can be stated:

$$V_1/T_1 = V_2/T_2$$

where "V_1" is the initial volume, "T_1" is the initial temperature, "V_2" is the
final volume, and "T_2" is the final temperature.

Because Gay-Lussac (1778-1850) was the first to publish the above
relationship, giving Charles credit for unpublished data, it was initially
known as the Gay-Lussac-Charles law. Eventually, Gay-Lussac's name
was dropped, and the relationship became known as Charles's law.
Charles died in Paris, France on April 7, 1823.

1. Gas Laws, ThinkQuest, http://library.thinkquest.org/10429/low/gaslaws/gasbody.htm#char..
2. Jacques Alexandre César Charles, http.www.centennialofflight.gov/essay/Dictionary/
    Charles/DI16.htm
3. Kelly, Judith, Jacques Alexandre César Charles , AS Chemists,
    http://pagead2.googlesyndication.com /pagead/ads?client=ca-pub.
4. Sears, Francis W., Mark W. Zemansky, and Hugh D. Young, College Physics, Seventh Edition,
    Addison Wesley, New York, 1991.

Benjamin Thompson (1753-1814)

Investigated the heat of friction

Benjamin Thompson, Count Rumford, was born in rural Woburn, Massachusetts on March 26, 1753. At the age of 13, he was apprenticed to John Appleton, a merchant of nearby Salem. There, he also occupied himself in chemical and mechanical experiments, and in engraving. Later, he began the study of medicine under Dr. John Hayin Worburn at Cambridge, but spent most of his time in manufacturing surgical instruments.

While working with the British armies in America, he conducted experiments concerning the force of gunpowder, the results of which were widely acclaimed. At the conclusion of the war he moved to London. For his services to England, King George III honored Thompson with a knighthood.

Thompson's most important scientific work took place in Munich, Germany, and centered on the nature of heat. While drilling out cannons in the Munich munitions works, he noticed that the canon became hot as long as the friction of boring continued. Furthermore, Thompson observed, the amount of heat released would be sufficient to completely melt the canon if it could be returned to the metal. Since more heat was being released than could have been originally contained in the metal, Thompson's observations were a contradiction to the accepted caloric theory. He therefore concluded that it was the mechanical process of boring which was producing the heat. He so argued in his 1798 publication *An Experimental Enquiry Concerning the Source of the Heat which is Excited by Friction*. This work subsequently became important in establishing the laws of conservation of energy in the 19th century.

Thompson was an active inventor, developing improvements for chimneys and fireplaces, and inventing the double boiler, a kitchen range, and a drip coffeepot. He is also credited with the invention of thermal underwear.

In 1791, Prince Maximilian of Bavaria invited Thompson to enter the civil and military service of that state. For his efforts, Thompson was made a Count of the Holy Roman Empire, and chose the title of Rumford. Thompson died in Paris on August 21, 1814, and is buried in the small cemetery of Auteuil.

1. Benjamin Thompson, Wikipedia, http://en.wikipedia.org/wiki/Benjamin_Thompson.
2. Rumford, Benjamin Thompson (1753-1814, Eric Wesstein's World of Biography, http://scienceworld.wolfram.com/biography/ Rumford.
3. Sir Benjamin Thompson, Count of Rumford, The Robinson Library, www.robinsonlibrary.com/ science/ science/ biography/b-thompson.

Ernst Chladni (1756-1827)

Studied the various modes of vibration in a mechanical surface

Ernst Florens Friedrich Chladni was a physicist and musician. He was born in Wittenberg, Germany on November 30, 1756. He came from an educated family of academics.

Chladni studied law and philosophy in Wittenberg and Leipzig, and obtained a law degree in 1782 from the University of Leipzig. When his father died in 1782, Chladni began his research in physics.

In 1787, Chladni published *Discovery of the Theory of Pitch*. In this and other works he laid the foundation for the study of music within physics. This led to what eventually became known as acoustics. For this, some call him the Father of Acoustics.

Chladni invented a technique to show the various modes of vibration in a mechanical surface. He extended the pioneering experiments of Robert Hooke (1635-1703), who had been able to see the nodal patterns associated with the modes of vibration of glass plates.

In 1787, Chladni published *Discoveries in the Theory of Sound*. In it, he described his technique of drawing a bow over a piece of metal whose surface was lightly covered with sand. The plate was bowed until it reached resonance and the sand formed a pattern showing the nodal regions.

Currently, a loudspeaker driven by an electronic signal generator is placed over or under the plate to achieve a more accurate adjustable frequency. Variations of this technique are commonly used in the design and construction of acoustic instruments, such as violins, guitars, and cellos.

Chladni discovered **Chladni's law**, a simple algebraic relation for approximating the modal frequencies of the free oscillations of plates and other bodies. He also estimated sound velocities in different gases by placing those gases in an organ pipe, playing it, and observing the sounds that emerged.

In 1794, Chladni proposed that meteorites have their origins in outer space. This was a very controversial statement at the time, since meteorites were thought to be of volcanic origin.

Chladni died on April 3, 1827 in Wrocław, Lower Silesia, an area that is now in southwestern Poland.

1. Ernst Chladni, Sacred Sound Tools, http://9waysmysteryschool.tripod.com/ sacredsoundtools/id19.
2. Ernst Chladni, Wikipedia, http://en.wikipedia.org/wiki/Ernst_Chladni.
3. Ernst Florens Friedrich Chladni, Answers.com, www.answers.com/topic/ernst-chladni?cat=technology.

John Dalton (1766-1844)

Showed that the total pressure exerted by a gaseous mixture is equal to the sum of the partial pressures of each individual component

John Dalton was a chemist, meteorologist, and physicist. He was born on September 6, 1766, the son of a Quaker weaver, in a small thatched cottage in the village of Eaglesfield, Cumberland, England.

In 1793, Dalton was appointed teacher of mathematics and physics at Manchester College. He remained in that position until 1800 when the college's worsening financial situation led him to resign his post and begin a new career in Manchester as a private tutor for mathematics and physics.

Dalton's color blindness led him to pursue the cause of the disorder. He incorrectly postulated that shortage in color perception was caused by discoloration of the liquid medium of the eyeball. Nevertheless, color blindness became synomamous with Daltanism.

In 1801, Dalton published a series of papers that culminated in what is now known as **Dalton's law** (or Dalton's law of partial pressures) for ideal gases. It states that the total pressure exerted by a gaseous mixture is equal to the sum of the partial pressures of each individual component in the mixture. Mathematically, this is expressed as

$$P_{total} = p_1 + p_2 + \ldots + p_n$$

The most important of Dalton's investigations are those concerned with the atomic theory in chemistry. In 1803, these investigations led him to the following conclusions: (1) Elements are made of tiny particles called atoms; (2) all atoms of a given element are identical; (3) the atoms of a given element are different from those of any other element; (4) atoms of one element can combine with atoms of other elements to form compounds; (5) a given compound always has the same relative numbers of types of atoms: (6) atoms cannot be created, divided into smaller particles, nor destroyed in the chemical process; and (5) a chemical reaction simply changes the way atoms are grouped together.

Dalton suffered a minor stroke in 1837, and a second one in 1838 left him with a speech impediment, though he remained able to do experiments. On July 27, 1844 Dalton fell from his bed and was found lifeless by his attendant. He was buried in Manchester in Ardwick cemetery.

1.   Dalton's Law of Partial Pressures, ThinkQuest, http://library.thinkquest.org/12596/dalton.
2.   John Dalton, Biographies/Images of Physicists and Astronomers, www.mlahanas.de/Physics/Bios/JohnDalton.
3.   John Dalton, Wikipedia, http://en.wikipedia.org/wiki/John_Dalton.

John Leslie (1766-1832)

Studied radiant heat

John Leslie was born in Largo in Fife, Scotland on April 10, 1766. He was raised and educated there during his early years. Due to his natural aptitude in mathematics Leslie was encouraged to attend the University of St. Andrews.

After his graduation, Leslie spent time working as a private tutor, and working on his first publication, *Natural History of Birds*, which was published in 1793.

In 1805, Leslie was elected to the chair of mathematics at the University of Edinburgh. In 1819, he was promoted to the chair of natural philosophy. He kept this position until his death in 1832. While there, he made a number of contributions to the field of physics, such as the differential thermometer. This type of thermometer usually has a U-shaped tube terminating in two air bulbs, and contains a colored liquid used for indicating the difference between the temperatures to which the two bulbs are exposed. The change of position of the colored fluid occurs due to the different expansions of the air in the bulbs. A graduated scale is attached to one leg of the tube. By adapting to this instrument various devices he was able to do a great variety of investigations, connected especially with photometry, hygroscopy, and the temperature of space.

Leslie also invented the atmometer (or evaporimeter), which is an instrument used for measuring the rate of evaporation to the atmosphere from a wet surface.

Leslie gave the first modern account of capillary action in 1802, and froze water using an air-pump in 1810—the first artificial production of ice.

Leslie is best known for his work involving research on heat. In 1804, he experimented with radiant heat using a cubical vessel filled with boiling water. One side of the cube was composed of highly polished metal, two sides of dull metal (copper), and one side painted black. He showed that radiation was greatest from the black side and negligible from the polished side. The apparatus is known as **Leslie's cube**.

Leslie was knighted in 1832. Later that same year, on November 3, he died at Coates, Scotland.

1.   Atmometer, Wikipedia, http://en.wikipedia.org/wiki/Atmometer.
2.   Horseburrgh, E. M., The Works of Sir John Leslie (1766-1832), www-groups.dcs.st-and.ac.uk/ ~history/ Extras/Leslie_works, April 2007.
3.   John Leslie, Wikipedia, http://en.wikipedia.org/wiki/John_Leslie_(physicist).
4.   Sir John Leslie (1766-1832) Online Encyclopedia, http://encyclopedia.jrank.org/ LEO_LOB/LESLIE_SIR_JOHN_1766_1832_.

Joseph Fourier (1768-1830)

Developed Fourier series for a theory of heat conduction

Jean Baptiste Joseph Fourier was born on March 21, 1768 in Auxerre, France. He was the ninth of twelve children born to a tailor. He was orphaned at age nine. Fourier was recommended to the Bishop of Auxerre, and through this introduction, he was educated by the Benvenistes of the Convent of St. Mark.

Fourier was trained for the priesthood, but instead turned to the life of mathematics. At the age of 16 he became a mathematics teacher at the military school in Auxerre.

Fourier took a prominent part in promoting the French Revolution, and was rewarded by an appointment in 1795 in the École Normale Supérieure, and subsequently by a chair at the École Polytechnique.

In 1798, Fourier followed Napoleon on his Eastern expedition and was given the position of governor of Lower Egypt. In 1801, Fourier returned to France and was named prefect of Grenoble.

Fourier moved to Paris in 1816. In 1822 he published his *Théorie Analytique de la Chaleur*, in which he based his reasoning on Newton's law of cooling, namely that the flow of heat between two adjacent molecules is proportional to the infinitely small difference of their temperatures. In this work he showed that any functions of a variable, whether continuous or discontinuous, can be expanded in a series of sine waves. This process is now known as **Fourier series**.

The **Fourier transform**, which transforms one function into another, is also named in his honor.

Fourier is also credited with the discovery in 1824 that gases in the atmosphere might increase the surface temperature of the Earth. This was the effect that would later be called the greenhouse effect.

Fourier's work provided the impetus for later work on trigonometric series and the theory of functions of a real variable. Fourier died in Paris on May 16, 1830.

Fourier left an unfinished work on determinate equations, which was edited by Claude-Louis Navier (1785-1836) and published in 1831. This work contained much original matter—in particular, a demonstration of Fourier's theorem on the position of the roots of an algebraic equation.

1.   Hewitt, Paul G., John Suchocki, and Leslie A. Hewitt, Conceptual Physical Science, Third Edition, Pearson/Addison Wesley, San Francisco, 2003.
2.   Jean Baptiste Joseph Fourier (1768 - 1830), www.maths.tcd.ie/pub/ HistMath/People/Fourier/ RouseBall/RB_Fourier.
3.   Biographies of mathematicians-Fourier, www.andrews.edu/~calkins/math/biograph/biofouri.

Thomas Seebeck (1770-1821)

Discovered the thermoelectric effect

Thomas Johann Seebeck was born in Reval (today Tallinn, Estonia) to a wealthy Baltic German merchant family on April 9, 1770. He studied at the University of Göttingen and the University of Berlin, and received a medical degree in 1802. He is better known as a physicist than a physician.

Seebeck returned to the University of Berlin around 1818 as a member of the faculty. There, he worked on the magnetization of iron and steel when electrical currents were passed through conductors. In numerous experiments on the magnetizability of various metals, he observed the anomalous reaction of magnetized red-hot iron, which resulted in the phenomenon now known as hysteresis.

In early 1820, Seebeck began to search experimentally for a relation between electricity and heat. In 1821, he joined two wires of dissimilar metals (copper wire and bismuth wire) to form a loop. Two junctions were formed where the ends of the wires were connected to each other. He then discovered that if he heated one junction to a high temperature than the other junction a magnetic field was observed around the circuit. He initially thought this was due to magnetism induced by the temperature difference. However, he quickly realized that an electrical current was being induced. The temperature difference was producing an electric potential which was driving an electric current in a closed circuit. He used the term thermomagnetic currents or thermomagnetism to express his discovery. Today, this phenomenon is known as the Seebeck effect, which is the direct conversion of temperature differences to electric voltage and vice versa.. It is the basis of thermocouples and thermopiles. A thermopile is a device consisting of a number of thermocouples connected in series or parallel, used for measuring temperature or generating current.

The voltage "V" produced in a thermocouple is proportional to the temperature difference $(T_h - T_c)$ between the two junctions:

$$V = \alpha(T_h - T_c)$$

The proportionality constant "α" is known as the **Seebeck coefficient**, and often referred to as the thermoelectric power or thermopower. Seebeck died in Berlin, Germany, on December 10, 1831.

1. Brief History of Thermocouples, www.thermoelectrics.caltech.edu/history_page.
2. Thomas Johann Seebeck, http://chem.ch.huji.ac.il/history/seebeck, June 13, 2003.
3. Thomas Johann Seebeck, Wikipedia, http://en.wikipedia.org/wiki/Thomas_Seebeck.
4. Thomas Johann Seebeck, Institute of Chemistry, http://chem.ch.huji.ac.il/history/seebeck.

Robert Brown (1773-1858)

Observed minute particles within vacuoles in pollen grains executing a continuous jittery motion

Robert Brown, the son of an Episcopalian clergyman, was born on December 21, 1773 in Montrose, Scotland. He studied medicine at the University of Edinburgh, but his real interest was in botony.

On a visit to London in 1798, Brown met the botanist, Sir Joseph Banks (1743-1820), who was president of the Royal Society. Banks allowed him the free use of his library and collections, and recommended Brown to the Admiralty for the post of naturalist aboard the 334 ton ship the Investigator. The ship was to embark on a surveying voyage along the northern and southern coasts of Australia.

For three and a half years he did intensive botanic research in Australia, collecting about 3400 species, of which about 2000 were previously unknown.

In 1805, Brown returned to England where he spent the next five years working on the material he had gathered. In 1810, he published the results of his collecting in *Prodromus Florae Novae Hollandiae et Insulae Van Diemen*, which was the first systematic account of Australian flora.

In 1827, while examining pollen grains and spores of mosses and Equisetum, which were suspended in water under a microscope, Brown observed minute particles within vacuoles in the pollen grains undergoing a continuous jittery motion. He then observed the same motion in particles of dust, allowing him to rule out the possibility that the motion was due to the pollen being a living organism. He did not provide a theory to explain the motion.

Although Jan Ingenhousz (1730-1799) had previously described the phenomenon in 1784 and 1785, it is nevertheless known as **Brownian motion** in honor of Robert Brown. We now understand that Brownian motion (the random movement of microscopic particles suspended in a fluid) is caused by collisions with molecules of the surrounding fluid.

After the division of the Natural History Department of the British Museum into three sections in 1837, Robert Brown became the first Keeper of the Botanical Department, remaining in the position until his death at Soho Square in London on June 10, 1858. He was buried in Kensal Green Cemetery.

1.   Brown, Robert, Microsoft Encarta Encyclopedia, 2003.
2.   Robert Brown (1773-1858), Council of Heads of Australasian Herbaria, www.anbg.gov.au/ biography/ brown-robert.
3.   Robert Brown (Botonist), Wikipedia, http://en.wikipedia.org/wiki/Robert_Brown_(botanist).

Thomas Young (1773-1829)

Demonstrated the wave nature of light

Thomas Young was born on June 13, 1773 in Milverton, Somerset, England. He became a physicist, physician, and Egyptologist. His lectures, while professor of natural philosophy from 1801 to 1803 at the Royal Institution in London, were published in 1807 as *A Course of Lectures on Natural Philosophy and the Mechanical Arts.* In this work, he introduced the modern physical concept of energy.

Young and Hermann von Helmholtz (1821-1894) proposed a trichromatic color theory of color vision, now known as the **Young-Helmholtz theory.** Young studied the structure of the eye and described the defect called astigmatism. He also contributed to the theory of tides, participated in the deciphering of the Rosetta Stone (which provided a key to understanding Egyptian hieroglyphic writings), and established a coefficient of elasticity known as **Young's modulus.** Young's modulus is calculated by dividing the tensile stress by the tensile strain.

Young, in his 1804 essay on the *Cohesion of Fluids*, proposed a theory of capillary phenomena based on the principle of surface tension. He also observed the constancy of the angle of contact of a liquid surface with a solid interface. From these two principles, he showed how to deduce the phenomena of capillary action, which is described by the **Young-Laplace equation,** independently discovered by Pierre-Simon de Laplace (1749-1847) in 1805.

In 1801, Young demonstrated the wave nature of light, as opposed to the corpuscular theory, with his interference experiment. He found that light directed through two closely spaced pin holes recombined to produce fringes of brightness and darkness on a screen behind. Bright fringes of light resulted from light waves from the two holes arriving peak to peak, and thus reinforcing each other. The dark areas resulted from light waves arriving trough to peak, and thus canceling each other.

In 1817, Young proposed that light waves are transverse (perpendicular to the direction of travel), rather than longitudinal (in the direction of travel). This explained polarization, which is the alignment of light waves to vibrate in the same plane. Young died in London on May 10, 1829.

1. Hewitt, Paul G., John Suchocki, and Leslie A. Hewitt, Conceptual Physical Science, Third Edition, Pearson/Addison Wesley, San Francisco, 2003.
2. Sears, Francis W., Mark W. Zemansky, and Hugh D. Young, College Physics, Seventh Edition, Addison Wesley, New York, 1991.
3. Thomas Young, www.geog.ucsb.edu/~jeff/115a/history/young.
4. Young, Thomas, Infoplease, www.infoplease.com/ce6/people/A0853151.

Jean-Baptiste Biot (1774-1862)

Helped develop the law which relates the strength of a magnetic field to the current flow in the wire that causes it

Jean-Baptiste Biot was born in Paris, France on April 21, 1774. After serving for a short time in the artillery, he was appointed in 1797 professor of mathematics at the École Centrale at Beauvais.

In 1800, Biout became Professor of Mathematical Physics at the Collège de France. His 1803 report on a meteorite fall convinced scientists for the first time that rocks fall from the sky.

In 1804, Biot ascended in a balloon with Joseph Louis Gay-Lussac (1778-1850) for the purpose of studying the magnetic, electrical, and chemical condition of the atmosphere at various elevations. They rose to a height of 13,000 feet. In 1809, Biot was appointed Professor of Physical Astronomy at the Faculty of Sciences.

Biot studied a wide range of mathematical topics, mostly in the applied mathematics area. He made advances in astronomy, elasticity, electricity and magnetism, heat, and optics. In pure mathematics, he did important work in geometry. He collaborated with François Arago (1786-1853) on refractive properties of gases.

Biot discovered the laws of rotary polarization by crystalline bodies, and applied these laws to the analysis of saccharine solutions. His fame rests chiefly on his work in polarization and double refraction of light.

Biot and Felix Savart (1791-1841) formulated the Biot-Savart law, which states that the intensity of a magnetic field set up by a current flowing through a wire varies inversely with the square of the distance from the wire. This is expressed mathematically as:

$$d\mathbf{B} = K_m \frac{I d\mathbf{l} \times \hat{\mathbf{r}}}{r^2}$$

where $K_m = \mu_0/4\pi$ and "μ_0" is the magnetic constant, "I" is the current, "dl" is the differential length vector of the current element, "$\hat{\mathbf{r}}$" is the unit displacement vector from the current element to the field point, and "r" is the distance from the current element to the field point. Biot died in Paris on February 3, 1862.

Biot was the first to discover the unique optical properties of mica, and therefore the mica-based mineral **biotite** was named after him.

1.   Jean-Babtiste Biot, NNDB, www.nndb.com/people/895/000100595/.
2.   Jean-Baptiste Biot, Catholic Encyclopedia, www.newadvent.org/cathen/02576a.
3.   Jean-Baptiste Biot, www.cartage.org.lb/en/themes/sciences/Physics/ Electromagnetism/ Magnetostatics/Currentsmagnetism/JeanBaptiste/JeanBaptiste.

André Ampère (1775-1836)

Founded the field of electromagnetism

André Marie Ampère was born in near Lyon, France on January 20, 1775. As a child, he voraciously consumed books of history, geography, literature, philosophy, and the natural sciences. His father, a successful businessman and a Lyon city official, encouraged Ampère to pursue his passion for mathematics.

In 1801, Ampère became professor of physics and chemistry at Bourg-en-Bresse. In 1809, he was appointed professor of mathematics at the Ecole Polytechnique, and held posts there until he was appointed to a chair at Université de France in 1826. He held that position until his death.

In Paris, Ampère worked on a wide variety of topics in mathematics, physics, and chemistry. He is best known for his important contributions to the study of what now is known as electromagnetism.

On September 11, 1820 Ampère learned of the discovery of Hans Ørsted (1777-1851) that a magnetic needle is acted on by a electrical current. A week later, on September 18, Ampère presented a far more complete paper to the Academy on the science of electromagnetism, or as he called it, electrodynamics. He formulated a combined theory of electricity and magnetism, and proposed that electric currents are the source of all magnetism.

Ampère's electrodynamic theory and his views on the relationship of electricity and magnetism were published in his *Recueil d'observations électrodynamiques* (Collection of Observations on Electrodynamics) in 1822, and in his *Théorie des phénomènes électrodynamiques* (Theory of Electrodynamic Phenomena) was published in 1826.

Ampère invented the astatic needle, a critical component of the modern astatic galvanometer, and was the first to demonstrate that a magnetic field is generated when two parallel wires are charged with electricity.

Ampère died June 10, 1836 in Marseilles, France and was buried in the Montmartre Cemetery in Paris. He is generally credited as one of the first to discover electromagnetism. The **ampere** (A), the unit of electric current, was named in honor of Ampère. One ampere is one coulomb/second.

1.   André-Marie Ampère, Encyclopedia Britannica Online, www.britannica.com/ eb/article-9007234/Andre-Marie-Ampere.
2.   André-Marie Ampère, www.rare-earth-magnets.com/magnet_university/ andre_marie_ampere.
3.   Hewitt, Paul G., John Suchocki, and Leslie A. Hewitt, Conceptual Physical Science, Third Edition, Pearson/Addison Wesley, San Francisco, 2003.

Étienne Malus (1775-1812)

Showed that intensity of transmitted light depends on the relative orientation between the polarization direction of the incoming light and the polarization axis of the filter

Étienne Louis Malus was born in Paris, France on July 23, 1775. His father, Louis Malus de Mitry, was Treasurer of France. Malus attended the engineering school, École Royale de Genie at Mézières, but he was dismissed without receiving a commission, and was obliged to enter the army as a private soldier.

While in the army, Malus was sent to École Polytechnique. where he was taught by Jean Baptiste Fourier (1768-1830). After three years at the École, he was admitted into the corps of engineers, and served in the army of the Sambre and Meuse. He was present at the passage of the Rhine in 1797 and at the affairs of Ukratz and Altenkirch. Then, from 1798 to 1801 he participated in Napoleon's expedition into Egypt.

Malus's mathematical work was almost entirely concerned with light. He studied ray systems, and conducted experiments to verify the theories of light as stated by Christiaan Huygens (1629-1695). He rewrote the theory in analytical form. Malus's discovery of the polarization of light by reflection was published in 1809, and his theory of double refraction of light in crystals in 1810.

For water, Malus identified the relationship between the polarizing angle of reflection and the refractive index of the reflecting material. He wasn't unable to do so for glass because of the low quality of the glass available to him. It wasn't until 1815 that David Brewster (1781-1868) was able to experiment with higher quality glass and arrived at Brewster's law.

Malus is best remembered for **Malus's law**, which states that when a perfect polarizer is placed in a polarized beam of light, the intensity "I" of the light that passes through is given by:

$$I = I_0 \cos^2 \theta_i$$

where "I_0" is the initial intensity and "θ_i" is the angle between the light's initial plane of polarization and the axis of the polarizer. Malus died of tuberculosis in Paris France on February 24, 1812.

1. Étienne-Louis Malus, Wikipedia, http://en.wikipedia.org/wiki/Etienne-Louis_Malus.
2. J O'Connor, J. J. and E F Robertson, Étienne-Louis Malus, MacTutor, ://www-history.mcs.st-andrews.ac.uk/Printonly/Malus, January 1997.
3. Malus's Law, Wikipedia, http://en.wikipedia.org/wiki/Malus%27s_law#Malus.27_law_and_other_properties.

Amedeo Avogadro (1776-1856)

Developed the hypothesis that gases at the same volume, pressure, and
temperature contain the same number of molecules

Lorenzo Romano Amedeo Carlo Avogadro was born in Turin, Italy on
August 9, 1776. His father, Philippe Avogadro, was a distinguished lawyer
and civil servant. In 1796, Amedeo earned his doctorate in ecclesiastical
law and began to practice.

In spite of a successful legal career, Avogadro developed an interest
mathematics and physics, and in 1800 began private studies.

In 1806, Avogadro was appointed demonstrator at the Academy of
Turin, and in 1809 became professor of natural philosophy at the College
of Vercelli.

In 1811, Avogadro published an article entitled *Essai D'une Manière
de Déterminer Les Masses Relatives des Molécules Élémentaires des
Corps, et Les Proportions Selon Lesquelles Elles Entrent Dans Ces
Combinaisons* (Essay on Determining the Relative Masses of the
Elementary Molecules of Bodies). In this article, Avogadro hypothesized
that equal volumes of gases at the same temperature and pressure contain
equal numbers of molecules.

Avogadro reasoned that simple gases were not formed of solitary
atoms, but were instead compound molecules of two or more atoms. He
did not actually use the word atom. He talked about three kinds of
"molecules," including an "elementary molecule"—what we now call an
atom. Avogadro's hypothesis was not accepted for half a century after it
was first published.

In 1820, when the first chair of mathematical physics in Italy was
established at the University of Turin, Avogadro was appointed to the
position. The post was short lived because political changes suppressed the
chair, and in 1822 Avogadro was out of a job.

The chair was at the University of Turin was eventually reestablished
in 1832, and Avogadro was reappointed to the position in 1834. He
remained there until his retirement in 1850.

In honor of Avogadro's contributions to the theory of molarity and
molecular weights, the number of molecules in one mole was renamed
Avogadro's number. It is approximately 6.0221415×10^{23}.

1.   Amedeo Avogadro, Chemical Achievers, www.chemheritage.org/classroom/chemach /periodic/
     avogadro.
2.   Hewitt, Paul G., John Suchocki, and Leslie A. Hewitt, Conceptual Physical Science, Third
     Edition, Pearson/Addison Wesley, San Francisco, 2003.
3.   Johnson, Chris, www.bulldog.u-net.com/avogadro/avoga, July 4, 2004.

Friedrich Gauss (1777 -1855)

Developed an alternative formulation to Coulomb's law of the relationship between electric charge and electric field

Johann Carl Friedrich Gauss was born on April 30, 1777 in Brunswick (now Germany). In 1795, Gauss left Brunswick to study at Göttingen University. There, he made one of his most important discoveries—the construction of a regular 17-gon by ruler and compasses. This was the most major advance in this field since the time of Greek mathematics. The work was published in *Disquisitiones Arithmeticae*. It was considered a masterpiece that took the theory of numbers far beyond its earlier state, and established Gauss as a mathematical genius.

In 1807, Gauss took the position of director of the Göttingen observatory, where he remained the rest of his life. He published his second book, *Theoria Motus Corporum Coelestium in Sectionibus Conicis Solem Ambientium*, in 1809. It was a major two volume treatise on the motion of celestial bodies. In the first volume Gauss discussed differential equations, conic sections, and elliptic orbits, while in the second volume he showed how to estimate and then to refine the estimation of a planet's orbit.

Gauss is best known for his development of an alternative formulation to Coulomb's law of the relationship between electric charge and electric field, now known as **Gauss's law**. Gauss's law is expressed mathematically as:

$$\Phi = \oint_S \mathbf{E} \cdot d\mathbf{A} = \frac{1}{\varepsilon_0} \int_V \rho \, dV = \frac{Q_A}{\varepsilon_0}$$

where "Φ" is the electric flux, "E" is the electric field, "dA" is a differential area on the closed surface "S," "Q_A" is the charge enclosed by the surface, "ρ" is the charge density at a point in the volume "V," "ε_0" is the permittivity of free space. E.dA is integrated over the entire surface "S" enclosing volume "V."

Gauss also invented the heliotrope, the magnetometer, the photometer, and, some 5 years before Samuel Morse, the telegraph.

Gauss died on died February 23, 1855 in Göttingen Hanover (now Germany).

1. Carl Friedrich Gauss, Biography, www.swlearning.com/quant/kohler/stat/ biographical_sketches/bio10.1.
2. Johann Carl Friedrich Gauss, MacTutor, www-groups.dcs.st-and.ac.uk/~history/Biographies/ Gauss.
3. Lectures 5 and 6 E-field and Gauss's Law, PHY205 Electromagnetism, www.shef.ac.uk/ physics/teaching/phy205/ lectures_5_and_6.

Hans Ørsted (1777-1851)

Discovered that a current in a wire can produce magnet effects

Hans Christian Ørsted was born on August 14, 1777 at Rudkøbing on the Danish island of Langeland, where his father practiced as an apothecary. In 1794, Ørsted entered the University of Copenhagen, where he earned his Ph.D. in 1799.

Soon after receiving his Ph.D., Ørsted became assistant to the professor of medicine at the same school. In 1806, he was appointed professor of natural philosophy in the University of Copenhagen.

Before 1820, the only magnetism known was that of iron magnets and lodestones, and the subjects of electricity and magnetism where taught separately. One evening, Ørsted arranged in his home a science demonstration to friends and students. He planned to demonstrate the heating of a wire by an electric current, and also to carry out demonstrations of magnetism, for which he had a compass needle mounted on a wooden stand.

While performing his electric demonstration, Ørsted noted that every time the electric current was switched on, the compass needle moved, thus showing, quite by accident, that electricity and magnetism are related.

During residence in Berlin, Ørsted wrote an essay in which he first developed the ideas on which were based his discovery of the connection existing between electricity and magnetism. As a result, he became regarded as the originator of the new science of electromagnetism. His discovery obtained for Ørsted the Copley Medal from the Royal Society of England and the principal mathematical prize in the gift of the Institute of Paris.

Five years after discovering electromagnetism, Ørsted became the first person to isolate the element aluminum. Although aluminum is one of the most plentiful elements on Earth, it is always combined with other elements. Many chemists who came before Ørsted thought it existed as a separate identity, but failed in their attempts to uncover it.

The CGS unit of magnetic induction (**oersted**) is named in honor of Ørsted's contributions to the field of electromagnetism. He died on March 8, 1851 in Copenhagen Denmark.

1. Hans Christian Oersted, www.rare-earth-magnets.com/magnet_university/hans_christian_orsted.
2. Hans Christian Ørsted, Biographies/Images of Physicists and Astronomers, www.mlahanas.de/Physics/Bios/HansChristianOersted.
3. Sears, Francis W., Mark W. Zemansky, and Hugh D. Young, College Physics, Seventh Edition, Addison Wesley, New York, 1991.
4. Stern, David P., Oersted and Ampere Link, electricity and magnetism, www.phy6.org/earthmag/oersted, November 25, 2001.

Joseph Gay-Lussac (1778-1850)

Discoved two laws related to gases

Joseph Louis Gay-Lussac was born in Saint Léonard on December 6, 1778. He was educated at the École Polytechnique and the École des Ponts et Chaussées in Paris. In 1809, he became professor of chemistry at the École Polytechnique. From 1808 to 1832 he was professor of physics at the Sorbonne, a post which he resigned for the chair of chemistry at the Jardin des Plantes.

In 1804, Gay-Lussac made two balloon ascensions to study magnetic forces and to observe the composition and temperature of the air at different altitudes. Jean-Baptiste Biot (1774-1862) accompanied him on the first ascent. On the second ascent, a height of about 7,016 meters was reached. Gay-Lussac collected samples of the air at different heights to record differences in temperature and moisture.

In 1805, Gay-Lussac and Alexander von Humboldt (1769-1859) discovered that the basic composition of the atmosphere does not change with increasing altitude. They also discovered that water is formed by two parts of hydrogen and one part of oxygen by volume.

Gay-Lussac is best known for his formulation of two laws, which relate to the properties of gases. The first law, known as **Gay-Lussac's law**, was discovered in 1802. It states that the pressure of a fixed amount of gas at fixed volume is directly proportional to its temperature:

$$P = kT$$

where "P" is the pressure of the gas, "T" is it's temperature in kelvins, and "k" is a proportionality constant.

Gay-Lussac's second law, known as the law of combining volumes, states that the ratio between the combining volumes of gases and the product, if gaseous, can be expressed in small whole numbers. He discovered this law in 1809.

In 1811, Amadeo Avogadro (1776-1856) used Gay-Lussac's Law to form Avogadro's hypothesis about the number of molecules in a mole of gas.

Gay-Lussac died in Paris on May 9, 1950, and was buried at the Père Lachaise cemetery in Paris.

1. Gay Lussac's Law, Wikipedia, http://en.wikipedia.org/wiki/Gay-Lussac's_law.
2. Gay-Lussac, Joseph Louis, Infoplease, www.infoplease.com/ce6/people/A0820366.
3. Gay-Lussac's Law, Wikipedia, http://en.wikipedia.org/wiki/Gay-Lussac's_law.
4. Joseph Louis Gay-Lussac, Chemical Achievers, ://www.chemheritage.org/classroom/ chemach/ gases/gay-lussac.

David Brewster (1781-1868)

Discovered the angle of incidence that produces completely polarized
reflected light

David Brewster was born in Jedburgh, an obscure country town in the
midst of the Scottish lowlands, on December 11, 1781. His family decided
he should study for the ministry of the Church of Scotland. So, at the age
of 12, he was consigned to the University of Edinburgh, and at age 19
awarded an honorary Master of Arts degree. This carried with it a license
to preach the gospel as a minister of the Scottish Established Church.

The first day Brewster mounted the pulpit was the last, for when he
saw a congregation eyeing him he vowed never to do that job again. So, in
1801 Brewster turned his talents to two of his life-long interests—the
study of optics and the development of scientific instruments.

Brewster's theory on the polarization of light is explained as follows:
Unpolarized light can be partially polarized by reflection. When it strikes a
reflecting surface between two optical materials, such as air and water,
preferential reflection, at one particular angle of incidence, called the
polarizing angle θ_B, no light is reflected except that in which the E vector
is perpendicular to the plane of incidence.

When light is incident at the polarizing angle, none of the component
parallel to the plane of incidence is reflected; this component is 100
percent polarized. Brewster noticed that when the angle of incidence is
equal to the polarizing angle "θ_B," the reflected and the refracted ray are
perpendicular to each other. This is known as **Brewster's law,** and is
expressed mathematically as:

$$\theta_B = \arctan\left(\frac{n_2}{n_1}\right)$$

where "n_1" and "n_2" are the indices of refraction of the two media.

For his theory, Brewster was admitted to the Royal Society of London,
and was later awarded the Rumford gold and silver medal.

Brewster is probably best known for his invention of the
Kaleidoscope, which means "beautiful form viewer."

In 1832, Brewster was knighted by William IV. He died on February
10, 1868.

1. Baker, Cozy, Sir David Brewster, Kaliedoscopes, www.brewstersociety.com/brewster_bio.
2. David Brewster, pourlemerite.org/peace, www.pourlemerite.org/peace/davidbrewster.
3. Sears, Francis W., Mark W. Zemansky, and Hugh D. Young, College Physics, Seventh Edition,
    Addison Wesley, New York, 1991.

Siméon Poisson (1781-1840)

Mathematically formulated the distribution of electric charges on the surface of conductors

Siméon Denis Poisson was born at Pithiviers, France on June 21, 1781. His father, Siméon Poisson, occupied a small administrative post at Pithiviers. Poisson studied at the École Polytechnique in Palaiseau, and was strongly influenced by the French mathematicians Joseph Louis Lagrange (1736-1813) and Pierre Simon Laplace (1749-1827).

In 1802, Poisson became the assistant of French mathematician Jean Baptiste Fourier (1768-1830), whose Chair he assumed in 1808. Later Poisson became the first professor of mechanics at the Sorbonne.

Poisson is best known for his contributions to theories of electricity and magnetism. However, he published extensively on other topics, including the calculus of variations, differential geometry, and probability theory.

The **Poisson distribution** is a special case of the binomial distribution in statistics. It expresses the probability of a number of events occurring in a fixed period of time if the events occur with a known average rate, and are independent of the time since the last event.

In his work on electricity, published in 1812, Poisson adopted the same two-fluid model of electricity that French physicist Charles Coulomb (1736-1806) had used (like fluids repel and unlike attract according to the inverse-square law). According to this model, a body became electrified, either positively or negatively, when the uniform distribution of both fluids is disturbed. He then used the potential function of Joseph Lagrange (1736-1813) to mathematically formulate the distribution of electric charges on the surface of conductors. In 1824, he demonstrated that these formulations were equally applicable to magnetism.

In astronomy, Poisson worked on the mathematics of the motion of the Moon and the dependence of the force of gravity on the distribution of mass within a planet. He also studied the theory of elasticity.

Also named for Poisson are **Poisson's integral**, **Poisson brackets** in differential equations, **Poisson's ratio** in elasticity, and **Poisson's constant** in electricity. Poisson published between 300 and 400 mathematical works in all. Despite this exceptionally large output, he was not highly regarded by other French mathematicians during his life. Poisson died on April 25, 1840 in Sceaux (near Paris), France.

1. O'Connor, J. J. and E F Robertson, Siméon Denis Poisson, MacTutor, www-groups.dcs.st-and.ac.uk/~history/Printonly/Poisson.
2. Poisson, Siméon Denis, Microsoft Encarta Encyclopedia, 2003.
3. Siméon Denis Poisson, NNDB, www.nndb.com/people/857/000093578.

William Sturgeon (1783-1850)

Made the first electromagnets and invented the first practical electric motor

William Sturgeon was born near Kirkby Lonsdale, North Lancashire, England on May 22, 1783. His father was a shoemaker who has been described as neglecting his family. After the death of his mother when he was 10 years old, William was apprenticed to a shoemaker. In 1802, he ran away and joined the army, and taught himself mathematics and physics. In 1820, at the age of 37, Sturgeon left the army and opened a boot-making business in Woolwich, where he studied physics at night and occasionally lectured on the subject.

In 1825, Sturgeon exhibited his first electromagnet, a seven-ounce piece of iron wrapped with wire through which a current from a single battery flowed. He displayed its power by having it lift a nine pound weight.

In 1832, Sturgeon invented the commutator for electric motors. That same year, he was appointed to the lecturing staff of the Adelaide Gallery of Practical Science in London, where he first demonstrated a DC electric motor incorporating his commutator. This device led to the invention of the telegraph, the AC electric motor, and numerous other devices basic to modern technology.

In 1836, Sturgeon made the first moving-coil galvanometer. He also improved the voltaic battery and worked on the theory of thermoelectricity. From more than 500 kite observations, he established that in serene weather the atmosphere is invariably charged positively with respect to the Earth, becoming more positive with increasing altitude.

In 1840, Sturgeon became superintendent of the Royal Victoria Gallery of Practical Science in Manchester. There he improved the solenoid by varnishing the iron to insulate it from the wound wires, and bent it into horseshoe shape so that each coil reinforced the next coil because they formed parallel wires with the current moving in the same direction.

Also in 1840, Sturgeon improved the cell devised by Alessandro Volta (1745-1827), and developed a longer lasting battery. The Gallery closed in 1842, and afterward Sturgeon earned a living by lecturing and demonstrating his inventions. He died in Prestwich, England on December 4, 1850, and is buried under a simple stone in the churchyard of the Parish Church of Saint Mary's.

1. The Electromagnet, Electricity, http://physics.kenyon.edu/EarlyApparatus/Electricity/ Electromagnet/ Electromagnet.
2. William Sturgeon, Institute of Chemistry, http://chem.ch.huji.ac.il/history/sturgeon.
3. William Sturgeon, Wikipedia, http://en.wikipedia.org/wiki/William_Sturgeon.

Leopoldo Nobili (1784-1835)

Invented a thermopile and the astatic galvanometer

Leopoldo Nobili was born in Trassilico, Garfagnana, Italy in 1784. After attending the military academy at Modena, he became an artillery officer. He was awarded the Légion d'honneur for his service in Napoleon's invasion of Russia.

Nobili carried out early research into electrochemistry and thermoelectricity, and is best known for inventing a number of instruments critical to investigating thermodynamics and electrochemistry. He is regarded as one of the Italian pioneers in the field of electromagnetism.

Nobili was appointed professor of physics at the Regal Museum of Physics and Natural History in Florence, where he worked with Vincenzo Antinori (1792-1865) on electromagnetic induction, which had been recently discovered by Michael Faraday (1791-1867).

In 1826, Nobili described the prismatically-colored films of metal, known as **Nobili's' rings**, deposited electrolytically from solutions of lead and other salts when the anode is a polished iron plate and the cathode is a fine wire placed vertically above it.

The first galvanometers, called simple galvanometers, were not shielded from terrestrial magnetic field, so when the electric current was running, two magnetic fields acted on the needle—the magnetic field produced by the instrument's electromagnet and the terrestrial one. The terrestrial magnetic field caused an error in measurement. Nobili provided the galvanometer with a system to remove the perturbation produced by the terrestrial magnetic field, thus developing the astatic galvanometer.

Nobili also invented a thermopile used to measure radiant heat in the form of infrared radiation. A thermopile is a device consisting of a number of thermocouples connected in series or parallel and used for measuring temperature or generating current. Nobili's thermopile made use of the thermoelectric effect, which is the direct conversion of temperature differences to electric voltage and vice versa.

Nobili connected his thermopile in series to the clips of the astatic galvanometer. He called it a thermomultiplier or electric thermoscope. For most of the 19th century, the thermomultiplier proved to be an irreplaceable instrument in the study of thermic radiation because of its high reactivity and quickness.

Nobili died in Florence, Italy in 1835.

1. Leopoldo Nobili, Institute of Chemistry, http://chem.ch.huji.ac.il/history/nobili.
2. Leopoldo Nobili, Wikipedia, http://en.wikipedia.org/wiki/Leopoldo_Nobili.

Louis Navier (1785-1836)

Developed the partial differential equations that describe the flow of non-turbulent, Newtonian fluids

Claude Louis Navier was born on February 10, 1785 in Dijon, France. In 1802, he enrolled at the École polytechnique, and in 1804 continued his studies at the École Nationale des Ponts et Chaussées, from which he graduated in 1806.

Navier eventually succeeded his uncle, Emiland Gauthey, as Inspecteur general at the Corps of Bridges and Roads. He directed the construction of bridges at Choisy, Asnières, and Argenteuil in the Department of the Seine, and built a footbridge to the Île de la Cité in Paris.

In 1830, Navier was appointed professor at the École Nationale des Ponts et Chaussées, and in the following year became professor of calculus and mechanics at the École polytechnique.

In 1819, Navier succeeded in determining the zero line of mechanical stress, finally correcting Galileo's incorrect results. In 1821, he formulated the general theory of elasticity in a mathematically usable form, making it available to the field of construction. And in 1826 he established the elastic modulus as a property of materials independent of the second moment of area. Navier is considered by many to be the founder of modern structural analysis.

Navier's major contribution is considered to be the **Navier-Stokes equations**, named for Navier and George Gabriel Stokes (1819-1903). The equations are the fundamental partial differential equations that describe the flow of non-turbulent, Newtonian fluids. They are derived by applying Newton's laws of motion. They establish that changes in momentum in infinitesimal volumes of fluid are simply the sum of dissipative viscous forces, changes in pressure, gravity, and other forces acting inside the fluid

The Navier-Stokes equations are central to fluid mechanics. They can be use to describe the physics of a large number of phenomena of academic and economic interest, including weather, ocean currents, water flow in a pipe, flow around an airfoil, and motion of stars inside a galaxy. They are used in the design of aircraft and cars, the study of blood flow, the design of power stations, and the analysis of the effects of pollution.

Navier died in Paris on August 21, 1836.

1. Claud-Louis Navier, Wikipedia, http://en.wikipedia.org/wiki/Claude-Louis_Navier.
2. Navier-Stokes Equations, Answers.com, ://www.answers.com/topic/navier-stokes-equations?cat=technology
3. Navier-Stokes equations, efunda, www.efunda.com/formulae/fluids/navier_stokes.cfm.

Joseph von Fraunhofer (1787-1826)

Invented the spectrascope and discovered what later was identified as atomic absorption spectra

Joseph von Fraunhofer was born in Straubing, Bavaria on March 6, 1787. He became an orphan at the age of 12, and began working as an apprentice to a harsh glassmaker. When the workshop physically collapsed, Joseph was resqued by an operation led by Maximilian IV, who then provided Joseph with books and forced his employer to allow him time to study.

After eight months of studying, Fraunhofer went to work at the Optical Institute at Benediktbeuern, a Benedictine monastery devoted to glass making. There, he learned how to make the world's finest optical glass and invented very precise methods for measuring dispersion. In 1818, he became the director of the Institute.

In 1814, Fraunhofer invented the spectroscope, which is an optical instrument for producing spectral lines and measuring their wavelengths and intensities. Using his new invention, he discovered 574 dark lines appearing in the solar spectrum. These were later shown to be atomic absorption lines. The lines are still called **Fraunhofer lines** in his honor.

Using the spectrscope, Fraunhofer found that the spectra of Sirius and other first-magnitude stars differed from each other and from the Sun, thus founding stellar spectroscopy. He labeled the most prominent spectral lines with letters, establishing a nomenclature that is still used to this day.

Fraunhofer also invented the diffraction grating, comprised of 260 close parallel wires. He used his diffraction grating to accurately measure wavelengths of specific colors and dark lines in the solar spectrum. In doing so, he transformed spectroscopy from a qualitative art to a quantitative science.

Fraunhofer diffraction is a form of wave diffraction which occurs when field waves are passed through an aperture or slit, causing only the size of an observed aperture image to change.

In 1817, Fraunhofer designed an achromatic objective lense. With minor modifications, his design is still in use today. In 1822, the University of Erlangen honored Fraunhofer by giving him an honorary doctorate.

Fraunhofer died of tuberculosis in Munich, Germany on June 7, 1826 at the age of 39.

1. Fox, William, Catholic Encyclopedia, www.newadvent.org/cathen/06250a.
2. Joseph von Fraunhofer (1787-1826), High Altitude Observatory, www.hao.ucar.edu/Public/ education/bios/ fraunhofer.
3. Joseph von Fraunhofer, Wikipedia, http://en.wikipedia.org/wiki/Joseph_von_Fraunhofer.

Jean Fresnel (1788-1827)

Showed the transverse nature of light waves

Augustin Jean Fresnel, the son of an architect, was born at Broglie near Bernay, Normandy on May 10, 1788. His early progress in learning was slow, and he still could not read when he was eight years old.

Fresnel was educated at the École Polytechnique, and then the École des Ponts et Chaussées.

Fresnel's chosen field of research was optics, and his analysis of interference, diffraction, and polarization turned the wave theory of light into an integral part of physical science.

At the onset of the 19th century, the most widely accepted belief among physicists was that light was a particle that traveled through ether, an invisible substance that made up the heavens. Soon after the turn of the century, however, came the revival of an old belief—that light was actually a wave that needed no medium for travel.

A question to be answered by the wave theory of light was the apparent failure of light waves to bend around a corner or edge, which is at complete variance with the behavior of water waves and sound waves. Fresnel set out to resolve this problem. According to his theory, the longitudinal light waves assumed by previous investigators, such as Thomas Young (1773-1828), were actually transverse waves. Using this information Fresnel showed that, under the wave theory, a slight bending does occur, and that it manifested itself in a succession of dark and bright bands at the edge of the shadow. Most importantly, Fresnel's mathematical formalism of the wave theory of light predicted the exact width of each of those bands. In honor of Fresnel, these patterns have been given the name **Fresnel diffraction**.

Fresnel also worked on the improvement of lighthouse lanterns. He replaced metal reflectors and thick lenses with lenses built from annular rings, and consequently eliminated spherical aberration. He also combined a fixed and a flashing light as a means of increasing the intensity of the light periodically.

A starting point for modern relativistic physics was Fresnel's prediction of the change in the speed of light in moving media.

Tuberculosis had plagued Fresnel throughout his career, and ultimately ended his life. He died at Ville d'Avray, near Paris, on July 14, 1827.

1. Augustin-Jean Fresnel, Encyclopedia of World Biography, www.bookrags.com/Augustin-Jean_Fresnel.
2. Brock, H.M., Augustin-Jean Fresnel, Catholic Encyclopedia, www.newadvent.org/ cathen/ 06280a.
3. Sears, Francis W., Mark W. Zemansky, and Hugh D. Young, College Physics, Seventh Edition, Addison Wesley, New York, 1991.

Georg Ohm (1789-1854)

Discovered that current flow is proportional to potential difference and inversely proportional to resistance

Georg Simon Ohm was born on March 16, 1789 in Erlangen, Bavaria, Germany. He was the oldest of seven children. His father, Johann Wolfgang Ohm, was a master mechanic and an avid reader of books on philosophy and mathematics. He cultivated Georg's obvious mathematical talents.

Ohm was educated at the University of Erlangen. Following the publication in 1817 of his first book, a textbook of geometry, he received an appointment as teacher of mathematics and physics at the Royal Prussian Konsistorium in Cologne. The well-equipped laboratory of the local Jesuit gymnasium was put at his disposal, and there he began his investigations on the characteristics of electric circuits.

In 1827, Ohm published his most renowned work, *The Galvanic Circuit Mathematically Treated*, in which he stated the very simple but important relationship between current "I," voltage "V," and resistance "R" for a metallic conductor at constant temperature. This relationship is known as **Ohm's law**, which is expressed mathematically as $I = V/R$.

A material that obeys Ohm's law is known as an ohmic or linear conductor. Some materials do not obey Ohm's law and are known as nonohmic or nonlinear. Ohm's law is an idealized model that describes the behavior of some materials quite well, but is not a general description of all matter.

Ohm's work was greeted with indifference and with some hostility. He withdrew from the academic world for six years. In 1833, he became professor of physics at the Polytechnic School in Nuremberg. The real turning point in his life came when the Royal Society of London awarded him the Copley Medal in 1841.

Ohm achieved his lifelong dream, a position with a major university, in 1849 as professor at the University of Munich. He was working on the manuscript of his textbook on optics when he died on July 6, 1854. The unit of electrical resistance, the **ohm** (Ω), was named in his honor.

1. Georg Simon Ohm, Biographies/Images of Physicists and Astronomers, www.mlahanas.de/ Physics/Bios/ GeorgSimonOhm.
2. Georg Simon Ohm, Bookrags.com, www.bookrags.com/Georg_Ohm.
3. Hewitt, Paul G., John Suchocki, and Leslie A. Hewitt, Conceptual Physical Science, Third Edition, Pearson/Addison Wesley, San Francisco, 2003.
4. Ohm, Georg Simon, Microsoft Encarta Encyclopedia, 2003.
5. Sears, Francis W., Mark W. Zemansky, and Hugh D. Young, College Physics, Seventh Edition, Addison Wesley, New York, 1991.

Felix Savart (1791-1841)

Co-discovered the fact that the intensity of magnetic field set up by a current flowing through a wire varies inversely with the distance from the wire

The son of an engineer, Felix Savart was born in Mézières, France on June 30, 1791. From 1808 to 1810 he studied medicine at a hospital in Metz. His studies were interrupted when he became a regimental surgeon in Napoleon's army. In 1814, Savart was discharged from the army and resumed his medical training. He received his medical degree from the University of Strasbourg in 1816.

After returning to Metz in 1817 Savart set up a medical practice. However, he spent more time studying physics than treating patients. He set up a physics laboratory to carry out experiments. He became fascinated with a study of sound, in particular the acoustics of musical instruments, such as the violin. He began to build violins to examine how vibrations were transmitted from the strings to the body of the violin. He became interested in trying to base the form of the instrument on mathematical principles. Jean-Baptiste Biot (1774-1862) was impressed with Savart's work, and found him a position teaching physics in Paris. In 1820, Savart and Biot began measuring the magnetic fields produced by a current. These experiments resulted in the **Biot-Savart law**, which is a quantitative relationship between an electric current and the magnetic field it produces. It is expressed mathematically as:

$$d\mathbf{B} = K_m \frac{I d\mathbf{l} \times \hat{\mathbf{r}}}{r^2}$$

where $K_m = \mu_0/4\pi$ and "μ_0" is the magnetic constant, "I" is the current, "dl" is the differential length vector of the current element, "$\hat{\mathbf{r}}$" is the unit displacement vector from the current element to the field point, and "r" is the distance from the current element to the field point.

Savart continued his studies on vibrations, developing methods for studying the vibrations of air, membranes, solids, and various other materials. He also studied the vocalizations of animals and humans and determined the lower frequency limits of hearing. To do this he used a toothed wheel, known as the **Savart wheel** that produced tones of various frequencies. He died at the age of 49 in Paris in on March 16, 1841.

1.   Felix Savart, Electro.Patent.Invent.com, www.electro.patent-invent.com/electricity/ inventors/ felix_savart.
2.   Felix Savart, World of Physics, www.bookrags.com/biography/felix-savart-wop
3.   Sears, Francis W., Mark W. Zemansky, and Hugh D. Young, College Physics, Seventh Edition, Addison Wesley, New York, 1991.

Michael Faraday (1791-1867)

Discovered electromagnetic induction

Michael Faraday was born on September 22, 1791 in Newington Butts, England, where his father worked as a blacksmith. Michael himself became apprenticed at the age of 14 to a bookbinder. He continued to work as a journeyman bookbinder until 1813, when he was appointed assistant in the laboratory of the Royal Institution of Great Britain.

In 1821, Faraday undertook a set of experiments which culminated in his discovery of electromagnetic rotation—the principle behind the electric motor.

It was not until nearly 10 years later that Faraday was able to resume his work on electro-magnetism, and in 1831 discovered electro-magnetic induction—the principle behind the electric transformer and generator.

Induction of current can be illustrated by moving a magnet back and forth close to a loop of wire connected to a galvanometer. It was this discovery that allowed electricity to be turned, during the 19th century, from a scientific curiosity into a powerful technology. Induction of an electric field by changing magnetic flux became known as **Faraday's law**.

Between 1832 and 1834 Faraday worked on a new theory of electro-chemical action, one of the results of which was that he coined, with William Whewell (1794-1866), the words electrode, electrolyte, anode, cathode, and ion. In 1833, Faraday was appointed Fullerian professor of chemistry at the Royal Institution.

In the later half of the 1830s, work on static electricity and electrical induction led Faraday to reject the traditional theory that electricity was a fluid. Instead he proposed that electricity was a form of force that passed from particle to particle of matter.

Faraday placed a piece of heavy glass on the poles of a powerful electro-magnet. Then he passed polarized light through the glass. When he turned the electro-magnet on he found that the state of polarization of the light changed. Thus, light had been affected by magnetic force, which later became known as the **Faraday Effect**.

After 54 years at the Royal Institution, Faraday died on August 25, 1867 in Hampton Court, London, England. He is buried in Highgate Cemetery in London.

1.   Heritage Faraday Page, The Royal Institution of Great Britain, www.rigb.org/heritage/faradaypage.jsp.
2.   Michael Faraday, Famous Physicists and Astronomers, www.phy.hr/~dpaar/fizicari/xfaraday.
3.   Michael Faraday, Microsoft Encarta Encyclopedia, 2003.
4.   Michael Faraday, NNDB, www.nndb.com/people/571/000024499.
5.   Sears, Francis W., Mark W. Zemansky, and Hugh D. Young, College Physics, Seventh Edition, Addison Wesley, New York, 1991.

Sadi Carnot (1796-1832)
Founded the field of thermodynamics

Nicolas Léonard Sadi Carnot was born in Paris, France on June 1, 1796. He was the oldest son of the French Revolutionary figure, Lazare Carnot. At the age of 18, Sadi graduated from École Polytechnique, after which he was commissioned as an engineer in Napoleon's army. After Waterloo, his father was exiled, and Carnot's army career and studies were interrupted for many years.

Following Carnot's retirement as an army officer, he began studying the processes involved in the operation of steam engines. In 1824, he published his theory of heat engines in *Réflexions sur la Puissance Motrice du Feu* (Reflections on the Motive Power of Fire).

Carnot was concerned with the relation between heat and mechanical energy. He devised an ideal engine in which a gas is allowed to expand to do work, absorbing heat in the process, and is expanded again without transfer of heat, but with a temperature drop. The gas is then compressed with heat being given off, and finally it is returned to its original condition by another compression, accompanied by a rise in temperature. This series of operations, now known as the **Carnot cycle**, showed that even under ideal conditions a heat engine cannot convert into mechanical energy all the heat energy supplied to it. Some of the heat energy must be rejected. This is an illustration of the second law of thermodynamics, which can be stated:

- Heat cannot spontaneously flow from a material at lower temperature to a material at higher temperature

Carnot's work was mostly ignored, largely because the British engineers were so highly regarded that no one believed that their work could be improved upon in France. However, in 1834 when Émile Clapeyron (1799-1864) began to publicize Carnot's work it was quickly accepted and became incorporated into the general theories of thermodynamics, developed in 1850 by Rudolf Clausius (1822-1888) in Germany and William Thomson (1824-1907) in Britain.

Carnot died of cholera in Paris, France on August 24, 1832. His work made him the world's first thermodynamacist.

1. Nicolas Léonard Sadi Carnot (1796-1832)http://library.thinkquest.org/C006011/ english/sites/ carnot_bio.php3?v=2.
2. Sadi Carnot (1796-1832), Eric Weisstein's World of Biography, Wolfram Research, http://scienceworld.wolfram.com/biography/CarnotSadi.
3. Sadi Carnot, NNDB, www.nndb.com/people/768/000082522.
4. Sears, Francis W., Mark W. Zemansky, and Hugh D. Young, College Physics, Seventh Edition, Addison Wesley, New York, 1991.

Joseph Henry (1797-1878)

Discovered self inductance and devised the first practical electric motor

Joseph Henry, the son of a day laborer, was born on December 17, 1797 in Albany, New York. At the age of 13 he was apprenticed to a watchmaker. After his interest in science was stimulated by reading George Gregory's 1808 lectures on experimental philosophy, astronomy, and chemistry, he decided to attend Albany Academy, and did so from 1819 to 1822.

In 1826, after a stint as a district schoolteacher and a private tutor, Henry was appointed Professor of Mathematics and Natural Philosophy at Albany Academy. While there, he was the first to wind insulated wires around an iron core to obtain powerful electromagnets. Before he left Albany, he built one for Yale that would lift 2,300 pounds, the largest in the world at that time.

In experimenting with electromagnets, Henry observed that a large spark was generated when he broke the circuit. He deduced that the phenomenon was due too self-inductance (an apposing electromotive force is induced in a coil of wire when the current in the wire changes). Self inductance is the inertial characteristic of an electric circuit.

While Henry was the doing the above experiments in the United States, Michael Faraday was doing similar work in England. Today Faraday is recognized as the discoverer of mutual inductance (a measure of the change in the electromotive force of a circuit caused by a change in the current flowing through an associated circuit.), while Henry is credited with the discovery of self-inductance.

Henry also invented the first practical electric motor, and his work on the electromagnetic relay was the basis of the electrical telegraph, jointly invented by Samuel Morse (1791-1872) and Charles Wheatstone (1802-1875).

In 1846, Henry was elected as the first secretary of the Smithsonian Institution in Washington D.C. He stayed there until his death on May 13, 1978. He was buried in Oak Hill Cemetery in northwest Washington, D.C.

During his lifetime, Henry was considered one of the greatest American scientists since Benjamin Franklin (1706-1790). The SI unit of inductance, the **henry** (H), is named after him.

1. Bailey, Herbert, S., Henry, Joseph, http://etcweb.princeton.edu/CampusWWW/ Companion/ henry_joseph.
2. Joseph Henry, www.ilt.columbia.edu/projects/bluetelephone/html/henry.
3. Sears, Francis W., Mark W. Zemansky, and Hugh D. Young, College Physics, Seventh Edition, Addison Wesley, New York, 1991.

Jean Poiseuille (1797-1869)

Developed an equation for the flow of a fluid in a narrow tube

Jean Louis Marie Poiseuille, a physician and physiologist, was born in Paris, France on April 22, 1797. From 1815 to 1816 he was trained in physics and mathematics at the École Polytechnique in Paris. In 1828, he earned his D.Sc. degree.

Poiseuille is best known for his research on the physiology of the circulation of blood through the arteries. This interest led him to study the flow rates of other fluids.

In 1840, Poiseuille formulated the law regarding the laminar flow of fluids in circular tubes. This law concerns the flow of an incompressible uniform viscous liquid (Newtonian fluid) through a cylindrical tube with constant circular cross-section. It can be successfully applied to blood flow in capillaries and veins, to air flow in lung airways, and to the flow of liquid through a drinking straw, catheter, or hypodermic needle.

Poiseuille's law states that the velocity of the steady flow of a fluid "ΔV/ΔT" through a narrow tube varies directly as the pressure "P" and the fourth power of the radius "r" of the tube and inversely as the length "L" of the tube and the coefficient of viscosity:

$$\frac{\Delta V}{\Delta t} = \frac{(P_1 - P_2)\,\pi r^4}{8\eta L}$$

where $(P_1 - P_2)$ is the difference in pressure and $(P_1 - P_2)/L$ is the pressure gradient. Gotthilf Hagen (1797-1884), a German hydraulic engineer, discovered Poiseuille's law independently in 1839.

Poiseuille also developed an improved method for measuring blood pressure. He used a mercury manometer in which he filled the connecting tubing to the artery with potassium carbonate to prevent coagulation. He used this instrument, known as a hemodynamometer, to show that blood pressure rises during expiration and falls during inspiration.

Poiseuille died in Paris on December 26, 1869. The unit of dynamic viscosity, **Poise (P)**, in the centimetre gram second system of units was named after him (1 P = 1 g/cm·s).

1. Jean Louis Marie Poiseuille, Wikipedia, http://en.wikipedia.org/wiki/ Jean_Louis_Marie_Poiseuille.
2. Poiseuille's+law, Merriam-Webster, http://medical.merriam-webster.com/medical/poiseuille.
3. Sears, Francis W., Mark W. Zemansky, and Hugh D. Young, College Physics, Seventh Edition, Addison Wesley, New York, 1991.

Franz Neumann (1798-1895)

Developed the first mathematical theory of electrical induction

Franz Ernst Neumann was born on September 11, 1798 in Joachimsthal, Prussia, near Berlin. He was raised by his grandparents. Neumann entered Berlin University as a student of theology, but soon turned to scientific subjects.

Neumann's earlier papers were mostly concerned with crystallography, and the reputation they gained him led to an eventual appointment as professor of mineralogy and physics at the University of Königsberg in 1829.

Neumann's 1831 study on the specific heats of compounds included what is now known as **Neumann's law**:

- The molecular heat of a compound is equal to the sum of the atomic heats of its constituents.

In a second study, Neumann investigated why, when equal volumes of hot water and cold water are mixed, the result does not have a temperature the average of the two temperatures. His concluded that the specific heat of water increases with temperature.

In 1832, Neumann investigated the wave theory of light and, using rigorous calculations, arrived at results agreeing with those of Augustin Louis Cauchy (1789-1857). He also succeeded in deducing laws of double refraction closely resembling those of Augustin-Jean Fresnel (1788-1827). Later, he condiered the conditions for a surface separating two crystalline media, and worked out the laws of double refraction in strained crystalline bodies.

In papers published in 1845 and 1847, Neumann established mathematically the laws of the induction of electric currents; that is, the process of converting mechanical energy to electrical energy.

In 1848, Neumann discovered fine patterns of parallel lines seen in cross-sections of many hexahedrite iron meteorites. The lines are indicative of a shock-induced deformation of the kamacite crystal, and are thought to be due to impact events on the parent bodies of the meteorites. The lines are named **Neumann's lines** in his honor.

Neumann retired in 1876, and died at Königsberg (now Kaliningrad), Russia in on May 23, 1895 at the age of 96.

1. O'Connor, J. J. and E F Robertson, Franz Ernst Neumann, MacTutor, www-groups.dcs.st-and.ac.uk/~history/Mathematicians/Neumann_Franz, May 2000.
2. Franz Ernst Neumann, Wikipedia, http://en.wikipedia.org/wiki/Franz_Ernst_Neumann.
3. Neumann Lines, The Internet Encyclopedia of Science, www.daviddarling.info/encyclopedia/N/Neumann_line.

Macedonio Melloni (1798-1854)

Demonstrated that radiant heat has similar physical properties to those of light

Macedonio Melloni was born in Parma, Italy on April 11, 1798. In 1824, he was appointed professor at the local University, but had to escape to France after taking part in the revolution of 1831. He studied at the Polytechnic in Paris.

On his return to Parma he was appointed professor of Physics at Ducale University. He devoted his time to meteorology, barometry, thermometry, and hygrometry. He also studied the magnetism of rocks, electrostatic induction, and photography.

In 1831, Melloni moved to Geneva to work with Leopoldo Nobili (1784-1835). Soon after the discovery of thermoelectricity by Johann Seebeck (1770-1831), Melloni and Nobili used the thermomultiplier to study the characteristics of black body radiation transmitted by various materials. The thermomultiplier is a combination of thermopile and astatic galvanometer.

Melloni used an optical bench fitted with thermopiles, shields, and light and heat sources (Locatelli's lamp, which had had a light and a heat source, or Leslie's cube) to show that radiant heat could be reflected, refracted, and polarized in the same way as light.

In 1839, Ferdinando II of the Bourbons asked Melloni to return to Naples as Honorary Professor of Physics and Director of the emerging Vesuvius Observatory.

In 1848, Melloni was removed from public duty after his involvement in the Republican demonstrations. However, because of his scientific merit, he was granted exile in Portici at Villa Moretta, and he devoted the long period he spent there to furthering his studies into the properties of radiant energy.

Melloni designed an electroscope for his studies, which had very different features from its predecessors. The instrument was described in detail in a manuscript that Melloni planned to present at the Congress of the Naples Academy of Science. He died of cholera at Portici, near Naples, Italy, on August 11, 1854, a few days before the Congress. His manuscript was read by Antonio Nobili, an eminent astronomer at the Capodimonte Observatory.

1. Macedonio Melloni (Parma, 1798 - Napoli, 1854), Directors of the Vesuvius Observatory, www.ov.ingv.it/inglese/storia/direttori.
2. Macedonio Melloni, NNDB, www.nndb.com/people/948/000100648.
3. Macedonio Melloni, Wikipedia, http://en.wikipedia.org/wiki/Macedonio_Melloni.

Benoît Clapeyron (1799-1864)
Co-founded thermodynamics

Benoît Paul Émile Clapeyron was born on February 26, 1799 in Paris, France. He studied at the École polytechnique and the École des Mines, before leaving for Saint Petersburg in 1820 to teach at the École des Travaux Publics.

Clapeyron returned to Paris only after the Revolution of July 1830, supervising the construction of the first railway line connecting Paris to Versailles and Saint-Germain.

In 1834, Clapyron made his first contribution to the creation of modern thermodynamics by publishing a report entitled the *Puissance motrice de la chaleur* (Driving force of the heat), in which he further developed the work of Nicolas Léonard Sadi Carnot (1796-1832). Though Carnot had developed an analysis of a generalized heat engine, he had employed the outmoded caloric theory that viewed heat as fluid called "caloric" that flowed from hotter to colder bodies. Clapeyron presented Carnot's work in a more analytic graphical form, showing the Carnot cycle as a closed curve on an indicator diagram (a chart of pressure against volume), which was named the **Clapeyron graph** in his honor.

In 1843, Clapeyron further developed the idea of a reversible process suggested by Carnot, and made a definitive statement of Carnot's principle, which is now known as the second law of thermodynamics.

These works allowed Clapeyron to make substantive extensions of the work of Rudolf Clausius (1822-1899), including the formula now known as the **Clausius-Clapeyron relation**, which characterizes the phase transition between two phases of matter. The Clausius-Clapeyron relation gives the slope "dP/dT" of the coexistence curve, and is expressed mathematically as:

$$\frac{\mathrm{d}P}{\mathrm{d}T} = \frac{L}{T\,\Delta V}$$

Where "L" is the latent heat, "T" is the Kelvin temperature, and "ΔV" is the volume change of the phase transition.

Clapeyron also worked on the characterization of perfect gases, the equilibrium of homogeneous solids, and calculations of the statics of beams. He died on January 28, 1864.

1.   Basic Principles, www.evitherm.org/default.asp?ID=459.
2.   Benoît Paul Émile Clapeyron, Wikipedia, http://en.wikipedia.org/wiki/%C3%89mile_Clapeyron.
3.   O'Connor, J. J. and E F Robertson, MacTutor, Benoît Paul Émile Clapeyron, October 1998.

19th Century Physics
(1800-1899)

H. Thomas Milhorn, MD, PhD

Christian Doppler (1803-1853)

Studied the apparent change in wave frequency due to relative motion between observer and source

Christian Andreas Doppler, the son of a successful stone-mason, was born in Salzburg, Austria on November 29, 1803. He could not work in his father's business because of his generally weak physical condition.

After completing high school, Doppler studied mathematics at the Vienna Polytechnic Institute. He excelled in his mathematical and other studies, and graduated in 1825. He returned to Salzburg, attended philosophy lectures at the Salzburg Lyceum, then went to the University of Vienna where he studied higher mathematics, mechanics, and astronomy. At the end of his studies at the University of Vienna in 1829, he was appointed to a temporary position of assistant to the professor of higher mathematics and mechanics at the University.

After a period of difficulty finding a university position, in 1841 Doppler was appointed professor of mathematics and physics at Prague Polytechnic in 1841. That same year, at the age of 39, he published his most notable work, which was on what would become know as the **Doppler effect**. We notice the Doppler effect today as the change of sound as an ambulance with it's siren going approaches us and then quickly passes and moves on. The pitch is higher as it approaches and lower as it moves away from us. The Doppler effect, as Doppler stated it, deals with apparent change in wavelength of sound caused by the motion of the source, observer, or both. Sound waves emitted by a moving object and received by an observer will be compressed if approaching and elongated if receding. Doppler's work dealing with sound was verified by John Russell (1802-1882) in 1848, and Armand Fizeau (1819-1896) showed that the Doppler effect also applies to light.

Doppler's research career in Prague was interrupted by revolutionary incidents, so in March 1848 he fled to Vienna. There, in 1850, he was appointed head of the Institute for Experimental Physics at the University of Vienna. He played an influential role in the development of young Gregor Mendel (1822-1884), who later became the founding father of genetics.

Doppler died of tuberculosis in Venus, Italy on March 17, 1853 at the age of 50. His tomb is located just inside the entrance of the Venetian island cemetery of San Michele.

1. C.J. Doppler, www.qerhs.k12.nf.ca/projects/physics/doppler.
2. Christian Doppler, NNDB, www.nndb.com/people/711/000061528.
3. Sears, Francis W., Mark W. Zemansky, and Hugh D. Young, College Physics, Seventh Edition, Addison Wesley, New York, 1991.

Heinrich Lenz (1804-1865)

Showed that an induced electric current flows in a direction such that the current opposes the change that induced it

Heinrich Friedrich Emil Lenz was born on February 12, 1804 in Dorpat (now Tartu, Estonia). He studied theology before switching to science, studying chemistry and physics at the University of Tartu.

From 1823 to 1826 Lenz traveled with Otto von Kotzebue (1787-1846) on his third expedition around the world. On the voyage, Lenz studied climatic conditions, and made accurate measurements of the salinity, temperature, and specific gravity of sea water. After a second voyage, Lenz joined the faculty at the University of St. Petersburg in 1840. He later served as Dean of Mathematics and Physics.

Lenz began studying electromagnetism in 1831. He is best-known for **Lenz's law**, which he discovered in 1834 while investigating magnetic induction. Lenz's law is stated as:

- An induced electric current flows in a direction such that the current opposes the change that induced it

According to Lenz's law, when the north pole of a bar magnet approaches a coil of wire, the induced current flows in such a way as to make the side of the coil nearest the pole of the bar magnet itself a north pole, so that it opposes the approaching bar magnet. Upon withdrawing the bar magnet from the coil, the induced current reverses itself and the near side of the coil becomes a south pole, producing an attracting force on the receding bar magnet. A small amount of work, therefore, is done in pushing the magnet into the coil and in pulling it out against the magnetic effect of the induced current. The small amount of energy manifests itself as a slight heating effect, the result of the induced current encountering resistance in the material of the coil. Lenz's law, therefore, upholds the general principle of the conservation of energy.

In 1833, Lenz showed that an increase in temperature increases the resistance of a metal. He also studied the relationship between heat and current and discovered, independently of James Joule (1818-1889), the law now known as Joule's law.

Lenz retired from the University of St. Petersburg in 1963. He died on February 10, 1865 in Rome after suffering a stroke.

1. Electricity, Microsoft Encarta Encyclopedia Online, 2003.
2. Heinrich Friedrich Emil Lenz, Institute of Chemistry, http://chem.ch.huji.ac.il/history/lenz.
3. World of Physics on Heinrich Lenz, BookRags, www.bookrags.com/Heinrich_Lenz.

Wilhelm Weber (1804-1891)

Determined the ratio between the electrodynamic and electrostatic units of charge

Wilhelm Weber was born on October 24, 1804 in Wittenberg, Germany, where his father, Michael Weber, was professor of theology. After receiving his doctorate at the University of Halle, Wilhelm was appointed Professor Extraordinary of natural philosophy there. Working with his brothers Ernst and Eduard, he published a large work on wave theory in 1825.

In 1831, on the recommendation of Carl Friedrich Gauss (1777-1855), Weber was called to Göttingen as professor of physics. There, in 1833, Gauss and Weber constructed the first electromagnetic telegraph, which connected the astronomy observatory, where Gauss spent most of his time, with Weber's laboratory in the Institute for Physics. The distance was nearly a mile. That same year, Weber published with brother Ernst a treatise on the mechanics of walking.

Working with Gauss, Weber made sensitive magnetometers to measure magnetic fields. He also made instruments to measure direct and alternating currents.

In 1837, Weber was dismissed from the university on political grounds. He served for a time as professor of physics at the University of Leipzig. In 1849, he was restored to his position at the University of Göttingen.

Weber's work on the ratio between the electrodynamic and electrostatic units of charge, which Weber's calculations indicated was approximately the speed of light, later become extremely important for the electromagnetic theory of James Maxwell (1831-1879).

In his later years, Weber's work on the electrical structure of matter led him to speculate that atoms contain positive charges around which negative particles rotate, and that applying a voltage to a conductor results in the movement of the negative particles between atoms.

Weber died at 86 years of age on June 23, 1891 in Göttingen, Germany. He is buried in the same cemetery as Max Planck (1858-1947) and Max Born (1882-1970). The SI unit of magnetic flux was named **weber** (Wb) in honor of Wilhelm Weber (1 Wb = 1 volt.second).

1. Wilhelm Weber, NNDB, www.nndb.com/people/097/000099797.
2. Sears, Francis W., Mark W. Zemansky, and Hugh D. Young, College Physics, Seventh Edition, Addison Wesley, New York, 1991.
3. Wilhelm Weber (1804-1891), National Magnetic Field Laboratory, www.magnet.fsu.edu/inhouseresearch

Thomas Graham (1805-1869)

Showed that the diffusion rate of gases is inversely proportional to the square root of their densities

Thomas Graham was born on December 21, 1805 in Glasgow, Scotland. His father, a successful textile manufacturer, wanted Thomas to enter into the Church of Scotland. Instead, at the age of 14 he became a student at the University of Glasgow in 1819, where he obtained his M.A. in chemistry in 1826.

Graham went on to become a professor of chemistry at a number of universities, including the Royal College of Science and Technology and the University of London.

One of the founders of physical chemistry, Graham showed that at a constant temperature and pressure the diffusion rate of a gas is inversely proportional to the square root of its density:

$$Rate_{diffusion} = k/(density)^{1/2}$$

This relationship has come to be known as **Graham's law**. Graham's work on the diffusion of gases was used in 1868 to discover the chemical formula for ozone (O_3).

Graham was the first to distinguish between colloids and crystalloids. His observation that certain substances, such as glue, gelatin, and starch, pass through a membrane more slowly than others, such as inorganic salts, led him to draw a distinction between the two groups. He called the slower-passing substances colloids and the faster-passing substances crystalloids. In this connection, he discovered dialysis, which is used in many medical facilities today in the form of the artificial kidney. Graham's colloid studies initiated the scientific field known today as colloid chemistry, of which Graham is credited as the founder.

Graham's investigation of phosphoric acid led to the present chemical concept of polybasic acids.

Graham died on September 16, 1869. His textbook, *Elements of Chemistry*, was widely used in both England and in Europe.

1. Graham, Thomas, Microsoft Encarta Encyclopedia Online, http://encarta.msn.com/ency clopedia_761553216/thomas_graham. 2007.
2. Lane, Dan and Jay Solon, Thomas Graham, Woodrow Wilson Organization, www.woodrow.org/ teachers/chemistry/institutes/1992/Graham.
3. Thomas Graham (Chemist), Wikipedia, http://en.wikipedia.org/wiki/Thomas_Graham_(chemist).

William Hamilton (1805-1865)
Developed the mathematical theory of quaternions

William Rowan Hamilton, the son of a solicitor, was born in Dublin, Ireland on August 3, 1805, and was put up for adoption. Before he was 13 years of age he had acquired, under the care of an uncle who was an extraordinary linguist, almost as many languages as he had years of age.

In 1823, at the age of 18, he entered Trinity College, in Dublin, where he excelled in mathematics and physics. He was also strong and active in gymnastics and swimming. In addition, he had interests in reading literature and poetry. Just before graduation, he was appointed Professor of Astronomy and Royal Astronomer of Ireland at Dunsink Laboratory. He was only 22.

In 1832, Hamilton presented to The Royal Irish Academy a paper entitled *Theory of Systems of Rays*, which included a theory of a single function that brought together mechanics, optics, and mathematics. This work helped to establish the wave theory of light.

In 1834, Hamilton published *General methods in Dynamics* in which the mathematical method allowed light to be treated as waves or particles equally, virtually predicting modern wave mechanics.

In 1835, Hamilton predicted the phenomenon of conical refraction in biaxial crystals. He was knighted in that same year.

The idea for his most famous discovery, Quaternions, came to him on October 16, 1943 when he was walking along the Royal Canal with his wife. In his excitement he carved the formula with his pin knife in the stone of Brougham Bridge:

$$i^2 = j^2 = k^2 = ijk = -1$$

This can still be seen today. Quaternions opened up a vast new field of mathematics. Today, quaternions are in use in computer graphics, control theory, signal processing, and orbital mechanics. It is common for spacecraft attitude-control systems to be commanded in terms of quaternions.

Hamilton's research was later significant for the development of quantum mechanics. He remained at Trinity College until his death on September 2, 1865.

1. Tribute to Sir William Hamilton, Sir William Rowan Hamilton, Biography, www.hamilton2005.ie.
2. Sir William Rowan Hamilton, Encyclopedia.com, www.encyclopedia.com/doc/1E1-HamiltWR.
3. Sir William Rowan Hamilton, NNDB, www.nndb.com/people/951/ 000101648.
4. William Rowan Hamilton, Biographies/Images of Physicists and Astronomers, www.mlahanas.de/Physics/Bios/WilliamRowanHamilton.

Robert von Mayer (1814-1878)

Proposed one of the first versions of the first law of thermodynamics

Julius Robert von Mayer, the son of a pharmacist, was born on November 25, 1814 in Heilbronn, Wurttemberg (modern day Germany). He enrolled at the University of Tubingen and, despite being expelled in 1837 for participation in a secret society, received his medical degree in 1838.

In 1841, Mayer stated one of the first versions of the first law of thermodynamics. This relation implies that, although work and heat are different forms of energy, they can be transformed into one another. This law led to the formulation of the general principle of conservation of energy.

Despite the fact that Mayer's work was received skeptically by the scientific community, he applied the principle to the explanation of numerous physical phenomena. Hermann von Helmholtz (1821-1894) definitively stated the law of conservation of energy in 1847, and was given credit for it.

Mayer determined a value for the mechanical equivalent of heat in 1842. Although his result was published five years before James Prescott Joule (1818-1889) published a similar result, Joule received credit.

The resistance from the scientific establilshment, and his failure to obtain credit for his work, undoubtedly was due to Mayer's inability to express himself scientifically because of lack of physics training.

In 1842, Mayer described the chemical process now referred to as oxidation as the primary source of energy for any living creatures. He also derived what is now known as **Mayer's Relation**, $C_p - C_v = R$, where "C_p" is the specific heat of a gas at constant pressure, "C_v" is specific heat of a gas at constant volume and "R" is the gas constant.

In 1848, two of his children died in rapid succession, and Mayer's mental health deteriorated. He attempted suicide on May 18, 1850 and was committed to a mental institution. After he was released in 1860, he was a broken man. However, during his absence his scientific fame had grown, and he had begun to be appreciated.

Mayer was awarded an honorary doctorate in 1859 by the philosophical faculty at the University of Tubingen. He worked vigorously as a physician the rest of his life.

Mayer died of tuberculosis on March 20, 1878 in Heilbronn, Germany. He is now recognized as one of the founders of thermodynamics.

1. Julius Robert Mayer, NNDB, www.nndb.com/people/013/000095725.
2. Julius Robert von Mayer, Wikipedia, http://en.wikipedia.org/wiki/Julius_Robert_von_Mayer.
3. Julius Robert von Mayer, World of Scientific Discovery, BookRags, www.bookrags.com/Julius_Robert_von_Mayer

Gustave Hirn (1815-1890)

Made important measurements of the mechanical equivalent of heat

Gustave Adolphe Hirn was an astronomer, mathematician, and engineer. His health was delicate as a child, a problem he never outgrew. He was self-taught, and never went to a University. He knew German, English, Latin, and Italian. He was an avid reader, and possessed a very good memory.

Hirn was born into a prosperous textile-manufacturing family in Logelbach, France on August 21, 1815. The owner of the company, Baron Haussmann, was a cousin. From an early age, Hirn worked in dye chemistry in the laboratory of the business.

Hirn made important measurements of the mechanical equivalent of heat and contributed to the early development of thermodynamics. He identified the mechanical equivalent of the calorie, based on the warming of the lead hitting a body.

In 1862, Hirn wrote *Escribió Exposición Analítica y Experimental de la Teoría Mecánica del Calor* (Exhibition Analytical and Experimental Mechanics of the Theory of Heat), a work that was one of the first treaties on systematic thermodynamics. It is regarded as a major work of the 19th century.

Hirn further did work on the practical development of steam engines. In 1889, he invented the pandynanometer, an instrument for measuring the degree of muscular power (also called ergometer). He also published a theory of the origin and chemical composition of Saturn's rings.

Hern was an excellent musician, and published articles dealing with music topics. In early life, he was a first-rate violin player

On the death of the owner of the textile company in 1880, the Logelbach business passed into other hands, and Hirn retired to his house in Colmar. There, he took an increased interest in art, especially music. He also occupied himself with meteorological and astronomical studies, which he had always been fascinated by.

Hirn died on January 14, 1890 in Logelbach, France, the result of an enfluenza epidemic. He had spent his entire life within a few miles of his birthplace.

1. Donkin, Bryan, Life and Work of Gustav Adolf Hirn (1815-1890), Cashier's Magazine, Volume IV, Pp 223-239, May 1893-October 1893.
2. Gustave Adolphe Hirn, Answers.com, www.answers.com/topic/gustave-adolphe-hirn.
3. Gustave-Adolphe Hirn (1815–1890), PARIS-PROVINCE: ENERGY PHYSICS IN MID-NINETEENTH-CENTURY FRANCE, Revue de la Maison Française d'Oxford Volume I, n° 2, 2003, www.mfo.ac.uk/Publications/Revue%20Fox/papanelopoulou.
4. Gustave-Adolphe Hirn, Wikipedia, http://en.wikipedia.org/wiki/Gustave-Adolphe_Hirn.

James Joule (1818-1889)

Discovered the mechanical equivalent of heat

James Prescott Joule was born into a wealthy brewing family in Salford, Manchester, England on December 24, 1818. He initially was educated at home before being tutored at the age of 16 by the eminent Manchester scientist, John Dalton (1766-1844). Joule soon began to conduct independent research at a laboratory built in the cellar of his father's home.

Joule was fascinated by electricity. He and his brother experimented by giving electric shocks to each other and to the family's servants

From 1837 to 1856 Joule worked in the family brewery. In his spare time he studied the relatively new subject of current electricity. He became interested in measuring the work done and the heat generated by electricity, and in 1840 discovered that the rate of generation of heat "Q" by an electric circuit was proportional to the square of the current "I" multiplied by the resistance "R" for a time "t":

$$Q = I^2Rt$$

Joule measured the heat produced by the action of falling weights that churned water inside a bucket, thereby slightly increasing the overall temperature of the water. He concluded that mechanical energy was being converted to heat energy. More generally, he concluded that all systems contain energy which can be converted from one form to another, but the total energy of a closed system remains constant. Thus, he established the principle of conservation of energy. This relationship served as the foundation to the first law of thermodynamics.

After working with physicist William Thomson (1824-1907) from 1852 to 1859, Joule described the **Joule-Thomson effect**, whereby an expanding gas is cooled as work is done to separate the molecules.

Joule also was an inventor. Among many of his inventions were electrical welding and the displacement pump. Much of his research was self-funded.

Joule died on October 11, 1889 in Sale, England. He was buried in Brooklands Cemetery. In Joule's honor, the unit of energy, the **joule** (J), is named after him (1 J = 1 N.m).

1. James Prescott Joule (1818 - 1889), Corrosion Doctors, www.corrosion-doctors.org/Biographies/ JouleBio.
2. James Prescott Joule, NNDB, www.nndb.com/people/275/000049128.
3. Sears, Francis W., Mark W. Zemansky, and Hugh D. Young, College Physics, Seventh Edition, Addison Wesley, New York, 1991.

H. Thomas Milhorn, MD, PhD

George Stokes (1819-1903)

Developed equations of fluid mechanics

George Gabriel Stokes, the son of a protestant minister, was born on August 13, 1819 in Skreen, County Sligo, Ireland. In 1835, at the age of 16, George moved to England and entered Bristol College where his talent for mathematics was shown. Then, in 1837 he enrolled at Pembroke College, Cambridge. He did so well that when he graduated he was immediately offered a fellowship. In 1845, Stokes published an important work on the aberration of light, the first of a number of important works on this topic. He also used his work on the motion of pendulums in fluids to consider the variation of gravity at different points on the Earth, publishing a work on geodesy of major importance in 1849.

Also in 1849, Stokes was appointed Lucasian Professor of Mathematics at Cambridge, and accepted an additional position as Professor of Physics at the Government School of Mines in London. His work on the motion of pendulums in fluids led to a fundamental paper on hydrodynamics in 1851, which described the velocity of a small sphere moving through a viscous fluid. A force "F" on acts on the sphere of radius "r" as it moves through a fluid of viscosity "η" with a velocity "v" is given by:

$$F = 6\pi\eta rv$$

This became known as **Stokes' law**.

The Navier-Stokes equations, named after Stokes and Claude-Louis Navier (1785-1836), describe the motion of liquids and gases. They establish that changes in momentum in infinitesimal volumes of fluid are simply the sum of dissipative viscous forces, changes in pressure, gravity, and other forces acting inside the fluid. The Navier-Stokes approach is an application of Newton's second law.

In 1852, Stokes described the phenomenon of fluorescence, as exhibited by fluorspar and uranium glass, materials, which he viewed as having the power to convert invisible ultra-violet radiation into radiation of longer wavelengths that are visible Stokes died on February 1, 1903 in Cambridge, Cambridgeshire, England.

1. George Gabriel Stoke, NNDB, swww.nndb.com/people/131/000097837.
2. George Gabriel Stokes (1819-1903), Corrosion Doctor, www.corrosion-doctors.org/Biographies/ StokesBio.
3. Sears, Francis W., Mark W. Zemansky, and Hugh D. Young, College Physics, Seventh Edition, Addison Wesley, New York, 1991.
4. Wood, Alistair, George Gabriel Stokes 1819 - 1903: An Irish Mathematical Physicist, www.cmde.dcu.ie/Stokes/GGStokes.

Jean Foucault (1819-1868)

Showed the relationship between mechanical energy, heat, and magnetism

Jean Bernard Leon Foucault, the son of a publisher, was born in Paris, France on September 18, 1819. He received his early schooling at home, and showed his mechanical skill by constructing a boat, a mechanical telegraph, and a working steam-engine. He enrolled in college and began to study medicine. Unable to bear the sight of blood, he abandoned medicine and worked for three years as an experimental assistant to Alfred Donné (1801-1878) in his course of lectures on microscopic anatomy.

In 1850, Foucault showed the relation between mechanical energy, heat, and magnetism.

In the 1850s, Foucault worked with the Armand Fizeau (1819-1896) in making determinations of the speed of light. He proved independently that the speed of light in air is greater than it is in water. With Fizeau, he took the first clear photograph of the Sun.

In 1851, Foucault gave a spectacular demonstration to prove that the Earth rotates on its axis. He suspended a pendulum on a long wire from the dome of the Pantheon in Paris. The daily movement of the pendulum duplicated the rotation of the Earth on its axis.

Foucault was one of the first to show the existence of the eddy currents generated by magnetic fields. These are now known as **Foucault currents**. He devised a method of measuring the curvature of telescope mirrors, developed a polarizing prism to orient and manipulate polarized light, and invented (and named) the gyroscope to demonstrate the earth's movement around its axis. The gyroscope is the basis of the modern gyrocompass.

In 1865, Foucault described a modification of Watt's governor to make the period of revolution constant and a new apparatus for regulating the electric light. In the following year, he showed how, by the deposition of a transparently thin film of silver on the outer side of the object glass of a telescope, the Sun could be viewed without injuring the eye.

Foucault died in Paris on February 11, 1868, and was buried in the Cimetière de Montmartre. He is considered one of the most versatile experimentalists of the 19ih century.

1. Foucault, Jean Bernard Leon, Microsoft Encarta Encyclopedia Online, http://ca.encarta.msn.com/encyclopedia_761573292/foucault,. 2007.
2. Jean Bernard Leon Foucault, Biographies/Images of Physicists and Astronomers, www.mlahanas.de/Physics/Bios/JeanBernardLeonFoucault.
3. Jean-Bernard-Leon Foucault, MOLECULAR Expressions, http://micro.magnet.fsu.edu/optics /timeline/people/foucault.
4. Léon Foucault, Wikipedia, http://en.wikipedia.org/wiki/L%C3%A9on_Foucault.

Louis Fizeau (1819-1896)

Made the first terrestrial measurement of the speed of light in air and water

Armand Hippolyte Louis Fizeau was born in Paris, France on September 23, 1819. His father was a distinguished physician and professor of medicine. On his death he left Louis an independent fortune, so that he was able to devote himself to scientific research without being encumbered by the need for employment.

Fizeau attended Stanislas College and began to study medicine, but had to abandon his studies because of ill-health. He soon began to study physics at the College de France.

Fizeau and Jean Foucault (1819-1868) began to collaborate. The two young men were particularly interested in the new photography method that had been developed by L. J. M. Daguerre (1787-1851). By making improvements, in 1845 they were able to obtain very detailed photographs of the sun's surface. Concentrating further on optics, Fizeau and Foucault studied the effects of interference on light waves, and later the interference patterns of heat waves. Their research helped to further the view that light acted as a wave rather than as a stream of particles.

In 1849, Fizeau was the first to determine experimentally the velocity of light on Earth. On the peak of a hill, he set up a light source and a spinning gear, arranged so that the light would shine through the gear's teeth. As it spun, the gear would alternately block and unblock the light, so that it would flash. On another hilltop eight kilometers away he positioned a mirror that reflected the light back to its source. He spun the gear very fast so that light passing through one gap in the gear's teeth would travel to the mirror, bounce back, and reenter through the next gap. By using a timer, he was able to determine the amount of time it took light to travel the 16 kilometers there and back. He arrived at the figure of 315,000 kilometers per second, which proved to be about five percent too high.

Fizeau's next step was to measure light's velocity as it passed through water. He began with a single beam of light and split it into two beams, one of which passed through water while the other passed through air. Then, using another rotating-gear-and-mirror system, he measured and compared the two velocities, finding that the speed of light was greater as it traveled through air.

1. Fox, William, Armand-Hippolyte-Louis Fizeau, Catholic Encyclopedia, www.newadvent.org/cathen/06088b.
2. Armand-Hippolyte-Louis Fizeau, World of Scientific Discovery, www.bookrags.com/biography/armand-hippolyte-louis-fizeau-wsd.
3. Sears, Francis W., Mark W. Zemansky, and Hugh D. Young, College Physics, Seventh Edition, Addison Wesley, New York, 1991.

Hermann von Helmholtz (1821-1894)

One of the founders of the law of conservation of energy

Hermann Ludwig Ferdinand Helmholtz, the son of a senior grammar school teacher in Potsdam, Germany, was born on August 31, 1821. As a young man, Helmholtz was interested in natural science, but his father wanted him to study medicine because there was financial support for medical students.

Helmholtz studied at the Berlin Military Academy and gained his Doctor of Medicine degree in 1842.

Initially, Helmholtz took a position as a military doctor in Potsdam, and subsequently taught anatomy at the Berlin Academy of Art. In 1849, he became professor of physiology in Königsberg. He remained there from 1849 to 1855, when he moved to the chair of physiology in Bonn. In 1858, he became professor of physiology in Heidelberg, and in 1871 he was called to occupy the chair of physics in Berlin. To this professorship was added in 1887 the directorship of the physico-technical institute at Charlottenburg, near Berlin.

Helmholtz's investigations involved the fields of physiological optics, physiological acoustics, chemistry, mathematics, electricity and magnetism, meteorology, and theoretical mechanics.

His numerous contributions to science included determination of the velocity of nerve impulses, invention of the ophthalmoscope, investigation of the mechanisms of sight and hearing, development of a theory of color vision, study of the vortex motion of fluids, application of the principle of least action to electrodynamics, development of the theory of electricity, and investigation of the motion of electricity in conductors.

Helmholtz is best known for being one of the founders of the law of the conservation of energy. He expressed the relationship between mechanics, heat, light, electricity, and magnetism by treating them all as manifestations of a single energy. In 1847, he published a mathematical proof for the law of conservation of energy in *Über die Erhaltung der Kraft*.

In 1894, shortly after a lecture tour of the United States, Helmholtz fainted and fell, suffering a concussion. He never completely recovered, and died of complications several months later on September 8, 1894 in Charlottenburg, Berlin, Germany.

1. Hermann von Helmholtz, NNDB, www.nndb.com/people/ 445/000072229.
2. HELMHOLTZ, Hermann Ludwig Ferdinand von, Biography, http://library.thinkquest.org/ 10170/voca/helmholtz.
3. Hermann von Helmholtz Biography (1821-1894), Free Health Encyclopedia, www.faqs.org/ health/bios/65/Hermann-von-Helmholtz.

Rudolf Clausius (1822-1888)

Developed the concept of entropy

Rudolf Julius Emanuel Clausius was born on January 2, 1822 in Köslin, Prussia, Germany. He studied at Berlin University from 1840 to 1844. In 1848 he took his doctorate from the University of Halle, where he studied the optical effects in the earth's atmosphere. In 1850, he was appointed professor of physics in the Royal Artillery and Engineering School at Berlin.

In 1855, Clausius became a professor at Zürich Polytechnic, accepting at the same time a professorship in the University of Zürich. In 1867, he moved to Würzburg as professor of physics, and two years later was appointed to the same chair at Bonn. His wife died in childbirth in 1875, leaving him to raise their six children.

Clausius stated formally the equivalence of heat and work (first law of thermodynamics). His restatement of the principle of Sadi Carnot (1796-1832) put the theory of heat on a sounder basis. He enunciated the second law of thermodynamics in a paper contributed to the Berlin Academy in 1850 in the well-known form, "Heat cannot of itself pass from a colder to a hotter body." Clausius is given credit for having made thermodynamics a science.

In 1865, Clausius developed the concept of entropy, which he named. Entropy can never decrease in a physical process, and can only remain constant in a reversible process—a restatement of the second law of thermodynamics. For a reversible isothermal process entropy can be stated mathematically as:

$$\Delta S = Q/T$$

where "ΔS" is the change in entropy, "Q" is the heat added, and "T" is the temperature at which the heat is added.

With James Maxwell (1831-1879), Clausius developed the kinetic theory of gases. His *Über die Art der Bewegung welche wir Wärme nennen*, published in 1857, was the first systematic treatment of the kinetic theory of molecular motion. In it he introduced the concept of mean-free path and correlated temperature and velocity. Clausius died in Bonn, German on August 24, 1888.

1. Clausius, Rudolf (1822- 1888), Wolfram Research, http://scienceworld.wolfram.com/biography/Clausius.
2. Rudolf Clausius, NNDB, www.nndb.com/people/951/000100651.
3. Rudolf Julius Emanuel Clausius, Biographies/Images of Physicists and Astronomers, www.mlahanas.de/Physics/Bios/RudolfClausius.

Gustav Kirchhoff (1824-1887)

Developed two rules of electric circuit analysis

Gustav Robert Kirchhoff was born in Königsberg, Prussia, Germany on March 12, 1824. He obtained his Ph.D. at the Albertus University of Königsberg in 1847, and became professor of physics at Breslau in 1850.

Kirchhoff contributed to the fundamental understanding of electrical circuits, spectroscopy, and the emission of black-body radiation by heated objects. He coined the term "black body" radiation in 1862.

In 1845, while still a student Kirchhoff formulated **Kirchhoff's circuit laws**, which are now ubiquitous in electrical engineering. These laws are:

Loop Rule: Around any closed loop in a circuit, the sum of the potential differences across all elements is zero. This law is a statement of energy conservation, in that any charge that starts and ends up at the same point with the same velocity must have gained as much energy as it lost.

Junction Rule: For a given junction or node in a circuit, the sum of the currents entering equals the sum of the currents leaving. This law is a statement of charge conservation.

Kirchhoff took a faculty position at the University of Heidelberg in 1854, where he collaborated in spectroscopic work with Robert Bunsen (1811-1899) of Bunsen burner fame. Together, in 1861 they discovered caesium and rubidium while studying the chemical composition of the Sun via its spectral signature.

Kirchhoff contributed greatly to the field of spectroscopy by formalizing three laws that describe the spectral composition of light emitted by incandescent objects

Kirchhoff demonstrated in the Sun many chemical elements already isolated on Earth; argued that the bulk of the Sun is comprised of a hot, incandescent liquid; and firmly established the hot, gaseous nature of the solar atmosphere. His work with Bunsen remains at the root of almost everything we know today about the Sun and stars.

Kirchhoff died on October 17, 1887, and was buried in the St. Matthäus Kirchhof Cemetery in Schöneberg, Berlin, only a few meters from the graves of the Brothers Grimm.

1. Gustav Kirchhoff, (1824-1887), High Altitude Observatory, www.hao.ucar.edu/Public/ education/bios/kirchhoff..
2. Gustav Kirchhoff, Biographies/Images of Physicists and Astronomers, www.mlahanas.de/ Physics/ Bios/GustavRobertKirchhoff.
3. Gustav Kirchhoff, Wikipedia, http://en.wikipedia.org/wiki/Gustav_Kirchhoff.
4. Gustav Robert Kirchhoff, NNDB, www.nndb.com/people/530/000097239.
5. Sears, Francis W., Mark W. Zemansky, and Hugh D. Young, College Physics, Seventh Edition, Addison Wesley, New York, 1991.

Johann Hittorf (1824-1914)

Computed the electricity-carrying capacity of charged atoms and ions and studied electrical discharge in gases

Johann Wilhelm Hittorf was born in Bonn, Germany on March 27, 1824. From 1852 to 1879 he was professor of physics and chemistry at Munster University, and from 1979 to 1989 he was also director of physical laboratories. Hittorf's early investigations were on the allotropes of phosphorus and selenium.

In 1853, Hittorf pointed out that some ions traveled more rapidly than others, an important factor in understanding electrochemical reactions. This observation led to the concept of transport number (the fraction of the electric current carried by each ionic species) and the first method for their measurements. He measured the changes in the concentration of electrolyzed solutions, computed from these the transport numbers of many ions, and in 1869 published his laws governing the migration of ions. .

Hittorf's work led to the development of X-ray and cathode ray tubes. The **Hittorf tube** consisted of a glass bulb, into which were sealed two electrodes. A partial vacuum was created in the tube, and its operation was dependant upon the presence of the residual gas. When a high voltage was applied across the tube, the gas became ionized, and current flowed between the electrodes. If the electrode at the narrow end of the rube was attached to the negative pole of the supply, and the electrode nearest the wide end of the tube attached to the positive pole, the tube was filled with a faint glow, but an area on the broad end of the tube would glow brightly. The Hittorf tube was used by Wilhelm Röntgen (1845-1923) in 1895 when he made the observations that led to the discovery of X-rays.

Hittorf determined that discharge in a vacuum tube was accomplished by the emission of rays (later termed cathode rays) capable of casting a shadow of an opaque body on the wall of the tube. He noted that the rays seemed to travel in straight lines and produce a fluorescent glow where they passed through the glass. His measurement of current in a vacuum tube was an important step towards the creation of the vacuum tube diode.

Hittorf also investigated the light spectra of gases and vapors, worked on the passage of electricity through gases, and discovered new properties of cathode rays. He died in Münster, Germany on November 28, 1914.

1. Hittorf Tube, B. A. Mansfield, www.xrayengineering.co.uk/glasstube.
2. Hittorf, Johann (1824-1914), Eric Weisstein's World of Biography, http://scienceworld.wolfram.com/biography/Hittorf.
3. Johann Wilhelm Hittorf, Institute of Chemistry, http://chem.ch.huji.ac.il/history/hittorf.
4. Johann Wilhem Hittorf, Wikipedia, http://en.wikipedia.org/wiki/Johann_Wilhelm_Hittorf.

William Thomson (1824-1907)

Developed the absolute temperature scale

William Thomson, the second son of James Thomson who was professor of mathematics in the University of Glasgow, was born in Belfast, Northern Ireland on June 26, 1824. In 1845, William graduated from Peterhouse, Cambridge University. In 1846, when only 22 years of age, he accepted the chair of natural philosophy in the university of Glasgow, which he filled for fifty-three years.

In 1848, Thomson proposed his absolute scale of temperature, which is independent of the properties of any particular thermometric substance. The absolute temperature scale sets zero temperature at the point where matter has its minimum possible kinetic energy. For comparison, absolute zero corresponds to -273.15° on the Celsius temperature scale. The absolute temperature increment is the same as the Celsius degree; hence the freezing point of water (0°C), and the boiling point of water (100°C) correspond to 273.15K and 373.15K.

In 1851, Thomson presented to the Royal Society of Edinburgh a paper on the dynamical theory of heat, which put the fundamental principle of the conservation of energy in a position to command universal acceptance.

The term "kinetic energy" was introduced by Thomson in 1856. He supported the mechanical equivalent of heat of James Joule ((1818-1889), and changed the view of heat as being a fluid to an understanding of the energy of motion of molecules. The names of these two scientists are linked with the **Joule-Kelvin effect,** which now makes refrigerators work. The Joule–Kelvin effect is the fall in temperature of a gas as it expands adiabatically through a narrow jet.

Thomson invented the mirror galvanometer as a long distance telegraph receiver which could detect extremely feeble signals, an improved gyro-compass, new sounding equipment, and a tide prediction with chart-recording machine. He also studied the electrical losses in cables. In 1866, he was knighted because of his achievements in submarine cable-laying, and in 1892 he was given the title Baron Kelvin of Largs. In his honor, the **Kelvin temperature scale** is named for him.

Thomson died in Netherhall, Largs, Scotland on December 17, 1907, and was buried next to Isaac Newton (1643-1727) in Westminster Abbey.

1. Lord Kelvin (William Thomson), www.todayinsci.com/K/Kelvin_Lord/Kelvin_Lord.
2. Lord Kelvin, Famous Physicists and Astronomers, www.phy.hr/~dpaar/fizicari/xkelvin.
3. Lord Kelvin, NNDB, www.nndb.com/people/607/000050457.
4. Sears, Francis W., Mark W. Zemansky, and Hugh D. Young, College Physics, Seventh Edition, Addison Wesley, New York, 1991.

Johann Balmer (1825-1898)

Developed an empirical formula to describe the hydrogen spectrum

Johann Jacob Balmer was born in Lausanne, Switzerland on May 1, 1825. His father was a Chief Justice. During his schooling he excelled in mathematics, and so decided to focus on that field when he attended university.

Balmer studied at the University of Karlsruhe and the University of Berlin, then completed his Ph.D. at the University of Basel in 1849. His dissertaion was on the cycloid.

Balmer then spent his entire life in Basel, where he taught at a secondary school for girls.

From 1865 until 1890 he was also a university lecturer in mathematics at the University of Basel where his main field of interest was geometry.

In 1885, at the age of 60, Balmer made his only significant scientific contribution. He discovered that there was a simple mathematical formula that gave the wavelengths of the spectral lines of the hydrogen atom, which are now called the **Balmer series**. He arrived at his result purely from empirical evidence, and was unable to explain why it yielded correct answers. Balmer's formula is:

$$\frac{1}{\lambda} = R_H \left(\frac{1}{2^2} - \frac{1}{n^2} \right), n = 3, 4, 5, \dots$$

where "λ" is wavelength, "R" is a constant called the Rydberg constant ($1.09678 \times 10^7\,\mathrm{m^{-1}}$), and "n" may have the values as shown.

In his paper of 1885, Balmer suggested that giving n other small integer values would give the wavelengths of other series produced by the hydrogen atom. Indeed this prediction turned out to be correct and these series of lines were later observed.

Balmer's formula proved to be of great importance in atomic spectroscopy and in developing the atomic theory. Not until the further development of the atomic theory by Niels Bohr (1885-1962) and others was this possible.

Balmer died in Basel on March 12, 1898.

4. John Jacob Balmer, Answers.com, www.answers.com/topic/johann-jakob-balmer?cat= technology.
5. John Jacob Balmer, Wikipedia, http://en.wikipedia.org/wiki/Johann_Jakob_Balmer.
6. O'Connor, J.J. and E. F. Robertson, Johann Jocob Balmer, MacTutor, www-history.mcs.st-and.ac.uk/history/Biographies/Balmer.
7. Sears, Francis W., Mark W. Zemansky, and Hugh D. Young, College Physics, Seventh Edition, Addison Wesley, New York, 1991.

George Stoney (1826-1911)

Conceived and calculated of the magnitude of the "atom of electricity"
which he proposed to call the electron

George Johnstone Stoney was born on February 15, 1826 at Oak Park, near Birr, County Offaly, in the Irish midlands. He attended Trinity College, Dublin and graduated with an M.A. in 1852. In 1848, he became an assistant to William Parsons (1800-1867), who had built the world's largest telescope—the 72-inch Leviathan of Parsonstown.

In 1852, Stoney became Professor of Natural Philosophy at Queen's College Galway, and in 1857 he moved to Dublin as Secretary of the Queen's University. He subsequently became superintendent of Civil Service Examinations in Ireland, a post he held until his retirement in 1893.

When the work of Rudolf Clausius (1822-1888) on the kinetic theory of gases was published in 1867, Stoney realized that it was possible to obtain the number of molecules in a given volume of gas under standard conditions, and therefore the actual mass of an atom of any chemical element. Sometime after the publication of his paper, William Thomson (1824-1907) gave a similar estimate, and a few months earlier another estimate had been made in Germany.

In 1891, Stoney showed that the waving about of the electrons inside a gaseous molecule in orbital motions with perturbations very similar to planetary perturbations must be the cause of lines in the spectrum.

Stoney's most important scientific work was the conception and calculation of the magnitude of the "atom of electricity." He showed that the real meaning of Faraday's laws of electrolysis is that electricity, like matter, consists of ultimately indivisible equal particles or atoms, and in 1891 he suggested the name "electron" for these atoms. His contributions to research in this area laid the foundations for the eventual discovery of the particle by J.J. Thomson (1856-1940) in 1897.

Stoney died in London on July 5, 1911 at the age of 84. His cremated ashes were buried in Dundrum, Dublin.

Stoney was the uncle of the Anglo-Irish physicist George FitzGerald (1851-1901). The two were in regular communication on scientific matters. In addition, both Stoney and FitzGerald were active opponents of Home Rule for Ireland.

1. George Johnstone Stoney 1826-1911, OHAS, www.offalyhistory.com/articles/264/1/George-Johnstone-Stoney-1826-1911/Page1, September 2, 2007.
2. George Johnstone Stoney, Irish Universities Promoting Science, www.universityscience.ie/pages/scientists/sci_georgestoney.php.
3. George Johnstone Stoney, Wikipedia, http://en.wikipedia.org/wiki/George_Johnstone_Stoney.

Joseph Swan (1828-1914)

Developed the first electric light bulb

Joseph Wilson Swan was born in the town of Bishopwearmouth in northeast England on October 31, 1828. He served an apprenticeship with a pharmacist there, and later became a partner in Mawson's, a firm in Newcastle upon Tyne that manufactured photographic plates.

From an early age, Swan had an interest in learning about new inventions. He used Sunderland library to read about Starr's electric lamp, patented in 1845, but unsuccessful because it blackened too quickly. Mawson encouraged him to pursue his scientific investigations, and introduced him to local chemical manufacturers.

At that time, one of the most difficult steps in the production of photographic plates involved adding chemicals in a liquid state to the plates. By 1871, Swan had developed the much simpler "dry-plate" photographic method. He also invented the first bromide paper, which is still widely used for printing black and white photos.

In 1850, Swan began working on a light bulb using carbonized paper filaments in a partially evacuated glass bulb. By 1860 he was able to demonstrate a working device, and obtained a UK patent for an incandescent lamp. However, the lack of a good vacuum and an adequate electric source resulted in an inefficient bulb.

Fifteen years later, in 1875, Swan returned to the light bulb problem with the aid of a better vacuum and a carbonized thread as a filament. Also, power generators had been improved to the point where they could supply cheaper, more reliable electricity than batteries. Soon Swan had invented a practical incandescent filament lamp

Swan received a British patent for his improved device in 1878, about a year before the work of Thomas Edison (1847-1931) in the United States.

Starting in February 1879 Swan began installing "glow bulbs" in homes in England. In 1881, he started his own company—The Swan Electric Light Company.

Swan agreed that Edison, who had developed a similar bulb, could sell the lights in America while he retained the rights in Britain. So, in 1883 the Edison & Swan United Electric Light Company was established.

Swan was knighted in 1904. He died on May 27, 1914 at Warlingham in Surrey.

1. Joseph Swan, Wikipedia, http://en.wikipedia.org/wiki/Joseph_Swan.
2. Joseph Wilson Swan, Timmonet, www.timmonet.co.uk/html/body_joseph_swan.
3. World of Invention on Joseph Swan, Bookrags, www.bookrags.com/biography /joseph-swan-woi.

James Maxwell (1831-1879)

Developed equations concisely stating the fundamentals of electricity and magnetism

James Clerk Maxwell was born in Edinburgh, Scotland on June 13, 1831. He was educated at the University of Edinburgh from 1847 to 1850, and in 1854 graduated from Cambridge University, where he was awarded a prestigious prize for original research for mathematically analyzing the stability of the rings around Saturn. Maxwell concluded that Saturn's rings could not be completely solid or fluid. Instead they must consist of small but separate solid particles—a conclusion that was corroborated more than 100 years later by the first Voyager space probe to reach Saturn.

Maxwell held the chair of Natural Philosophy in Marischal College, Aberdeen from 1856 until the fusion of the two colleges in 1860. For eight years subsequently he held the chair of Physics and Astronomy in King's College, London.

In 1871, Maxwell assumed the newly founded professorship of Experimental Physics in Cambridge; and it was under his direction that the plans of the Cavendish Laboratory were prepared.

In 1865, Maxwell left London and moved to the estate in Scotland he had inherited from his father. There, he devoted himself to research on electricity and magnetism. His formulation of electricity and magnetism was published in *A Treatise on Electricity and Magnetism* in 1873, which included the formulas today known as the **Maxwell equations**. These equations concisely state the fundamentals of electricity and magnetism, from which one can develop most of the working relationships in the field. Maxwell also showed that these equations implicitly required the existence of electromagnetic waves traveling at the speed of light.

Maxwell's electromagnetic theory and its associated field equations paved the way for special theory of relativity of Albert Einstein (1879-1955), which established the equivalence of mass and energy.

Maxwell also made significant advances in the area of optics and color vision, and with Rudolf Clausius (1822-1888) developed the kinetic theory of gases.

Maxwell died November 5, 1879 in Cambridge, England.

1. James Clerk Maxwell, Famous Physicists and Astronomers, www.phy.hr/~dpaar/fizicari/xmaxwell.
2. Lamont, Ann, James Clerk Maxwell (1831-1879), www.answersingenesis.org/Home/Area/bios/jc_maxwell.asp.
3. Maxwell, James (1831-1879), ScienceWorld.Wolfram.com, http://scienceworld.wolfram.com/biography/Maxwell.
4. Sears, Francis W., Mark W. Zemansky, and Hugh D. Young, College Physics, Seventh Edition, Addison Wesley, New York, 1991.

William Crookes (1832-1919)
Investigated the properties of cathode rays

William Crookes was born in London on June 17, 1832. He was the eldest son of Thomas Crookes, who was a tailor. At the age of 16 William entered the Royal College of Chemistry in London. Then, from 1850 to 1854 he filled the position of assistant in the college. In 1854, he became superintendent of the meteorological department at the Radcliffe Observatory in Oxford, and in 1855 he was appointed lecturer in chemistry at the Chester training college.

In chemistry, Crookes' first important discovery, using spectroscopy, was the element thallium in 1861. The work firmly established his reputation. He then invented the Crookes radiometer, in which a system of vanes, each blackened on one side and polished on the other, in a vacuum is set in rotation when exposed to light, with faster rotation with more intense light. He did not provide and explanation of the apparent attraction and repulsion resulting from radiation.

Crookes published numerous papers on spectroscopy, including investigation of the rare-earth metals. A spectroscope is an optical instrument for producing spectral lines and measuring their wavelengths and intensities.

He investigated the properties of the forerunner of the cathode ray tube (an evacuated glass tube known as a Crookes tube). He showed that the rays in the tube traveled in straight lines, caused phosphorescence in objects upon which they impinge, and by their impact produced great heat. He mistakenly believed the rays to consist of streams of particles of ordinary molecular magnitude. J. J. Thomson (1855-1941) later showed that cathode rays consist of streams of electrons. Nevertheless, Crookes's experimental work in this field was the foundation of discoveries which eventually changed the whole of chemistry and physics.

In 1903, Crookes achieved the separation from uranium of its active transformation product, which later turned out to be protactinium. At about the same time, he observed that when particles ejected from radio-active substances impinge upon zinc sulfide, each impact is accompanied by a minute scintillation—an observation which forms the basis of one of the most useful methods in radioactivity techniques.

Crookes was knighted in 1897. He died in London on April 4, 1919, and is buried in London's Brompton Cemetery.

1.   Sir William Crookes, NNDB, www.nndb.com/people/965/000100665.
2.   William Crookes, Sir : 1832 – 1919, Adventures in Cybersound, www.acmi.net.au/AIC/CROOKES_BIO.
3.   William Crookes, Wikipedia, http://en.wikipedia.org/wiki/William_Crookes.

Joseph Stefan (1835-1893)

Related radiation to absolute temperature

Joseph Stefan, the son of an illiterate shopkeeper, was born at St. Peter near Flagenfurt on March 24, 1835. His parents, both ethnic Slovenes, were married when Joseph was 11. His father was a milling assistant and his mother was a maid servant.

After graduating at the top of his class in the gymnasium, Stefan briefly considered joining the Benedictine order, but his great interest in physics prevailed.

Stefan graduated in mathematics and physics from the University of Vienna in 1857, and became Professor of Physics at Vienna in 1863.

In 1866, Stefan was appointed Director of the Institute for Experimental Physics in Vienna. He remained there for the rest of his life.

Stefan's wide-ranging work included investigations into electromagnetic induction, thermomagnetic effects, optical interference, thermal conductivity of gases, diffusion, capillarity, and the kinetic theory of gases. Based on his calculations of diffusion, modern meteorologists have found a flow between the droplets of water and ice crystals—today named **Stefan's flow**.

Stefan is best known for his work in 1879 on heat radiation. By analyzing measurements of a glowing platinum wire, he demonstrated that the rate of radiation of energy from a hot body is proportional to the fourth power of its temperature:

$$P/A = \delta T^4$$

where power "P" is in watts, area "A" is in meters squared, temperature "T" is kelvin, and the Stefan-Boltzmann constant "δ" is equal to 5.67×10^{-8} W/m^2K^4.

Subsequently, in 1884 Stefan's pupil, Ludwig Boltzmann (1844-1906), gave this relationship a theoretical foundation, which became the basis of the theory of gases proposed by James Clerk Maxwell (1831-1879). This relationship is know known as the **Stefan-Boltzmann Law**, and was used to make the first satisfactory estimate of the sun's surface temperature.

Stefan died in Vienna, Austria-Hungary on January 7, 1893.

1. Joseph Stefan, Answers.com, www.answers.com/topic/joseph-stefan?cat=technology.
2. Joseph Stefan, Microsoft Encarta Encyclopedia, 2003.
3. Stefan-Boltzmann law, The Free Dictionary by Farlex, http://encyclopedia.farlex.com/Stefan-Boltzmann+law.
4. Stephan Joseph (1835-1893), Biographies/Images of Physicists and Astronomers, www.mlahanas.de/Physics/Bios/StefanJosef.

Johannes van der Waals (1837-1923)

Developed an equation of state for non-ideal substances

Johannes Diderik van der Waals, the son of a carpenter, was born in Leiden, the Netherlands on November 23, 1837. From 1862 to 1865 he attended the University of Leiden to obtain certification to teach mathematics and physics in high schools. In 1873, he obtained his doctor's degree at Leiden University. His thesis work involved developing a model in which the liquid and the gas phase of a substance merge into each other in a continuous manner. In deriving the equation of state, he assumed not only the existence of molecules, but also that they were of finite size and attracted each other. This equation is now known as the **van der Waals equation**. It is based on a modification of the ideal gas law. The intermolecular attractive forces are known as van **der Waals forces**.

Van der Waals forces include momentary attractions between molecules, diatomic free elements, and individual atoms. They differ from covalent and ionic bonding in that they are not stable, but are caused by momentary polarization of particles.

The **van der Waals radius** of an atom is the radius of an imaginary hard sphere which can be used to model the atom for many purposes. Van der Waals radii are determined from measurements of atomic spacing between pairs of unbonded atoms in crystals.

Van der Waals published *The Law of Corresponding States* in 1880. This law shows that, after scaling temperature, pressure, and volume by their respective critical values, a general form of the equation of state is obtained which is applicable to all substances. This law served as a guide during the experiments that led to the liquefaction of helium by Heike Kamerlingh Onnes (1853-1926).

In 1877, van der Waals was appointed the first professor of physics at the newly-established University of Amsterdam.

In 1890, van der Waals suggested the notion of binary solutions—states in which a substance exists as both a gas and a liquid at the same time. The calculations that van der Waals made on binary solutions later proved crucial in the fledgling field of cryogenics.

For his pioneering work on the equation of state of liquids and gases van der Waals won the 1910 Nobel Prize in physics. He died in Amsterdam on March 8, 1923.

1. Johannes Diderik van der Waals, BookRags, www.bookrags.com/ Johannes_Diderik_van_der_Waals.
2. Johannes Diderik van der Waals, Infoplease.com, www.infoplease.com/ce6/people/A0850432.
3. Johannes Diderik van der Waals, Wikipedia, http://en.wikipedia.org/wiki/ Johannes_Diderik_van_der_Waals.

Ernst Mach (1838-1916)

Studied conditions that occur when an object moves through fluid at high speed

Ernest Mach was born in Brno, Czechia on February 8, 1838. He studied mathematics, physics, and philosophy at the University of Vienna, and received a doctorate in physics in 1860. In 1864, he took a job as professor of mathematics in Graz, and in 1866 he was appointed professor of physics. In 1867, he took the chair of Professor of Experimental Physics at Charles-Ferdinand University in Prague, where he stayed for 28 years.

Most of Mach's studies in the field of experimental physics were devoted to interference, diffraction, polarization, and refraction of light in different media under external influences.

Mach is best known for his studies of supersonic velocity. In 1887, he described the sound that occurred during the supersonic motion of a projectile. He experimentally confirmed the existence of a cone-shaped shock wave, which has the projectile at its apex. The ratio of the speed of projectile to the speed of sound, v_p/v_s, is now called the **Mach number**. It plays a crucial role in aerodynamics and hydrodynamics.

In his book *The Science of Mechanics,* which was published in 1893, Mach put forth the idea that it did not make sense to speak of the acceleration of a mass relative to absolute space. Rather, one would do better to speak of acceleration relative to the distant stars. The phrase, **Mach principle**, was coined by Albert Einstein (1879-1955) in 1918. It is the name given by Einstein to a vague hypothesis first supported by Mach. The broad notion is that "mass there influences inertia here." This concept was a guiding factor in Einstein's development of the general theory of relativity.

Mach bands are an optical illusion discovered and named after Ernst Mach. The term refers to bands adjacent to a light to dark gradient that appear lighter or darker than justified by the underlying light. It is usually supposed that this effect is caused by lateral inhibition of the receptors in the eye.

In 1897, Mach suffered a stroke, and in 1901 he retired from the University and was appointed to the upper chamber of the Austrian parliament. He died in Haar, Germany on February 19, 1916.

1. Ernst Mach, Wikipedia, http://en.wikipedia.org/wiki/Ernst_Mach.
2. Lateral inhibition and adaptation, Vision, www.siggraph.org/education/materials/ HyperVis/ vision/latinib..
3. Mach's Principle, Wolfram Research, http://scienceworld.wolfram.com/physics/ MachsPrinciple.
4. Sears, Francis W., Mark W. Zemansky, and Hugh D. Young, College Physics, Seventh Edition, Addison Wesley, New York, 1991.

Josiah Gibbs (1839-1903)

Founder of chemical thermodynamics and physical chemistry

Josiah Willard Gibbs was born in New Haven, Connecticut on February 11, 1839. His father was a professor of sacred literature at the Yale Divinity School, and Josiah was educated at Yale University. In 1863, Yale awarded him the first American Ph.D. in engineering.

In 1866, Gibbs went to Europe to study, spending a year each at Paris, Berlin, and Heidelberg, where he was influenced by Kirchhoff (1824-1887) and Helmholtz (1821-1894). In 1871, he was appointed professor of mathematical physics at Yale, where he remained until his death.

In 1873, at 34 years of age, Gibbs published his first work, titled *Graphical Methods in the Thermodynamics of Fluids*. The paper included the thermodynamics formula for which he is most famous:

$$dU = TdS - PdV$$

where "U" is the internal energy, "T" is Kelvin temperature, "S" is entropy, and "P" is pressure.

Between 1876 and 1878 Gibbs wrote a series of papers collectively titled *On the Equilibrium of Heterogeneous Substances*. In these papers, he applied thermodynamics to physicochemical phenomena. He developed mathematical techniques for predicting the outcome of chemical reactions before making a physical trial. He also derived the laws of thermodynamics from the basic concept of heat as movement of particles.

Gibbs also did work in statistical mechanics, providing a mathematical framework for quantum theory. He contributed to crystallography, the determination of planetary and comet orbits, and electromagnetic theory.

Gibbs was also interested in the practical side of science. He received a patent in 1866 for an improved type of railroad brake.

Gibbs never married, living all his life in his childhood home with a sister and his brother-in-law, the Yale librarian. Gibbs died in New Haven on April 28, 1903, and is buried in Grove Street Cemetery.

Gibbs is recognized as the founder of chemical thermodynamics and physical chemistry.

1. Gibbs (Josiah) Willard, Answers.com, www.answers.com/topic/willard-gibbs?cat=technology.
2. Gibbs J(osiah) Willard, Microsoft Encarta Encyclopedia Online, http://uk.encarta.msn.com/encyclopedia_761573954/Gibbs_J(osiah)_Willard. 2007.
3. Josiah Willard Gibbs, Biographies/Images of Physicists and Astronomers, www.mlahanas.de/ Physics/Bios/ WillardJGibbs.
4. Josiah Willard Gibbs, Wikipedia, http://en.wikipedia.org/wiki/Willard_Gibbs.
5. Josiah Willard Gibbs (1839-1903), Corrosion Doctors, www.corrosion-doctors.org/Biographies/ GibbsBio.

Ernst Abbe (1840-1905)

Made major inprovements to the microscope

Ernst Karl Abbe was born in Eisenach, Germany on January 23, 1840. He attended both the University of Jena and the University of Göttingen, earning his doctorate from the latter in 1861. He was appointed to a teaching position at Jena in 1863, and eventually became professor of physics and mathematics and the director of both the astronomical and meteorological observatories.

In 1866, Carl Zeiss (1816-1888) appointed Abbe research director of Zeiss Optical Works, a manufacturer of microscopes. Abbe remained in his teaching and directorship positions at the University, and used his combined resources to develop major advances in optical science.

In 1868, Abbe invented the apochromatic lens system for the microscope. This important breakthrough eliminates both the primary and secondary color distortion of microscopes.

Abbe created the mathematical foundation of microscope design. In particular, he discovered the **Abbe sine condition**, a breakthrough in lens design. It is a condition that must be fulfilled by a lens or other optical system for it to produce sharp images of off-axis as well as on-axis objects.

Abbe designed the first refractometer, an optical instrument that is used to determine the refractive index of a substance. A refractometer can be used to determine the identity of an unknown substance, based on its refractive index; to assess the purity of a particular substance; or to determine the concentration of one substance dissolved in another.

Abbe discovered the **Abbe number**, also known as the V-number or constringence, which is a measure of any transparent material's variation of refractive index with wavelength. He also invented the **Abbe condenser**, mounted below the stage of the microscope, which concentrates and controls the light that passes through the specimen and enters the objective. It has two controls, one which moves the Abbe condenser closer to or further from the stage, and another, the iris diaphragm, which controls the diameter of the beam of light. The controls can be used to optimize brightness, evenness of illumination, and contrast.

In 1876, Zeiss made Abbe a partner in his business, which made him a wealthy man. Zeiss died in 1888, leaving the entire Zeiss Works to Abbe. Abbe died in Jena on January 14, 1905.

1. Ernst Carl Abbe, BookRags, :www.bookrags.com/Ernst_Karl_Abbe.
2. Ernst Carl Abbe, Wikipedia, http://en.wikipedia.org/wiki/Ernst_Abbe.
3. O'Connor, J. J. and E F Robertson Ernst Abbe, MacTutor, www-history.mcs.st-andrews.ac.uk/history/Printonly/Abbe.

Alfred Cornu (1841-1902)

Made a more accurate measurement of the speed of light

Marie Alfred Cornu was born on March 6, 1841 in Orléans, France. After being educated at the École polytechnique and the École des mines, in 1867 he became professor of experimental physics at École polytechnique, where he remained the rest of his life.

A proponent of the wave theory of light, Cornu was also interested in the relationship between electricity and optics and the understanding of weather phenomena. He played a considerable part in the creation of the Observatory of Nice.

Cornu made a wide variety of contributions to the fields of optics and spectroscopy, but is most noted for significantly increasing the accuracy of the calculation of the speed of light.

In 1878, Cornu made adjustments to an earlier method of measuring the velocity of light developed by Armand Fizeau (1819-1896) in the 1840s. The changes and improved equipment resulted in the most accurate measurement taken up to that time—2, 299, 990 kilometers per second. This achievement won for him, in 1878, the Prix Lacaze and membership of the Academy of Sciences in France, and the Rumford medal of the Royal Society in England.

Knowledge of the exact speed of light in a vacuum is important for a number of reasons, one of them being that the meter is now defined as the distance light travels through a vacuum during a time interval of 1/299,792,458 seconds.

Other significant accomplishments of Cornu include a photographic study of ultraviolet radiation and the establishment of a graphical approach known as the **Cornu spiral**.

The Cornu spiral is a graphical aid for evaluating the Fresnel integrals, which show up in the evaluation of the diffraction intensities for the Fresnel diffraction of the light from a slit. It lends itself to the calculation of diffraction from slits, barriers, and opaque strips.

In 1899, at the jubilee commemoration of Sir George Stokes (1819-1903), Cornu was Rede lecturer at Cambridge. His subject was *The Udulatory Theory of Lght and its Influence on Modern Physics*. On that occasion, the University conferred on him the honorary degree of D.Sc.

Cornu died in Paris, France on April 12, 1902.

1.   Alfred Cornu, NNDB, www.nndb.com/people/962/000100662.
2.   Marie Alfred Cornu, Classic Encyclopedia, www.1911encyclopedia.org/Marie_Alfred_Cornu.
3.   Marie Alfred Cornu, Molecular Expressions, Science, Optics and You,
      http://micro.magnet.fsu.edu/optics/timeline/ people/cornu.
4.   Marie Alfred Cornu, Wikipedia, http://en.wikipedia.org/wiki/Marie_Alfred_Cornu.

James Dewar (1842-1923)
Investigated low-temperature phenomena

James Dewar was born in Kincardine-on-Forth, Scotland on September 29, 1842, and was educated at the University of Edinburgh. In 1869, he was appointed lecturer in chemistry at the Royal Veterinary College, Edinburgh, and from 1873 also held the post of assistant chemist to the Highland and Agricultural Society of Scotland.

In 1875, Dewar was elected Jacksonian professor of natural experimental philosophy at Cambridge, becoming a fellow of Peterhouse, and in 1877 became Fullerian professor of chemistry in the Royal Institution, London.

With the British chemist, Sir Frederick Abel (1827-1902), Dewar invented cordite, a smokeless gunpowder, which for a long time was the British standard military propellant.

In 1867, Dewar developed structural formulas for benzene. And using charcoal as an absorbing agent, he separated hydrogen, neon, and helium from the air, and created the highest known vacuum of his day.

Dewar also carried out work in spectroscopy, particularly concerned with the absorption spectra of metals.

Dewar is best known for his work with low-temperature phenomena. He studied the specific heat of hydrogen. In 1898, he was the first person to produce hydrogen in liquid form and to solidify it in 1899. In 1891, Dewar constructed a machine for producing liquid oxygen in quantity.

Dewar's discovery in 1905 that cooled charcoal can be used to help create high vacuums later proved useful in atomic physics.

It was Dewar's work at low temperatures which led to the idea of the thermos or vacuum, known as the **Dewar flask**. It had double walls with a vacuum between and silvering on the inner walls helped to reflect heat. The common, modern-day thermos bottle is an adaptation of the Dewar flask.

The vacuum flask was not manufactured for commercial or home use until 1904, when two German glass blowers formed Thermos GmbH. They held a contest to rename the vacuum flask, and a resident of Munich submitted "Thermos," which came from the Greek word "Therme" meaning "hot."

Dewar was knighted in 1904. He died in London, England on March 27, 1923, and his remains were cremated at Golders Green.

1. James Dewar, Microsoft Encarta Encyclopedia Online, http://encarta.msn.com/encyclopedia_761562847/dewar_sir_james, 2007.
2. Sir James Dewar (1842-1923), Famous Scots www.rampantscotland.com/famous/blfamdewar.
3. Sir James Dewar, Inventers, About.com, iinventors.about.com/library/inventors/blthermos.

Osborne Reynolds (1842-1912)

Developed a mathematical framework for turbulence

Osborne Reynolds was born into a family of Anglican clerics in Belfast, Ireland on August 23, 1842. He did not go straight to university after his secondary education, but took an apprenticeship with the engineering firm of Edward Hayes in 1861. After gaining experience in the engineering firm, he studied mathematics at Cambridge, graduating in 1867. He again took up a post with an engineering firm, this time the civil engineers John Lawson of London, spending a year as a practicing civil engineer. In 1868, he became the first professor of engineering in Manchester (and the second in England), a position he held until his retirement.

Reynolds' earliest professional research dealt with electromagnetic properties of the Sun and of comets. He soon began to concentrate on fluid mechanics, and it was in this area that his contributions were of world-leading importance. He studied the change in a flow along a pipe when the flow goes from laminar to turbulent.

A paper Reynolds published in 1883 on the motion of water in parallel channels introduced what is now known as the **Reynolds number**, which is commonly used in modeling fluid flow. The Reynolds number is ratio of inertial forces to viscous forces. Laminar flow occurs at lower Reynolds numbers and turbulent flow occurs at higher ones.

In 1889, Reynolds produced an important theoretical model for turbulent flow, which has become the standard mathematical framework used in the study of turbulence. In 1886, he formulated a theory of lubrication.

Reynolds' studies of condensation and heat transfer between solids and fluids brought radical revision in boiler and condenser design, and his work on turbine pumps permitted their rapid development. He also studied wave engineering and tidal motions in rivers and made pioneering contributions to the concept of group velocity.

Among Reynolds' other contributions were the explanation of the radiometer and an early absolute determination of the mechanical equivalent of heat.

By the beginning of the 1900s Reynolds' health began to fail and he retired in 1905. He died in Watchet, Somerset, England in February 21, 1912.

1. Osborne Reynolds (1842-1912), Corrosion Doctors, www.corrosion-doctors.org/Biographies/ ReynoldsBio.
2. Osborne Reynolds, Encyclopedia Britannica Online, www.britannica.com/ eb/article-9063387/ Osborne-Reynolds.
3. Reynolds number, Wikipedia, http://en.wikipedia.org/wiki/ Reynolds_number.

William Strutt (1842-1919)

Discovered the element argon

John William Strutt was a physicist and mathematician who worked in many disciplines, including electromagnetics, physical optics, and sound wave theory. He was born on November 12, 1842 in Langford Grove, Essex, England. Upon the death of his father, the second Lord Rayleigh, Strutt became the third Lord Rayleigh. In his early years, he suffered poor health, catching smallpox, and then whooping cough.

In 1861, Strutt entered Trinity College, Cambridge where he exhibited strong promise in mathematics and an avocation for photography. In 1865, he obtained his B.A. and obtained his M.A. in 1868 from the same institution. In 1879, Strutt suceeded James Clerk Maxwell (1831-1879) as professor of physics at the University of Cambridge.

An attack of rheumatic fever in 1872 almost cost him his life. It was on a recuperative trip to Egypt that he began his work on the theory of sound. The first volume was released in 1877 and a second in 1878. Their discussion of vibration, resonance, and acoustics remains one of the principal accomplishments in the field.

Another undertaking by Strutt was a mathematical explanation for the light scattering that gives the sky its blue appearance. The **Rayleigh scattering** law evolved from this theory and has become a landmark in the study of wave propagation. Rayleigh also examined the precision of electrical measures and standardized the ohm.

In optics, the **Rayleigh Criterion** defines the limit of resolution of a diffraction-limited optical instrument. The criterion is defined as the condition that arises when the center of one diffraction pattern is superimposed with the first minimum of another diffraction pattern produced by a point or line source equally bright as the first.

Strutt discovery of argon was due to the detection of a minute inconsistency in density between atmospheric nitrogen and chemical nitrogen. He deduced that there must be a heavier gas than nitrogen in air. Isolating the element proved extremely difficult: however, Strutt eventually succeeded. The gas was named argon, the Greek word for inactive, because it refused to combine chemically with other substances. For this work, Strutt received the Nobel Prize in 1904. He died on June 30, 1919 in Terling Place, Witham, Essex, England.

1. John Strutt 3rd Baron Rayleigh, Wikipedia, http://en.wikipedia.org/wiki/John_Strutt,_3rd_Baron_Rayleigh.
2. Lord Rayleigh (John William Strutt), Optical Expressions, http://micro.magnet.fsu.edu/optics/timeline/people/rayleigh.
3. Lord Rayleigh, History of Ultrasound in Obstetrics and Gynecology, www.ob-ultrasound.net/rayleigh.

Ludwig Boltzmann (1844-1906)

Made founding contributions in the fields of statistical mechanics and statistical thermodynamics.

Ludwig Eduard Boltzmann was born in Vienna, then capital of the Austrian Empire, on February 20, 1844. His father, George Boltzmann, was a tax official. Soon after Ludwig's birth the family moved to the Upper Austria, first to Wels and later to Linz.

Boltzmann received his Ph.D. in physics in 1866, working under the supervision of Joseph Stefan (1835-1893). In 1869, he was appointed full Professor of Mathematical Physics at the University of Graz. Then, after faculty positons at the University of Vienna and the University of Munich, he succeeded Stefan at the University of Vienna.

Boltzmann's most important contributions were in kinetic theory. He established the relationship between entropy and the statistical analysis of molecular motion in 1877, founding the branch of physics known as statistical mechanics. The **Boltzmann distribution** predicts the distribution function for the fractional number of particles occupying a set of states, which each respectively possess a given energy. Boltzmann's work paved the way for the development of quantum mechanics, which is inherently a statistical theory.

Boltzmann's constant (1.3807×10^{-23} joules per kelvin) defines the relation between absolute temperature and the kinetic energy contained in each molecule of an ideal gas, and is equal to the ratio of the gas constant "R" to the Avogadro constant 6.022×10^{23}.

The **Boltzmann's equation** describes the statistical distribution of particles in a fluid. It is used to study how a fluid transports physical quantities, such as heat and charge, and thus to derive transport properties, such as electrical conductivity, Hall conductivity, viscosity, and thermal conductivity.

Boltzmann was subject to depressed moods alternating with elevated moods—almost certainly bipolar disorder. On September 5, 1906, while on a summer vacation in Duino, Boltzmann committed suicide. He is buried in the Central Cemetery in Vienna. The equation S = klogW, where "S" is entropy, "k" is Boltzmann's constant, and "W" is the number of possible states of a system, is written on his tombstone.

1. Boltzmann, Ludwig (1844-1906), Wolfram Research, http://scienceworld.wolfram.com/biography/Boltzmann.
2. Boltzmann's Constant, Whais.com, http://whatis.techtarget.com/definition/0,,sid9_gci858320,00.
3. Ludwig Boltzmann, Biographies.Images Physicists and Astronomers, www.mlahanas.de/Physics/Bios/LudwigBoltzmann.
4. Ludwig Boltzmann, Wikipedia, http://en.wikipedia.org/wiki/Ludwig_Boltzmann.

Gabriel Lippmann (1845-1921)

Reproduced colors photographically based on the phenomenon of interference

Gabriel Jonas Lippmann was born to Franco-Jewish parents in Hollerich, Luxemburg on August 16, 1845. When Gabriel was three, his family moved back to France to live in Paris, where he was home schooled. He entered the École Normale Supérieure in 1868, and received a D.Sc. degree from the Sorbonne in 1875.

In 1878, Lippmann joined the Faculty of Science in Paris and became director of the research laboratory in 1886.

Lippmann invented the first process for making color photographs based on the phenomenon of interference—the combining of different light waves arriving simultaneously at the same point. To receive the image, Lippmann used a glass plate coated on one side with a light-sensitive emulsion—a mixture of gelatin, grains of silver nitrate, and potassium bromide. In the camera, the emulsion side of the plate faced a plate holder coated with mercury, which acted as a mirror. When the camera lens was opened, light was reflected from the objects in the lens's field of view through the lens to the emulsion-coated plate and through the plate to the mirror; the various wavelengths of this light corresponded to the various colors of the objects in the field of view. The incoming light was then reflected back into the emulsion by the mirror, where they created interference patterns in the silver grains of the emulsion. These patterns were then fixed on the plate by chemical baths. When the plate dried, the interference patterns reflected light in various wavelengths corresponding to the original colors of the photographic objects.

Lippmann invented the capillary electrometer, which could detect small electric currents by the movement of mercury in the apparatus's tube. For many years it was used in electrocardiograph machines. He also did research in piezoelectricity and seismology.

In addition, Lippmann invented the coelostat, an astronomical tool that compensated for the earth's rotation, and allowed a region of the sky to be photographed without apparent movement.

In 1908, Lippmann was awarded the Nobel Prize in Physics for reproducing colors photographically based on the phenomenon of interference. Lippmann died on July 13, 1921.

1. Gabriel Lippmann, Jewish Virtual Library, www.jewishvirtuallibrary.org/jsource/ biography/ lippmann.
2. Gabriel Lippmann, Microsoft Encarta Encyclopedia Online, http://encarta.msn.com/ encyclopedia_761583250/lippmann_gabriel_jonas, 2007.
3. Gabriel Lippmann, Wikipedia, http://en.wikipedia.org/wiki/Gabriel_Lippmann.

Wilhelm Röntgen (1845-1923)

Produced and detected electromagnetic radiation in a wavelength range today known as X-rays or Röntgen rays

Wilhelm Conrad Röntgen was born on March 27, 1845 at Lennep in the Lower Rhine Province of Germany. He was the only child of a merchant and manufacturer of cloth. In 1848, the family moved to Apeldoorn and Wilhelm was raised in the Netherlands. He attended the Federal Polytechnic Institute in Zurich as a student of mechanical engineering. In 1869, he graduated with a Ph.D. After a number of appointments at other universities, in 1900 he obtained the physics chair at the University of Munich by special request of the Bavarian government.

Röntgen worked on elasticity, capillary action of fluids, specific heats of gases, conduction of heat in crystals, absorption of heat by gases, and piezoelectricity. However, he is best known for the discovery of X-rays.

In 1895, Röntgen was repeating an experiment with one of the tubes of Philipp Lenard (1862-1947) in which a thin aluminium window had been added to permit the cathode rays to exit the tube. A cardboard covering was added to protect the aluminium from damage by the strong electrostatic field that was necessary to produce the cathode rays. Röntgen observed that the invisible cathode rays caused a fluorescent effect on a small cardboard screen painted with barium platinocyanide when it was placed close to the aluminium window. Using a Hittorf-Crookes tube, he covered it with cardboard and attached electrodes to a Ruhmkorff coil to generate an electrostatic charge. As he passed the charge through the tube, he turned to prepare the next step of the experiment and noticed a faint shimmering from a bench a meter away from the tube. It was coming from the barium platinocyanide screen he had been intending to use next. Röntgen speculated that a new kind of ray, which he called X-rays, might be responsible for the effect. Further investigation revealed that paper, wood, and aluminum, among other materials, are transparent to this new form of radiation. Eventually, X-rays came to bear Röntgen's name— Röntgen Rays. Röntgen's discovery of X-rays heralded the age of modern physics and revolutionized diagnostic medicine.

In 1901, Röntgen was awarded the first Nobel Prize in Physics. He died on February 10, 1923 of carcinoma of the intestine, not believed to be due to his work with X-rays.

1. Wilhelm Conrad Roentgen, Microsoft Encarta Encyclopedia Online, http://encarta.msn.com/ encyclopedia_761555545/roentgen, 2007.
2. Wilhelm Conrad Rontgen, Famous Scientists, Crystalinks, www.crystalinks.com/rontgen.
3. Wilhelm Conrad Röntgen, Wikipedia, http://en.wikipedia.org/wiki/ Wilhelm_Conrad_R%C3%B6ntgen.

Thomas Edison (1847-1931)

Invented the carbon microphone and phonograph

Thomas Alva Edison was born in Milan, Ohio on February 11, 1847. He was raised in Port Huron, Michigan. He was partially deaf from a young age. The cause was uncertain. Early on, Edison became a telegraph operator His first telegraphy job away from Port Huron was at Stratford Junction, Ontario on the Grand Trunk Railway. In 1866, at the age of 19, he moved to Louisville, Kentucky, where he was an employee of Western Union. He requested the night shift, which allowed him plenty of time to read and experiment. One night in 1867, he was working with a battery when he spilled sulphuric acid. It ran between the floorboards and onto his boss's desk below. The next morning Edison was fired.

Edison began his career as an inventor in Newark, New Jersey with the automatic repeater and improved telegraphic devices, but the invention which first gained him fame was the 1877 phonograph, which initially consisted of recording on tinfoil wrapped around a grooved cylinder.

During 1877 and 1878, Edison invented and developed the carbon microphone, which was used in all telephones, along with the Bell receiver, until the 1980s. His microphone was also used in radio broadcasting and public address work through the 1920s.

Edison's first successful test of his light bulb was in 1877. He continued to improve the design, and in 1880 he was granted a patent for an electric lamp using a carbon filament or strip which was coiled and connected to platina contact wires. To avoid a possible court battle with Joseph Swan (1828-1914), whose British patent had been awarded a year before Edison's, Swan and Edison formed a joint company called Ediswan to market the invention in Britain.

Edison patented an electric distribution system in 1880, which was essential to capitalize on the electric lamp, and in that same year he founded the Edison Electric Illuminating Company. On September 4, 1882, Edison switched on his Pearl Street generating station's electrical power distribution system, which provided 110 volts direct current to 59 customers in lower Manhattan.

Edison is also credited with inventing the fluoroscope and improving the telephone of Alexander Graham Bell (1847-1922). He died on October 18, 1931 at 84 years of age.

1.   Bellis, Mary, The Inventions of Thomas Edison, Inventers, About.com, http://inventors.about.com/ library/ inventors/bledison..
2.   Edison, Thomas Alva, Microsoft Encarta Encyclopedia Online, http://encarta.msn.com/ encyclopedia_761563582/edison_thomas_alva, 2007.
3.   Thomas Edison, Wikipedia, http://en.wikipedia.org/wiki/Thomas_Edison.

Roland von Eötvös (1848-1919)

Demonstrated the equivalence of gravitational and inertial mass

Roland von Eötvös was born in Hungary on July 27, 1848. To satisfy the wishes of his family, he entered the University of Budapest in 1865 in pursuit of a law degree. He terminated his law studies in 1867 and went to study physics at Heidelberg, where he was taught by Gustav Kirchhoff (1824-1887), Hermann Von Helmholtz (1821-1894), and Robert Bunsen (1811-1899). He received a doctorate in physics at Heidelberg in 1870.

Eötvös is primarily remembered for his experimental work on gravity, in particular his study of the equivalence of gravitational and inertial mass, and his study of the gravitational gradient on the earth's surface.

Eötvös's weak equivalence principle plays a prominent role in relativity theory, and the Eötvös experiment was cited by Albert Einstein (1879-1955) in his 1916 paper *The Foundation of the General Theory of Relativity*. Measurements of the gravitational gradient are important in applied geophysics, such as the location of petroleum deposits.

In the early 1900s, a German team from the Institute of Geodesy in Potsdam carried out gravity measurements on moving ships in the Atlantic, Indian, and Pacific Oceans. While studying their results the Eötvös noticed that the readings were lower when the boat moved eastwards and higher when it moved westward. He identified the cause as a consequence of the rotation of the Earth. The phenomenon has been named the **Eötvös effect**.

In fluid dynamics, the **Eötvös number** is a dimensionless number, which may be regarded as proportional to buoyancy forces divided by surface tension forces. The **Eötvös rule** enables the prediction of the surface tension of an arbitrary liquid pure substance at all temperatures. The density, molar mass, and the critical temperature of the liquid have to be known.

From 1886 until his death on April 8, 1919, Eötvös researched and taught at the University of Budapest, which in 1950 was renamed after him (Eötvös Loránd University). The CGS unit for gravitational gradient is named the **eotvos** in his honor.

1. Eötvös Effect, Wikipedia, http://en.wikipedia.org/wiki/E%C3%B6tv%C3%B6s_effect.
2. Eötvös Number, Wikipedia, http://en.wikipedia.org/wiki/E%C3%B6tv%C3%B6s_number.
3. Eötvös Rule, Wikipedia, http://en.wikipedia.org/wiki/E%C3%B6tv%C3%B6s_rule.
4. Loránd Eötvös, Wikipedia, http://en.wikipedia.org/
    wiki/Lor%C3%A1nd_E%C3%B6tv%C3%B6s.
5. O'Conner, J .J. and E. F. Robertson, Loránd Baron von Eötvös, MacTutor Biography, www-groups.dcs.st-and.ac.uk/~history/Mathematicians/Eotvos.
6. Roland von Eötvös, Biographies/Images of Physicists and Astronomers, www.mlahanas.de/
    Physics/Bios/RolandE%f6tv%f6s..

Ferdinand Braun (1850-1918)

Contributed to the development of wireless telegraphy and built the first
cathode ray tube oscilloscope

Karl Ferdinand Braun was born on June 6, 1850 in Fulda, Germany. He
was educated at the University of Marburg, and received a Ph.D. from the
University of Berlin in 1872. His dissertation was on the vibrations of
elastic rods and strings. He began his career as a research assistant at the
University of Würzburg, and later held positions at universities in Leipzig,
Marburg an der Lahn, Karlsruhe, and Tübingen, where he founded the
Physics Institute.

In 1874, Braun discovered the rectifying action that occurs when a
crystal of galena is probed by a metal point. This discovery led to the
development of crystal radio detectors in the early days of wireless
telegraphy and radio.

Braun became director of the Physical Institute and professor of
physics at the University of Strasbourg in 1895.

In 1897, Braun built the first cathode-ray tube (CRT) oscilloscope.
CRT technology is to this day used by most television sets and computer
monitors. The CRT is still called the **Braun tube** in German-speaking
countries and in Japan.

Around 1898, Braun invented a crystal diode rectifier known as the
cat's whisker diode.

Studying the transmitter of Guglielmo Marconi (1874-1937), which
used a sparking apparatus to produce periodic waves that traveled through
the air, Braun learned that attempts to increase the power output by
increasing the length of the spark gap eventually reached a limit at which
the spark caused a decrease in output. Braun solved this problem by
producing a sparkless antenna circuit. He magnetically coupled the power
from the transmitter to the antenna circuit using a transformer effect
instead of placing the antenna directly in the power circuit.

In 1909, Braun shared the Nobel Prize for physics with Marconi for
his contributions to the development of wireless telegraphy.

Braun was called to the United States in 1914 to testify in litigation
involving radio broadcasting. He was still in the country in 1917 when the
United States entered World War I, and he was not allowed to return to
Germany. Braun died in a Brooklyn, New York hospital in 1918.

1. Karl Ferdinand Braun, Institute of Chemistry, http://chem.ch.huji.ac.il/history/braun, May 10, 2003.
2. Karl Ferdinand Braun, Microsoft Encarta Encyclopedia Online, http://encarta.msn.com/encyclopedia_761583126/Braun_Karl_Ferdinand , 2007.
3. Karl Ferdinand Braun, Wikipedia, http://en.wikipedia.org/wiki/1872.

Oliver Heaviside (1850-1925)

Reformulated Maxwell's field equations in terms of electric and magnetic forces and energy flux

Oliver Heaviside was born in London's Camden Town on May 18, 1850. He was a self-taught electrical engineer, mathematician, and physicist. In his youth, scarlet feaver left him partially deaf. At age 16 he left school and began learning about telegraphy and electromagnetism. He became a telegraph operator, and worked for Great Northern Telegraph Company.

Heaviside showed mathematically that uniformly distributed inductance in a telegraph line would diminish both attenuation and distortion, and that if the inductance were great enough and the insulation resistance not too high, the circuit would be distortionless, while currents of all frequencies would be equally attenuated. Heaviside's equations helped further the implementation of the telegraph. In 1880, he patented, in England, the co-axial Cable.

In 1884, Heaviside recast the mathematical analysis of James Clerk Maxwell (1831-1879) from its original cumbersome form to its modern vector terminology, thereby reducing the original equations to four differential equations we now know as Maxwell's equations.

Between 1880 and 1887, Heaviside developed the operational calculus involving the "D" notation for the differential operator, a method of solving differential equations by transforming them into ordinary algebraic equations.

In two papers of 1888 and 1889, Heaviside calculated the deformations of electric and magnetic fields surrounding a moving charge, as well as the effects of it entering a denser medium. This included a prediction of what is now known as Cherenkov radiation, and inspired George FitzGerald (1851-1901) to suggest what now is known as the **Lorentz-Fitzgerald contraction**.

In 1902, Heaviside proposed the existence of an electrically conductive layer in the upper atmosphere that reflects radio waves around the earth's curvature.. The existence of the ionosphere was confirmed in 1923 and named the **Kennelly-Heaviside Layer**.

In 1905, Heaviside was given an honorary doctorate by the University of Göttingen. He died at Torquay in Devon, England on February 3, 1925, and is buried in Paignton cemetery.

1. Oliver Heaviside (British Physicist), Encyclopedia Britannica Online, www.britannica.com/eb/topic-258889/Oliver-Heaviside.
2. Oliver Heaviside, Biographies/Images of Physicists and Astronomers, www.mlahanas.de/Physics/Bios/OliverHeaviside.
3. Oliver Heaviside, MacTutor, www-history.mcs.st-andrews.ac.uk/Biographies/Heaviside.
4. Oliver Heaviside, Wikipedia, http://en.wikipedia.org/wiki/Oliver_Heaviside.

George FitzGerald (1851-1901)

Proposed a method of producing radio waves and predicted the
foreshortening of objects in the direction of their motion

George Francis FitzGerald was born in Dublin, Ireland on August 3,
1851. His father, William FitzGerald, was Professor of Moral Philosophy
at Trinity College and vicar of the Irish Protestant Church, St Anne's. At
the age of 16, George entered Trinity College. He graduated in 1871, and
then studied on his own for six years before obtaining a fellowship at
Trinity College and becoming a tutor in the Department of Experimental
Physics.

In 1881, FitzGerald was appointed to the Erasmus Smith Chair of
Natural and Experimental Philosophy.

At a meeting of the British Association in Southport in 1883,
FitzGerald, following from Maxwell's equations, suggested a device for
producing rapidly oscillating electric current to generate electromagnetic
(radio) waves, a phenomenon later shown experimentally by Heinrich
Hertz (18571894) in 1888.

Beginning in 1876, FitzGerald published articles on the equilibrium of
an elastic surface, the rotation of the plane of polarization of light by
reflection from the pole of a magnet, and the electromagnetic theory of the
reflection and refraction of light.

FitzGerald is best known for his proposal in 1889 that if all moving
objects were foreshortened in the direction of their motion, it would
account for the result of the Michelson-Morley experiment, which is
considered to be the first strong evidence against the theory of a
luminiferous ether, which is a hypthetical medium thought necessary for
the propagation of light. FitzGerald based this idea in part on the way
electromagnetic forces were known to be affected by motion. He drew on
equations that had been derived a short time before by Oliver Heaviside
(1850-1925). Hendrik Lorentz (1853-1928), came up with a very similar
idea in 1892 and developed it more fully in connection with his theory of
electrons. The FitzGerald-Lorentz contraction later became an important
part of the special theory of relativity proposed by Albert Einstein (1879-
1955) in 1905.

FitzGerald died from a perforated ulcer at home on February 22, 1901.
He was 49 years old.

1.   Coey, J. M. D., George Francis FitzGerald, School of Physics, University of Dublin Trinity
     College, School of Physics, www.tcd.ie/Physics/history/fitzgerald/fitzgerald_introduction.php.
2.   George Fitzgerald, Wikipedia, http://en.wikipedia.org/wiki/George_Francis_FitzGerald.
3.   O'Connor, J. J. and E F Robertson, George Francis FitzGerald, www-groups.dcs.st-and.ac.uk/
     ~history/Printonly/FitzGerald.html.

Jacques d'Arsonval (1851-1940)

Invented the moving-coil galvanometer

Jacques Arsène d'Arsonval was born in La Porcherie, France on June 8, 1851. He came from a family of France's ancient nobility. He studied classics at the Lycée Imperial de Limoges and later at the Collège St.-Barbe. After obtaining a baccalaureate degree from the Université de Poitiers in 1869, he decided on a career in medicine.

A physician and physicist, d'Arsonval was an important contributor to the emerging field of electrophysiology, which is the study of the effects of electricity on biological organisms.

Among d'Arsonval's inventions were dielectric heating and various measuring devices, including the thermocouple ammeter and the moving-coil galvanometer. These tools helped establish the field of electrical engineering. The d'Arsonval galvanometer, used for measuring weak electric currents, became the basis for almost all panel-type pointer meters.

After the Franco-Prussian war of 1870, d'Arsonval went to Paris where he worked under Claude Bernard (1813-1878) from 1873 to 1878. His first projects involved animal heat and body temperature. After Bernard's death, d'Arsonval assisted Charles-Édouard Brown-Séquard (1817-1894). As Brown-Séquard's assistant he worked on endocrine extract experiments. Their investigations of the therapeutic properties of animal extracts revealed clues to the later controversial hormone theory of wound healing. They found that testicular extracts from guinea pigs had definite antiseptic properties. D'Arsonval succeeded Brown-Séquard in 1897 at the Collège de France. He was professor at the Sorbonne from 1894 to 1932.

D'Arsonval's most outstanding scientific contributions involved the biological and technological applications of electricity. Much of this work concerned muscle contractions.

In 1881, d'Arsonval proposed tapping the thermal energy of the ocean. But it was his student, Georges Claude (1870-1960), who built the first Ocean Thermal Energy Conversion (OTEC) plant in Cuba in 1930. OTEC is a method for generating electricity which uses the temperature difference that exists between deep and shallow waters to run a heat engine. D'Arsonval was also involved in the industrial application of electricity. He died on December 13, 1940.

1. Arsonval, Arsène d', InfoPlease.com, www.infoplease.com/ce6/people/A0804845.
2. Jacques-Arsène d'Arsonval, Wikipedia, http://en.wikipedia.org/wiki/Jacques-Ars%C3%A8ne_d'Arsonval.
3. Jaques-Arsène d' Arsonval, Institute of Chemistry, http://chem.ch.huji.ac.il/history/arsonval.
4. Jaques-Arsène d' Arsonval, WhoNamedIt.com, www.whonamedit.com/doctor.cfm/289.

Albert Michelson (1852-1931)

Disproved ether; determined the speed of light

Albert Abraham Michelson, the son of a Jewish merchant, was born on December 19, 1852 in what is today Strzelno, Poland. He moved to the United States with his parents in 1855 at two years of age. He grew up in the rough mining towns of Murphy's Camp, California and Virginia City, Nevada, where his father continued in the merchant business.

In 1869, President Ulysses S. Grant awarded Michelson a special appointment to the U.S. Naval Academy. After his graduation in 1873 and two years at sea, he returned to the Academy in 1875 to become an instructor in physics and chemistry.

From 1880 to 1882, Michelson undertook postgraduate study in Berlin and Paris.

In 1883, Michelson accepted a position as professor of physics at the Case School of Applied Science in Cleveland, Ohio. In 1887, he and Edward Morley (1838-1923) carried out the famous Michelson-Morley experiment to measure the velocity of the Earth through the ether, a hypothetical substance that scientists believed filled the universe and which they believed the Earth moved through. In their experiment, a single source of white light was split into two beams traveling at right angles to one another. After leaving the splitter, the beams traveled out to the ends of long arms where they were reflected. The beams then recombined on the far side of the splitter producing a pattern of constructive and destructive interference based on the length of the arms. Any slight change in the amount of time the beams spent in transit was then observed as a shift in the positions of the interference fringes. Their results helped disprove the presence of ether.

In 1889, Michelson became a professor at Clark University at Worcester, Massachusetts, and in 1892 was appointed professor and the first head of the department of physics at the newly organized University of Chicago. There, he determined the velocity of light with a high degree of accuracy, using instruments of his own design.

In 1907, Michelson was the first American to receive a Nobel Prize in Physics—for for his optical precision instruments and the spectroscopic and metrological investigations carried out with their aid. He died on May 9, 1931 in Pasadena, California at the age of 78.

1. ALBERT ABRAHAM MICHELSON 1852-1931, Selected Papers of Great American Physicists, www.aip.org/history/gap/Michelson/Michelson.
2. Albert Abraham Michelson, Wikipedia, ttp://en.wikipedia.org/wiki/Albert_Abraham_Michelson.
3. Albert Michelson, Microsoft Encarta Encyclopedia Online, http://encarta.msn.com/ encyclopedia_761555191/albert_michelson., 2007.

H. Thomas Milhorn, MD, PhD

Henri Becquerel (1852-1908)

Discovered Radioactivity

Antoine Henri Becquerel was born into a family of physicists in Paris France on December 15, 1852. He studied science at the École Polytechnique and engineering at the École des Ponts et Chaussées.

In 1892, Becquerel became the third in his family to occupy the physics chair at the Muséum National d'Histoire Naturelle. In 1894, he became chief engineer in the Department of Bridges and Highways.

Becquerel undertook to investigate whether there was some fundamental connection between X-rays and visible light, such that all luminescent materials, however stimulated, would also yield X-rays. So, in 1896, while investigating phosphorescence in uranium salts, he wrapped a fluorescent substance (potassium uranyl sulfate) in photographic plates and black material in preparation for an experiment requiring bright sunlight. However, prior to actually performing the experiment, he found that the photographic plates were exposed. He went on to discover that all uranium compounds, not just specific salts, fogged the plates. This discovery led Becquerel to investigate the spontaneous emission of nuclear radiation.

Becquerel made three more important scientific contributions. From the charge to mass value he showed that the beta particle was the same as the recently identified electron of Joseph John Thompson (1856-1940).

Another discovery was the circumstance that the allegedly active substance in uranium lost its radiating ability in time, while the uranium, though inactive when freshly prepared, eventually regained its lost radioactivity.

Becquerel's last major achievement concerned the physiological effect of the radiation. His report in 1901 of the burn caused when he carried an active sample of radium in his vest pocket inspired investigations by physicians, leading ultimately to medical use.

Becquerel was awarded a share of the 1903 Nobel Prize in physics for his discovery of spontaneous radioactivity. He shared the Prize with Pierre Curie (1859-1906) and Marie Curie (1867-1934) for their joint researches on the radiation phenomena discovered by Becquerel.

Becquerel died on August 25, 1908. The SI unit for radioactivity, the **becquerel** (Bq), is named in his honor.

1. Antoine-Henri Becquerel (1852 - 1908), Radiation History, www.rtstudents.com/radiology/antoine-henri-becquerel.
2. Eric Weisstein's World of Biography, Wolfram Research, http://scienceworld.wolfram.com/biography/BecquerelHenri.
3. Henri Becquerel, Wikipedia, http://en.wikipedia.org/wiki/Henri_Becquerel.

John Poynting (1852-1914)

Developed an equation for calculating the energy flow of electromagnetic waves

John Henry Poynting was born in Monton, near Manchester, England on September 9, 1852. From 1872 to 1876 he was educated at the universities of Manchester and Cambridge. He was a professor of physics at Mason Science College (now the University of Birmingham) from 1880 until his death.

Poynting is best known for the **Poynting vector**, introduced in his paper *On the Transfer of Energy in the Electromagnetic Field* in 1884. In this paper he showed that the flow of energy at a point can be expressed by a simple formula in terms of the electric and magnetic forces at that point. In the MKS system, the Poynting vector "S" is given by:

$$S = \frac{1}{\mu_0} E \times B$$

where "μ_0" is the permeability of free space, "**B**" is the magnetic field, and "**E**" is the electromagnetic field. The Poynting vector is used in the **Poynting theorem**, which is a statement about energy conservation for electric and magnetic fields.

In 1891, Poynting determined the mean density of the Earth, and in 1893 he made a determination of Newton's gravitational constant through the accurate use of torsion balances. His results were published in *The Mean Density of the Earth* in 1894 and *The Earth; Its Shape, Size, Weight and Spin* in 1913.

In 1903, Poynting suggested the existence of the effect of radiation from the Sun that causes smaller particles in orbit about the Sun to spiral close and eventually plunge into it. This concept was further developed by Howard Robertson (1903-1961), and was related to the theory of relativity in 1937, becoming known as the **Poynting-Robertson effect**.

Poynting also devised a method for measuring the radiation pressure from a body. His method can be used to determine the absolute temperature of celestial objects. He also did research in various areas of physical chemistry, including phase changes and osmotic pressure.

Poynting died on March 30, 1914. In his honor, the main physics building at the University of Birmingham was named for him.

1. John Henry Poynting, Answers.com, www.answers.com/topic/john-henry-poynting?cat=technology
2. John Henry Poynting, Wikipedia, http://en.wikipedia.org/wiki/John_Henry_Poynting.
3. Poynting Vector, Eric Weisstein's World of Physics, Wolfram Research, Eric Weisstein's World of Physics, http://scienceworld.wolfram.com/physics/PoyntingVector.

Heike Onnes (1853-1926)

Made discoveries of the properties of matter at very low temperatures
which led to the production of liquid helium

Heike Kamerlingh Onnes, the son of a brickworks owner, was born on September 21, 1853 in Groningen, Netherlands. He studied under Robert Bunsen (1811-1899) and Gustav Kirchhoff (1824-1887) from 1871 to 1873 at the University of Heidelberg. In 1879, he obtained his doctorate from the University of Groningen. His disertation was entitled *Nieuwe bewijzen voor de aswenteling der aarde* (New proofs of the rotation of the earth). In this, he gave both theoretical and experimental proofs that the pendulum experiment of Jean Foucault (1819-1868) should be considered as a special case of a large group of phenomena which can be used to prove the rotation of the Earth.

Onnes taught at Polytechnic School at Delft in the Netherlands in 1880-1882, during which time he was in close contact with Johannes Diderik van der Waals (1837-1923), professor of physics in Amsterdam. From 1882 to 1924 Onnes served as professor of experimental physics at the University of Leiden.

Onnes is best known for his work in cryogenics—the study of the production and effects of extremely low temperatures. He established the Cryogenic Laboratory at Leyden University in 1894, and in 1908, using the Joule-Thomson effect, he produced liquid helium for the first time. The Joule-Thompson effect is a process in which the temperature of a gas is either decreased or increased by letting the gas expand freely with no heat transferred to or from the gas.

Onnes studied the properties of materials at liquid helium temperatures, and in 1911 discovered that metals, such as lead, tin, and mercury, lost all resistance when cooled to such temperatures, a phenomenon known as superconductivity.

Onnes was awarded the 1913 Nobel Prize in Physics his investigations on the properties of matter at low temperatures, which led to the production of liquid helium.

Onnes was said to be A man of great personal charm and philanthropic humanity. He was very active during and after the First World War in smoothing out political differences between scientists.

Onnes died, after a short illness, on February 21, 1926.

1. Heike Kamerlingh Onnes, Microsoft Encarta Encyclopedia Online, http://encarta.msn.com/ encyclopedia_761560151/kamerlingh_onnes_heike, 2007.
2. Heike Kamerlingh Onnes, Wikipedia, http://en.wikipedia.org/wiki/Heike_Kamerlingh_Onnes.
3. Kamerlingh-Onnes, Heike (1853-1926, Eric Weissman's World of Biography, Wolfram Research, http://scienceworld.wolfram.com/biography/Kamerlingh-Onnes.

Hendrik Lorentz (1853-1928)

Formulated the relations between electricity, magnetism, and light

Hendrik Antoon Lorentz was born on July 18, 1853 in Arnhem, Netherlands. He was the son of a shopkeeper. Lorentz studied physics and mathematics at the University of Leiden, and received a doctoral degree in 1875. In 1878, at 24 years of age, he was appointed to the newly established chair in theoretical physics at the University of Leiden.

Lorentz made contributions to mechanics, thermodynamics, hydrodynamics, kinetic theories, solid state theory, light, and propagation. However, he was primarily interested in the theory of electromagnetism to explain the relationship of electricity, magnetism, and light.

Lorentz theorized that atoms might consist of charged particles, and suggested that the oscillations of these charged particles were the source of light. When colleague and former student, Pieter Zeeman (1865-1943), discovered the Zeeman effect (a change in spectrum lines in a magnetic field) in 1896, Lorentz supplied its theoretical interpretation.

Lorentz' name is now associated a number of things in physics: The **Lorentz-Lorenz equation** relates the refractive index of a dilute gas to its temperature, pressure, and molar refractivity. The **Lorentz transformation** converts between two different observers' measurements of space and time, where one observer is in constant motion with respect to the other. The **Lorentz force** is the force exerted on a charged particle in an electromagnetic field. The **Lorentz factor** appears in several equations in special relativity, including time dilation, length contraction, and the relativistic mass formula. The **Lorentzian distribution**, also called the Cauchy distribution, is a continuous distribution describing resonance behavior. The **Fitzgerald-Lorentz** contraction is the physical phenomenon of a decrease in length detected by an observer in objects that travel at any non-zero velocity relative to that observer. These contractions only become noticeable at substantial fractions of the speed of light.

Lorentz and Zeeman shared the Nobel Prize in physics in 1902 for their researches into the influence of magnetism upon radiation phenomena.

In 1912, Lorentz retired early to become director of research at Teylers Museum in Haarlem, although he remained external professor at Leiden and gave weekly lectures there. He died on February 4, 1928 in Haarlem, Netherlands.

1. Hendrik Lorentz, Wikipedia, http://en.wikipedia.org/wiki/Hendrik_Lorentz.
2. Lorentz, Hendrik Antoon Infoplease, www.infoplease.com/ce6/people/A0830307.
3. O'Connor, J. J. and E F Robertson Hendrik Antoon Lorentz, MacTutor, www-groups.dcs.st-and.ac.uk/~history/Printonly/Lorentz, 2003.

Henri Poincaré (1854-1914)

Formulated a preliminary version of the special theory of relativity

Jules Henri Poincaré was born on April 29, 1854 in Nancy, France. His father, Leon Poincaré, was a professor of medicine at the University of Nancy. Poincaré's brother, Raymond, became president of the French Republic during World War I.

During the Franco-Prussian War of 1870 Poincaré served alongside his father in the Ambulance Corps.

Poincaré entered the École Polytechnique in 1873. There, he studied mathematics. On graduation he went on to study at the École des Mines, continuing to study mathematics in addition to mining engineering. He received the degree of ordinary engineer.

Poincaré went on to the University of Paris, and obtained his doctorate in 1879.

Beginning in 1881, Poincaré taught at the University of Paris, eventually holding the chairs of Physical and Experimental Mechanics, Mathematical Physics and Theory of Probability, and Celestial Mechanics and Astronomy.

Poincaré is considered the originator of the theory of analytic functions of several complex variables. He sketched a preliminary version of the special theory of relativity in which he stated that the velocity of light is a limit velocity, and that mass depends on speed. He formulated the principle of relativity, according to which no mechanical or electromagnetic experiment can discriminate between a state of uniform motion and a state of rest, and he derived the Lorentz transformation.

Poincaré found that a three body system is often chaotic in the sense that a small perturbation in the initial state, such as a slight change in one body's initial position, might lead to a radically different later state than would be produced by the unperturbed system. If the slight change isn't detectable by our measuring instruments, then we wouldn't be able to predict which final state will occur. So, Poincaré's research proved that the problem of determinism and the problem of predictability are distinct problems.

Poincaré died on July 17, 1912 in Paris, France.

1. Henri Poincaré, Biographies/Images of Physicists and Astronomers, www.mlahanas.de/Physics/Bios/HenriPoincare.
2. Henri Poincaré, Wikipedia, http://en.wikipedia.org/wiki/Henri_Poincar%C3%A9.
3. Jules Henri Poincaré (1854-1912), The Internet Encyclopedia of Philosophy, www.iep.utm.edu/p/poincare.
4. Poincaré, Henri (1854-), Eric Weisstein's World of Biography, Wolfram Research, http://scienceworld.wolfram.com/biography/Poincare.

Johannes Rydberg (1854-1919)

Discovered line spectra series could be described by a universal constant

$\mathbf{J}$ohannes (Janne) Robert Rydberg was born in Halmstad, Sweden on November 8, 1854. His father Sven, a local merchant and minor ship owner, died when Janne was four years old. Rydberg received his bachelor's degree in 1875 from the University of Lund.

In 1980, Rydberg was appointed to the post of lecturer in mathematics at Lund, but his interests were now turning towards mathematical physics rather than to pure mathematics.

In 1882, Rydberg moved from a lectureship in mathematics to become a lecturer in physics at Lund. In 1879, he was promoted to a professorship in physics. From that time until his retirement in 1919 he held the chair of physics at Lund.

Rydberg is known mainly for devising in 1988 the **Rydberg formula**, which is used to predict the wavelengths of photons of light and other electromagnetic radiation emitted by changes in the energy level of an electron in an atom. For a hydrogen atom it is:

$$\frac{1}{\lambda_{\text{vac}}} = R_{\text{H}} Z^2 \left(\frac{1}{n_1^2} - \frac{1}{n_2^2} \right)$$

where "λ_{vac}" is the wavelength of the light emitted in a vacuum, "R_{H}" is the **Rydberg constant** for this element, "Z" is the atomic number of this element, and "n_1" and "n_2" are integers such that $n_1 < n_1$. By setting n_1 to 1 and letting n_2 run from 2 to infinity, the spectral lines known as the Lyman series converging to 91nm are obtained. Similarly, by setting n_1 to 2 and letting n_2 run from 3 to infinity the spectral lines known as the Balmer series converging to 365nm is obtained. By following the same procedure for higher values of n_1 and n_2, other series are obtained.

Excited atoms with very high values of the principal quantum number, represented by n in the Rydberg formula, are called **Rydberg atoms**. The **Rydberg unit** is a wave number characteristic of the wave spectrum of each element.

Rydberg died of a brain hemorrhage on December 28, 1919 in Lund, Sweden.

1. Janne Rydberg (1854-1919), Rydberg in Atomic Physics, http://w3.msi.vxu.se/~pku/Rydberg/ LifeWork.
2. Johannes Rydberg, Wikipedia, http://en.wikipedia.org/wiki/Johannes_Rydberg.
3. O'Connor, J. J. and E F Robertson, Johannes Robert Rydberg, MacTutor, www-groups.dcs.st-and.ac.uk/~history/Printonly/Rydberg.
4. Rydberg Formula, Answers.com, www.answers.com/topic/rydberg-formula.
5. Sears, Francis W., Mark W. Zemansky, and Hugh D. Young, College Physics, Seventh Edition, Addison Wesley, New York, 1991.

Edwin Hall (1855-1938)

Discovered the development of a transverse electric field in a solid material when it carries an electric current and is placed in a magnetic field that is perpendicular to the current

Edwin Herbert Hall was born in Great Falls, Maine on November 7, 1855. He did his undergraduate work at Bowdoin College in Brunswick, Maine, graduating in 1875. Hall did his graduate work at Johns Hopkins University, obtaining a Ph.D. in 1880.

Hall discovered what has become known as the **Hall effect** in 1879 while working on his doctoral thesis under experimental physicist Henry Augustus Rowland (1848-1901).

Rowland stimulated Hall's interest in electricity, and encouraged him to question a statement that James Clerk Maxwell (1831-1879) had made in his 1873 *Treatise on Electricity and Magnetism*. Maxwell had said that the force acting on an electric conductor in a magnetic field acts on the conductor directly and not on the electric current. Hall discovered that an electric current flowing through a gold conductor in a magnetic field produced an electric potential (**Hall voltage**) that was perpendicular to both the current and the field. The strength of the potential was directly related to the strength of the magnetic field, and created a transverse current in the conductor. Today, the Hall effect is used in magnetic field sensors, which are made in the millions.

The strength of the electric field "E_x" is given by $E_x = R_H B_y I_z$, where E_x is perpendicular to both the current "I_z" and the magnetic field "B_y." "R_H" is known as the **Hall coefficient**.

In a classical model of the Hall effect, R_H is simply $1/nq$, where "n" is the number of charge carriers per unit volume and "q" is their charge. The Hall effect is a characteristic of the material from which the conductor is made.

Hall took a job as instructor at Harvard and was appointed as Harvard's professor of physics in 1895.

Hall retired in 1921, and died of heart failure on November 20, 1938 in Cambridge, Massachusetts.

In 1966, it was discovered that the Hall effect influenced the electrons located at the interface between a semiconductor and an electric insulator. In the realm of quantum theory, the interaction is known as the **quantum Hall effect**.

1. Edwin Hall, Wikipedia, http://en.wikipedia.org/wiki/Edwin_Hall.
2. Edwin Herbert Hall Biography, The World of Physics on Edwin H. Hall, BookRags, World of Physics on Edwin Herbert Hall, www.bookrags.com/biography/edwin-herbert-hall-wop.
3. Hall Effect, Wikipedia, http://en.wikipedia.org/wiki/Hall_effect.

J. J. Thomson (1856-1940)
Discovered the electron

Joseph John (J. J.) Thomson was born of Scottish parentage on December 18, 1856 at Cheetham Hill, Manchester in England,. His father was a book seller. In 1870, Joseph studied engineering at the University of Manchester, and moved on to Trinity College, Cambridge in 1876, where he obtained an M.A. in mathematics in 1883. In 1884, Thomson became Cavendish Professor of Physics at cambridge.

Beginning in 1895, Thomson used the cathode ray tube in three experiments. In his first, he determined that the negative charge of cathode rays was inseparable from the rays themselves. In the second experiment, he found that cathode rays bent under the influence of an electric field. In the third experiment, he measured the charge-to-mass ratio of the cathode rays by measuring how much they were deflected by a magnetic field and how much energy they carried. He found that the charge to mass ratio was over a thousand times higher than that of a proton, suggesting either that the particles were very light or were very highly charged.

Thomson concluded that cathode rays were made of particles, which he called corpuscles, and that these corpuscles came from within the atoms of the electrodes themselves, meaning that atoms are in fact divisible.

Thomson imagined the atom as being made up of these corpuscles swarming in a sea of positive charge. This model was later proved incorrect by Ernest Rutherford (1871-1937).

In 1906, Thomson demonstrated that hydrogen has only a single electron, whereas previous theories allowed various numbers of electrons.

Thomson was awarded the 1906 Nobel Prize in Physics for his theoretical and experimental investigations on the conduction of electricity by gases. He was knighted in 1908.

In other work, Thomson in 1913 channeled a stream of ionized neon through a magnetic and an electric field and measured its deflection by placing a photographic plate in its path. He observed two patches of light on the photographic plate which suggested two different parabolas of deflection. He concluded that the neon gas was composed of atoms of two different atomic masses—neon-20 and neon-22.

Thomson was chosen Master of Trinity in 1918, and guided the college until shortly before his death on August 39, 1940. He was buried in Westminster Abbey, close to Isaac Newton (1879-1955).

1.   J. J. Thomson, Wikipedia, http://en.wikipedia.org/wiki/J._J._Thomson.
2.   Joseph John Thomson, Chemical Achievers, www.chemheritage.org/classroom/ chemach/atomic/ thomson.
3.   Sir J.J. Thomson (1856 - 1940), AtomicArchive.com, www.atomicarchive.com/Bios/Thomson.

Heinrich Hertz (1857-1894)

Discovered radio waves and the photoelectric effect

Heinrich Rudolf Hertz was born in Hamburg, Germany on February 22, 1857. His father was a prominent lawyer and legislator. Hertz obtained his Ph.D. at the Universtiy of Berlin in 1880. He was a student of Gustav Kirchhoff (1824-1887) and Hermann von Helmholtz (1821-1894).

In 1883, Hertz took a post as a lecturer in theoretical physics at the University of Kiel. In 1885, he became a full professor at the University of Karlsruhe.

In the 1880s, physicists were trying to obtain experimental evidence of electromagnetic waves. Their existence had been predicted in 1873 by the mathematical equations of James Clerk Maxwell (1831-1879). Hertz clarified and expanded the electromagnetic theory of light and set out to find the experimental evidence for electromagnetic waves. He used an oscillator made of polished brass knobs, each connected to an induction coil and separated by a tiny gap over which sparks could leap. Hertz reasoned that, if Maxwell's predictions were correct, electromagnetic waves would be transmitted during each series of sparks. To confirm this, he made a simple receiver of looped wire. At the ends of the loop were small knobs separated by a tiny gap. The receiver was placed several yards from the oscillator. According to theory, if electromagnetic waves were spreading from the oscillator sparks, they would induce a current in the loop that would send sparks across the receiver gap. This indeed happened when Hertz turned on the oscillator, producing the first transmission and reception of electromagnetic waves.

Hertz also discovered the photoelectric effect, which refers to the ejection of electrons from the surface of a metal in response to incident light. Albert Einstein (1879-1955) received the Noble Prize in physics in 1921 for explaining the mechanism of the photoelectric effect.

Hertz died of blood poisoning in Bonn, Germany on January 1, 1894. He was 36 years old. His experiments with electromagnetic waves led to the development of the wireless telegraph, the radio, radar, and television. The SI unit of frequency, the cycle per second, is called the **hertz** (Hz) in honor of Heinrich Hertz.

1. Heinrich Hertz, Microsoft Encarta Encyclopedia Online, http://encarta.msn.com/ encyclopedia_761568022/heinrich_hertz, 2007.
2. Heinrich Hertz, Wikipedia, http://en.wikipedia.org/wiki/Heinrich_Hertz.
3. Heinrich Rudolph Hertz (1857 - 1894), Corrosion Doctors, www.corrosion-doctors.org/ Biographies/ HertzBio.
4. Heinrich Rudolph Hertz, Biographies/Images of Physicists and Astronomers, www.mlahanas.de/ Physics/Bios/HeinrichRudolfHertz.

Nikola Tesla (1857-1943)

Developed alternating current

Nikola Tesla was born to Serbian parents in Smiljan, Croatia (then part of Austria–Hungary) on July 10, 1857. He studied engineering at the Technical University in Graz, Austria from 1877 to 1880. He then went to the University of Prague to continue his studies, but the death of his father caused him to leave without graduating.

In 1881, Tesla went to Budapest as an engineer for a telephone company, and a year later took a similar position in Paris. In 1884, he went to the United States and worked for Thomas Edison (1847-1931) for a year before setting up his own workshop.

To generate electricity, Tesla developed the idea of spinning a piece of iron between stationary coils of wire electrified by two alternating currents not quite in step with each other. The current through the coil produced a rotating magnetic field, which induced an alternating current in the piece of iron, called a rotor. The current induced in the rotor could then be used to power electrical appliances. To counter fears of alternating-current, Tesla gave exhibitions in his laboratory in which he lighted lamps without wires by allowing electricity to flow through his body.

George Westinghouse (1846-1914) quickly bought Tesla's patents. And in 1893, he won the contract to generate electricity at Niagara Falls, New York. He used Tesla's system to supply electricity to local industries and deliver alternating current to the town of Buffalo, New York.

The **Tesla coil** is one of Tesla's most famous inventions. It is a combination of two circuits, each having a coil of wire wound together around a hollow tube. One of the coils is made of heavy wire and has just a few turns around the tube. The other circuit's coil is made of finer wire wound many times around the tube. When an alternating current passes through the coil of heavy wire, it produces a magnetic field. The magnetic field induces current in the fine wire. The frequency of the current in the finer coil is much higher. The voltage is also higher in the finer coil.

Among Tesla's other inventions are the induction motor, the rotating magnetic field, and wireless technology.

Tesla lived his last years as a recluse, and died impoverished in the New Yorker Hotel in New York City on January 7, 1943. After his death the Supreme Court credited him as being the inventor of the radio. The SI unit measuring the magnetic field, the **tesla** (T), was named in his honor.

1. Nicola Tesla, Wikipedia, http://en.wikipedia.org/wiki/Nikola_Tesla.
2. Nikola Tesla, A Biographical Sketch, The Life, www.mercury.gr/tesla/lifeen.
3. Tesla, Nikola, Microsoft Encarta Encyclopedia Online, http://uk.encarta.msn.com/ encyclopedia_761567992/tesla_nikola, 2007.

H. Thomas Milhorn, MD, PhD

Max Planck (1858-1947)

Originator of quantum theory

Max Karl Ernst Ludwig Planck was born in Kiel, Germany on April 23, 1858. His father was a law professor. Planck was gifted when it came to music; he took singing lessons and played the piano, organ and cello, and composed songs and operas, but chose physics as a profession.

Planck studied at the Universities of Munich and Berlin, where his teachers included Gustav Kirchhoff (1824-1887) and Heinrich Helmholtz (1821-1894), and received his doctorate of philosophy at Munich in 1879.

In 1885, the University of Kiel appointed Planck an associate professor of theoretical physics. Work on entropy and its treatment, especially as applied in physical chemistry, followed.

Within four years Planck was named the successor to Kirchhoff's position at the University of Berlin.

In 1894, Planck turned his attention to black-body radiation. The problem had been stated by Kirchhoff in 1859—how does the intensity of the electromagnetic radiation emitted by a black body depend on the frequency of the radiation and the temperature of the body? The question had been explored experimentally, but no theoretical treatment gave sufficient agreement with the experimental results. The central assumption behind Planck's final derivation was the supposition that the electromagnetic energy could be emitted only in quantized form, in other words, the energy could only be a multiple of an elementary unit $E = hv$, where "h" is **Planck's constant** and "v" is the frequency of the radiation. This relationship has come to be known as **Planck's law**.

The discovery of Planck's constant allowed Planck to define a new universal set of physical units, such as the **Planck length** and the **Planck mass**, all based on fundamental physical constants.

Planck's discoveries, which were later verified by other scientists, were the basis of an entirely new field of physics known as quantum mechanics. For his work, Planck received the Nobel Prize for physics in 1918.

Planck retired from Berlin in 1926, and died on October 4, 1947 in Göttingen, Germany.

1. Max Planck(1858-1947), Eric Weisstein's World of Biography, http://scienceworld.wolfram.com/ biography/Planck.
2. Max Planck, Microsoft Encarta Encyclopedia Online, http://encarta.msn.com/encyclopedia_761566401/max_planck, 2007.
3. Max Planck, Wikipediea, http://en.wikipedia.org/wiki/Max_Planck.
4. O'Connor, J. J. and E F Robertson, Max Karl Ernst Ludwig Planck, MacTutor, www-groups.dcs.st-and.ac.uk/~history/Printonly/Planck.

Pierre Curie (1859-1906)

Co-isolated polonium and radium

Pierre Curie was born in Paris, France on May 15, 1859. He was educated at home by his physician father. By the age of 18 he had completed the equivalent of a master's degree, but did not proceed immediately to a doctorate due to lack of money. Instead he worked as a laboratory instructor. In 1880, Curie and his older brother, Jacques (1856-1941), demonstrated piezoelectricity—the generation of an electric potential when crystals are compressed. In 1881, they demonstrated the reverse effect; that is, that crystals can be made to deform when subjected to an electric field. Almost all digital electronic circuits today rely on this phenomenon in the form of crystal oscillators.

Working on his Ph.D., Curie studied ferromagnetism, paramagnetism, and diamagnetism, and discovered the effect of temperature on paramagnetism, which is now known as **Curie's law**. The material constant in Curie's law is known as the **Curie constant.** He also discovered that ferromagnetic substances exhibit a critical temperature transition, above which the substances lose their ferromagnetic behavior. This is now known as the **Curie point.**

Curie worked with his wife, Marie, in isolating polonium and radium by fractionation of pitchblende. They were the first to use the term 'radioactivity." Curie made the first discovery of nuclear energy by identifying the continuous emission of heat from radium particles. He also investigated the radiation emissions of radioactive substances, and through the use of magnetic fields showed that some of the emissions were positively charged, some were negatively charged, and others were neutral. These particles correspond to alpha, beta, and gamma radiation.

Curie shared the 1903 Nobel Prize in physics with his wife, Marie Curie (1867-1934) and Henri Becquerel (1852-1908) in recognition of their joint researches on radiation phenomena. The Curie's daughter Irène Joliot-Curie (1897-1956) and son-in-law Frédéric Joliot-Curie (1900-1958) would receive the 1934 Nobel Prize for chemistry.

Pierre Curie died as a result of a carriage accident in a snow storm while crossing the *Rue Dauphine* in Paris on April 19, 1906. His head was crushed under the carriage wheel. The **curie** is a unit of radioactivity (3.7×10^{10} decays per second), named in honor of Pierre Curie.

1. Pierre Curie (1859 - 1906), AtomicArchive.com, www.atomicarchive.com/Bios/PierreCurie.
2. Pierre Curie (1859-1906), Marie Curie and the Science of Radioactivity, www.aip.org/history/curie/pierre.
3. Pierre Curie, Microsoft Encarta Encyclopedia Online, http://encarta.msn.com/encyclopedia_761555255/curie_pierre, 2007.
4. Pierre Curie, Wikipedia, http://en.wikipedia.org/wiki/Pierre_Curie.

Charles Guillaume (1861-1938)

Made precision measurements in physics by his discovery of anomalies in nickel steel alloys

Charles Édouard Guillaume, the son of a clock maker, was born in Fleurier, Switzerland to French parents on February 15, 1861. In 1878, he entered the Zurich Federal Institute of Technology, gaining his doctorate in physics in 1882, and then spending a brief time as an artillery officer in the Swiss Army.

In 1883, Guillaume became an assistant at the newly established International Bureau of Weights and Measures at Sèvres, near Paris. He was appointed director in 1915, and held this post until his retirement in 1936.

Guillaume's first assignment at the Bureau of Weights and Measures was to study mercury thermometers, and his published research in precision thermometry became the standard text. In other research, he redetermined the exact volume of the liter at 1000.028 cubic centimeters. He was also involved in developing the international standards for the meter and kilogram. His many publications communicated progress in metrology to the international scientific community.

Guillaume's research on thermal expansion of possible standards materials led him to investigate various alloys. He is best known for his discovery of nickel-steel alloys he named "invar" and "elinvar." Invar has a near-zero coefficient of thermal expansion, making it useful in constructing precision instruments whose dimensions need to remain constant in spite of varying temperature. Elinvar has a near-zero thermal coefficient of the modulus of elasticity, making it useful in constructing instruments with springs that need to be unaffected by varying temperature, such as the marine chronometer. Elinvar is also non-magnetic, which is a secondary useful property for antimagnetic watches.

Guillaume served at the Observatoire de Paris—Section de Meudon. He conducted several experiments with thermostatic measurements at the observatory. His treatise of 1889 on this subject became a standard text for metrologists.

In 1920 Guillaume received the Nobel Prize for physics for his researches into nickel–steel alloys.

Guillaume died on May 13, 1938 in Sevres, France.

1. Charles Édouard Guillaume, Answers.com, www.answers.com/topic/ charles-edouard-guillaume?cat=technology.
2. Charles Édouard Guillaume, Wikipedia, http://en.wikipedia.org/ wiki/Charles_Edouard_Guillaume.
3. Charles-Édouard Guillaume, Microsoft Encarta Encyclopedia Online, http://encarta.msn.com/encyclopedia_761583200/charles_edouard_guillaume, 2007.

Philipp Lénárd (1862-1947)

Discovered many of the properties of cathode rays

Philipp Eduard Anton von Lénárd, the son of a wealthy winemaker, was born on June 7, 1862 in Pressburg, Austria-Hungary. He studied under Robert Bunsen (1811-1899) and Hermann von Helmholtz (1821-1894) at the University of Heidelberg, obtaining his doctoral degree in 1886. After a number of posts, in 1907 he became head of the Philipp Lenard Institute in Heidelberg (formerly Heidelberg's physics institute).

Lénárd's early work included studies on the size and shape distributions of raindrops. In 1892, he was the first person to study the separation of electric charges accompanying the aerodynamic breakup of water drops. This has become known as the **Lenard effect**.

Lénárd's major contributions were in the study of cathode rays, which he began in 1888. Prior to his work, cathode rays were produced in primitive tubes, which made cathode rays difficult to study because they were inside sealed glass tubes. Lénárd overcame this problem by devising a method of making small metallic windows in the glass that were thick enough to withstand the pressure differences, but thin enough to allow passage of the rays. This window is known as a **Lenard window**. He detected the rays and measured their intensity by means of paper sheets coated with phosphorescent materials.

Lénárd observed that the absorption of the cathode rays was proportional to the density of the material they had to pass through. He also showed that the rays could pass through several inches of air, appearing to be scattered by it. This implied that they must be particles that were smaller than the molecules in air. He ultimately arrived at the understanding that cathode rays were streams of energetic electrons. The general realization that electrons were parts of the atom enabled Lénárd to claim correctly that atoms consist mainly of empty space.

Lénárd also studied the photoelectric effect, showing that the energy of the rays produced by radiating metals in a vacuum with ultraviolet light was independent of the light intensity, but was greater for shorter wavelengths of light. This observation was later explained by Albert Einstein (1879-1955) as a quantum effect. In 1937, Lénárd left academia to serve as an advisor to Adolf Hitler

Lénárd was awarded the 1905 Nobel Prize in physics for his investigation of cathode rays, which contributed to the knowledge of atoms. He died on May 20, 1947 in Messelhausen, Germany.

1. Philipp Lénárd, Wikipedia, http://en.wikipedia.org/wiki/Philipp_Lenard.
2. Phillip Lenard, Microsoft Encarta Encyclopedia Online, http://encarta.msn.com/encyclopedia_761583243/lenard_philipp_eduard_anton, 2007.

William Bragg (1862-1942)

Invented the X-ray spectrometer and with his son, William Lawrence
Bragg, founded the new science of X-ray analysis of crystal structure

William Henry Bragg, the son of a sea captain who had become a farmer,
was born in North Adelaide, South Australia on July 2, 1862. His mother
died when he was 7, and he was raised by his uncle, William Bragg, at
Market Harborough, Leicestershire.

Bragg was educated at King William's College, Isle of Man, and
Trinity College, Cambridge, graduating in 1884. In 1885, he was
appointed Professor of Pure and Applied Mathematics at the University of
Adelaide in Australia.

At Adelaide, Bragg showed clearly the well-defined ranges of alpha-
particles, and distinguished the four groups of alpha-particles emitted by—
radium, radon, RaA and RaC. He also showed the stopping power of
substances was approximately proportional to the square roots of the
atomic weights.

In 1908, Bragg became Cavendish professor at Leeds University.
There, Bragg continued the work on X-rays he had started at Adelaide,
inventing the X-ray spectrometer. With with his son, William Lawrence
Bragg, he founded the new science of X-ray analysis of crystal structure.
An X-ray spectrometer is a divice for determining the electronic structure
of materials by using X-ray excitation.

In 1915, father and son were jointly awarded the Nobel Prize in
Physics for their studies of the X-ray spectrometer, X-ray spectra, X-ray
diffraction, and crystal structure.

Bragg was appointed Quain Professor of physics at University College
London in 1915, but did not take up his duties there until after World War
I. He did much work for the government at this time, largely connected
with submarine detection. He returned to London in 1918 as consultant to
the admiralty.

While Quain professor at London he continued his work on crystal
analysis. He was knighted in 1820.

From 1923, Bragg was Fullerian professor of chemistry at the Royal
Institution and director of the Davy Faraday Research Laboratory.

Bragg died in London on March 10, 1942 after a period in which heart
trouble reduced his activity.

1.   Bragg, Sir William Henry, Microsoft Encarta Encyclopedia Online, http://encarta.msn.com/
     encyclopedia_761579323/bragg_sir_william_henry, 2007.
2.   Tomlin, S. G., Bragg, Sir William Henry (1862 - 1942), Australian Dictionary of Biography,
     www.adb.online.anu.edu.au/biogs/A070396b.
3.   William Henry Bragg, Wikipedia, http://en.wikipedia.org/wiki/William_Henry_Bragg.

Paul Drude (1863-1906)

Developed a model to explain the thermal, electrical, and optical
properties of matter

Paul Karl Ludwig Drude, the son of a physician, was born on July 12, 1863 in Braunschweig, Germany. His scientific work covered thermodynamics and statistical physics as well as optics and transport theory.

Drude began his studies in mathematics at the University of Göttingen, but later changed his major to physics. His dissertation, covering the reflection and diffraction of light in crystals, was completed in 1887.

Drude's first experiments involved the determination of the optical constants of various solids, which he measured to unprecedented levels of accuracy. He then worked to derive relationships between the optical and electrical constants and the physical structure of substances.

In 1894, Drude was responsible for introducing the symbol "c" for the speed of light in a perfect vacuum.

Toward the end of his tenure at Leipzig, Drude wrote a book on optics. The book, *Lehrbuch der Optik*, published in 1902, brought together the formerly distinct subjects of electricity and optics.

In 1900, Drude developed a model to explain the thermal, electrical, and optical properties of matter. The **Drude model** was further advanced in 1933 by Arnold Sommerfeld (1868-1951) and Hans Bethe (1906-2005) to become the **Drude-Sommerfeld model**.

In 1900, Drude became the editor for the scientific journal *Annalen der Physik*, the most respected physics journal at that time From 1901-1905,

Drude was ordinarius professor of physics at Giessen University. And in 1905 he became the director of the Physics Institute of the University of Berlin.

At the height of his career, Drude became a member of the Prussian Academy of Sciences. A few days after his inauguration lecture, on July 5, 1906, for inexplicable reasons, he committed suicide.

The Paul-Drude-Institut für Festkörperelektronik (Paul Drude Institute for Soldid State Electronics), in Berlin, was named in his honor. The Institute performs materials research, solid state physics, and nanotechnology on low-dimensional structures.

1. Kuzemsky, A.L, Biography of Paul Drude (1863-19006), Bogoliubov Laboratory of Theoretical Physics, http://theor.jinr.ru/~kuzemsky/drudbio, 2006.
2. Paul Drude, Paul-Drude-Institut für Festkörperelektronik www.pdi-berlin.de/drude, 2006.
3. Paul Drude, Wikipedia, http://en.wikipedia.org/wiki/Paul_Karl_Ludwig_Drude

Wilhelm Wien (1864-1924)

Developed a formula for determining the energy density associated with particular wavelengths for any given temperature of a radiating body

Wilhelm Carl Werner Otto Fritz Franz Wien, the son of landowner, was born on January 13, 1864 at Fischhausen (Rybaki), Province of Prussia (now Primorsk, Russia). In 1866, his family moved to Drachstein in Rastenburg (Rastembork).

In 1886, Wien received his Ph.D. with a thesis on the diffraction of light by metals and on the influence of various materials on the color of refracted light.

In 1890, Wien became an assistant to the German physicist Hermann von Helmholtz (1821-1894) at the Imperial Physical Technical Institute at Charlottenburg.

From 1896 to 1899, Wien lectured at the Aachen University of Technology.

In 1893, Wien used theories about heat and electromagnetism to compose **Wien's displacement law**, which relates the maximum emission of a blackbody to its temperature.

In 1896, Wien developed an approximation, now called **Wien's approximation** or **Wien's law**, to describe the spectrum of thermal radiation. Wien's equation accurately describes the high frequency spectrum of thermal emission from objects, but fails to accurately fit the experimental data for low frequency emission.

Wien's contributions to the field of radiation laid the foundation for the development of the quantum theory.

In 1898, while studying streams of ionized gas, Wien identified a positive particle equal in mass to the hydrogen atom. The particle was later named the proton. Wien, with this work, laid the foundation of mass spectroscopy.

In 1900, Wien went to the University Würzburg and became the successor of Wilhelm Röntgen (1845-1923).

For his work on thermal radiation, Wien received the Nobel Prize in Physics in 1911.

In 1913, Wien visited the United States to lecture at Columbia University.

Wien died on August 30, 1928.

1. Wilhelm Carl Werner Otto Fritz Franz Wien, MacTutor, www-groups.dcs.st-and.ac.uk/~history/Printonly/Wien.
2. Wilhelm Wien, Microsoft Encarta Encyclopedia Online, http://encarta.msn.com/encyclopedia_761553093/wien_wilhelm, 2007.
3. Wilhelm Wien, Wikipedia, http://en.wikipedia.org/wiki/Wilhelm_Wien.

Pieter Zeeman (1865-1943)

Discovered that a spectral line is split into several components in the presence of a magnetic field

Pieter Zeeman was born on May 25, 1865 in Zonnemaire, a small town on the island of Schouwen-Duiveland, Netherlands. His father was a minister of the Dutch Reformed Church. Zeeman studied physics at the University of Leiden under Kamerlingh Onnes (1853-1926) and Hendrik Lorentz ((1853-1928)). His doctoral thesis was devoted to the Kerr effect— a change in the refractive index of a material in response to an electric field.

After obtaining his doctorate, Zeeman became Privatdozent in mathematics and physics in Leiden. In 1896, three years after submitting his thesis on the Kerr effect, he made the discovery of what is now known as the **Zeeman effect**. As an extension of his thesis research, he began investigating the effect of magnetic fields on a light source. He discovered that a spectral line is split into several components in the presence of a magnetic field. Lorentz presented Zeeman with an explanation of his observations, based on the Lorentz theory of electromagnetic radiation.

Because of Zeeman's work, according to Lorentz theory, it appeared that the oscillating particles were the source of light emission, were negatively charged, and were a thousand-fold lighter than the hydrogen atom. This conclusion was reached well before Thomson's discovery of the electron. The Zeeman effect thus became an important tool for elucidating the structure of the atom.

In 1923, a new laboratory was built in Amsterdam, which in 1940 was renamed Zeeman Laboratory. This new facility allowed Zeeman to pursue refined investigation of the Zeeman effect.

For the remainder of his career, Zeeman remained interested in research in Magneto-Optics. He also investigated the propagation of light in moving media.

Zeeman shared the 1902 Nobel Prize in Physics with Hendrik Lorentz for their researches into the influence of magnetism upon radiation phenomena.

Zeeman died in Amsterdam on October 9, 1943, and was interred in Haarem.

1. Pieter Zeeman - Biographhy, http://shaila-legends.blogspot.com/2008/01/pieter-zeeman-1902-nobel-prize-in.
2. Pieter Zeeman, Wikipedia, http://en.wikipedia.org/wiki/Pieter_Zeeman.
3. Pieter Zeeman-1902 Nobel Prize in Physics, Famous People, http://shaila-legends.blogspot.com/2008/01/pieter-zeeman-1902-nobel-prize-in.

Marie Curie (1867-1934)

Co-isolated the elements radium and polonium

Marie Curie was born Maria Skłodowska in Warsaw, Poland on November 7, 1867. Both parents were teachers and ardent Polish nationalists. Soon after Maria was born they lost their teaching posts and had to take in boarders. Higher education was not available to women in Poland at that time. In 1891, Maria went to join her sister in Paris, where she entered the University of Paris. There, she studied mathematics, physics, and chemistry. In 1903, under the supervision of Henri Becquerel (1852-1908), she received her D.Sc. degree, becoming the first woman in France to complete a doctorate.

Maria married Pierre Curie (1859-1906) in 1895 and changed the spelling of her first name to Marie. At the time, Pierre was an instructor in the School of Physics and Chemistry at the École Supérieure de Physique et de Chimie Industrielles de la Ville de Paris. Maria was a student at the University of Paris. Their mutual interest in magnetism drew them together.

Eventually, the Curies began studying radioactive materials, particularly pitchblende, which is the ore from which uranium was extracted. By April 1898, they deduced that pitchblende must contain traces of an unknown substance far more radioactive than uranium. In July 1898, they published an article announcing the existence of a new element, which they named polonium in honor of Marie's native Poland. In 1898, they announced the discovery of a second element, which they named radium for its intense radioactivity.

In 1903, Marie Curie, Pierre Curie, and Becquerel were awarded the Nobel Prize in physics in recognition of their joint research on radiation phenomena. Marie Curie was the first woman to be awarded a Nobel Prize. Eight years later, she received the 1911 Nobel Prize in Chemistry in recognition of her discovery of the elements radium and polonium.

In her later years, Curie headed the Pasteur Institute and a radioactivity laboratory created for her by the University of Paris.

Curie died on July 4, 1934 from aplastic anemia, almost certainly due to exposure to radiation. She was 67 years old. Much of her work had been carried on with no safety measures. She was interred at the cemetery in Sceaux, where Pierre lay. Sixty years later, in 1995, in honor of their work, the remains of both were transferred to the Panthéon in Paris.

1. Marie Curie, Microsoft Encarta Encyclopedia Online, http://encarta.msn.com/encyclopedia_762505345/curie_marie, 2007.
2. Marie Curie, Wikipedia, http://en.wikipedia.org/wiki/Marie_Curie.
3. Marie Sklodowska Curie Physicist, LucidCafe, www.lucidcafe.com/library/95nov/curie.

Arnold Sommerfeld (1868-1951)

One of the founders of quantum mechanics

Arnold Johannes Wilhelm Sommerfeld, the son of a physician, was born in Königsberg, East Prussia on December 5, 1868. He studied mathematics and physics at Albertina University, and received his Ph.D. there in 1891. He passed the national teaching exam in 1892, and then began a year of military service, which was done with the reserve regiment in Königsberg.

In 1992, Sommerfeld joined the faculty of the University of Göttingen. In 1900, he was appointed to the Chair of Applied Mechanics at the RWTH Aachen University as extraordinarius professor. There, he developed the theory of hydrodynamics.

From 1906, Sommerfeld was professor of physics and director of the new Theoretical Physics Institute at the University of Munich. While at Munich, Sommerfeld came in contact with the special theory of relativity by Albert Einstein (1879-1955), which was not yet widely accepted. Sommerfeld's mathematical contributions to the theory helped its acceptance by the skeptics.

In 1914, Sommerfeld worked with Léon Brillouin (1889-1969) on the propagation of electromagnetic waves in dispersive media. As one of the founders of quantum mechanics, he co-discovered the **Sommerfeld-Wilson quantization rules** (1915), generalized Bohr's atomic model, introduced the **Sommerfeld fine-structure constant** (1916), and co-discovered with Walther Kossel (1888-1956) the **Sommerfeld-Kossel displacement law** (1919).

In 1918, Sommerfeld succeeded Einstein as chair of the Deutsche Physikalische Gesellschaft. In 1927, he applied Fermi-Dirac statistics to the Drude model of electrons in metals. The new theory solved many of the problems predicting thermal properties the original model failed to do. The model then became known as the **Drude-Sommerfeld model**.

Sommerfeld proposed a solution to the problem of a radiating hertzian dipole over a conducting Earth, which over the years led to many applications. His **Sommerfeld identity** and **Sommerfeld integrals** are still the most common way to solve this kind of problem.

Sommerfeld died on April 26, 1951 in Munich from injuries after a traffic accident while walking with his grandchildren.

1. Arnold Sommerfeld, Microsoft Encarta Encyclopedia Online, http://encarta.msn.com/encyclopedia_761580361/Sommerfeld, 2007.
2. Arnold Sommerfeld, Wikipedia, http://en.wikipedia.org/wiki/Arnold_Sommerfeld.
3. O'Conner, J. J. and E. F. Robertson, Arnold Johannes Wilhelm Sommerfeld, www-history.mcs.st-and.ac.uk/history/Printonly/Sommerfeld, October 2003.

Robert Millikan (1868-1953)

Measured the charge on the electron and showed that the charge exists
only as a whole number of units of that charge

Robert Andrews Millikan, the son of a minister, was born in Morrison,
Illinois on March 22, 1868. He received his Ph.D. in physics from
Columbia University in 1895. After a year in Germany at the Universities
of Berlin and Göttingen he joined the faculty of the University of Chicago
in 1896.

Starting in 1909, Millikan worked on an oil-drop experiment in which
he measured the force on tiny charged droplets of oil suspended against
gravity between two metal electrodes. Knowing the electric field, the
charge on the droplet could be determined. Repeating the experiment for
many droplets, Millikan showed that the results could be explained as
integer multiples of a common value (1.592×10^{-19} coulomb), the charge
on a single electron. This value is slightly lower than the modern value of
$1.602\ 176\ 53 \times 10^{-19}$ coulomb.

When Albert Einstein (1879-1966) published his 1905 paper on the
particle theory of light, Millikan was convinced that it had to be wrong
because of the vast body of evidence that had shown light to be a wave.
He undertook a decade-long experimental program to test Einstein's
theory. His results confirmed Einstein's predictions in every detail.

In 1917, Millikan went to Washington to be executive officer of the
National Research Council of the National Academy of Sciences, charged
with war research on the detection of submarines and other essential
problems.

In 1921, Millikan became director of the Norman Bridge Laboratory
of Physics of the California Institute of Technology (Caltech). He served
in that position until 1945. At Caltech, most of his scientific research
focused on the study of cosmic rays (a term he coined), X-rays, and the
experimental determination of Planck's constant..

Millikan was awarded the 1923 Nobel Prize in physics for his oil-drop
experiments, which had measured the charge on an electron and shown
that the charge exists only as a whole number of units of that charge.

Millikan died of a heart attack at his home in San Marino, California
on December 19, 1953 at age 85, and was interred in the Court of Honor at
Forest Lawn Memorial Park Cemetery in Glendale, California.

1. Robert Andrews Millikan (1868-1953), Corrosion Doctors, www.corrosion-
    doctors.org/Biographies/MillikanBio.
2. Robert Andrews Millikan, Wikipedia, http://en.wikipedia.org/wiki/Robert_Andrews_Millikan.
3. Robert Millikan, Microsoft Encarta Encyclopedia Online, http://encarta.msn.com/
    encyclopedia_761561973/robert_millikan, 2007.

Charles Wilson (1869-1959)

Developed a method of making the paths of electrically charged particles visible by condensation of vapor

Charles Thomson Rees (C. T. R.) Wilson, the son of a farmer, was born in the parish of Glencorse in Scotland on February 14, 1869. At the age of four, he lost his father, and his mother moved with the family to Manchester.

Wilson was educated at Owen's College, studying biology with the intent to become a physician. He then went to Sidney Sussex College, Cambridge, where he became interested in physics and chemistry.

In 1893, Wilson began to study clouds and their properties. He worked for some time at the observatory on Ben Nevis, the highest mountain in the British Isles. There, he made observations of cloud formation. He then tried to reproduce this effect on a smaller scale in the laboratory in Cambridge, expanding humid air within a sealed container. Wilson showed that water droplets can form around charged particles in the absence of dust. He found that if he exposed the air in a chamber to X-rays, many more droplets formed. He concluded that X-rays give the air molecules an electrical charge.

As an ion moves through a cloud chamber, drops of water form around it. Because the ion moves very quickly, the string of drops of water in the air appears as a continuous path, marking the movement of the ion. A strong light directed at the cloud makes the path more apparent. When a magnetic field is applied to the cloud chamber, the ions follow curved paths, depending on the strength and nature of the charge, the mass of the ion, and the strength and direction of the magnetic field.

In 1911, Wilson was the first person to see and photograph the tracks of individual alpha-particles, beta-particles, and electrons.

For the invention of the cloud chamber, Wilson shared the Nobel Prize for physics in 1927 with Arthur Compton (1892-1962). Compton's share was for showing wavelengths of an X-rays or gamma rays increase when they interact with matter (the Compton effect).

After retirement, Wilson moved back to Scotland, where he finished his long-promised manuscript on the theory of thundercloud electricity. He died on November 15, 1959.

1. C. T. R. Wilson, Biography, NobelPrize.org, http://nobelprize.org/nobel_prizes/ physics/laureates/1927/wilson-bio.
2. Charles T. R. Wilson, Microsoft Encarta Encyclopedia Online, http://encarta.msn.com/ encyclopedia_761583406/wilson_charles_thomson_rees, 2007.
3. Charles Wilson Rees Wilson, Wikipedia, http://en.wikipedia.org/wiki/ Charles_Thomson_Rees_Wilson.

Nils Dalén (1869-1937)

Developed automatic valves designed to be used in combination with gas
accumulators in lighthouses and light-buoys

Nils Gustaf Dalén was born in Stenstorp in present day Falköping
municipality, Sweden on November 30, 1869. He took care of the family
farm, but also started a market garden, a seed merchant, and operated a
dairy. In 1892, he invented a milk-fat tester for checking the quality of
milk. Dalén sold the farm to attended Chalmers University of Technology,
where he earned his Master's degree and then a Doctorate in 1896.

His studies completed, Dalén initially worked with acetylene—an
extremely explosive hydrocarbon gas. This work involved a new way of
producing and using acetylene gas, mainly for lighting. He invented
Agamassan (Aga), a substrate used to absorb the gas, allowing safe storage
and hence commercial use.

In 1906, Dalén became Chief Engineer at the Gas Accumulator
Company (AGA), a manufacturer and distributor of acetylene. There, he
introduced welding using acetylene gas in Sweden in 1902.

Because acetylene produces an ultra-bright white-light, it immediately
superseded the duller-flamed liquid petroleum gas as the fuel of choice in
lighthouses. Dalen exploited the new fuel, developing the Dalén light,
which incorporated a further invention, the sun valve, which could both
save on the expensive acetylene gas in light houses and provide beacons of
varying character. The sun valve automatically switched on lighthouse
beacons when darkness fell and switched them off at dawn. This
prolonged their service life of the light. The Dalén light used a flashing
apparatus so that, except for a small pilot light, the light only consumed
gas during the flash stage. As a result, it used one-tenth as much gas as its
predecessor. Dalén became president of AGA in 1909, and the company
began to produce unmanned, reliable, low-energy lighthouses, which
improved safety at sea. It was an immediate international success.

In 1912, Dalén was blinded in an acetylene explosion. Later the same
year he was awarded the Nobel Prize for physics for the development of
automatic valves designed to be used in combination with gas
accumulators in lighthouses and light-buoys.

Despite his blindness, Dalén remained in control of AGA until 1937.
He died on December 9 of that year.

1. Gustaf Dalén , Wikipedia, http://en.wikipedia.org/wiki/AGA_cooker.
2. Gustaf Dalén, History, www.aga.com/web/web2000/com/WPPcom.nsf/pages/
   History_GustafDalen.
3. Nils Gustaf Dalén, Microsoft Encarta Encyclopedia Onlilne, http://encarta.msn.com/
   encyclopedia_761583155/dal%C3%A9n_nils_gustaf, 2007.

Jean Perrin (1870-1942)

Verified Einstein's explanation of Brownian motion and thereby
the atomic nature of matter

Jean Baptiste Perrin was born in Lille, France on September 30, 1870. Educated at the Ecole Normale Superieure in Paris, Perrin received his Ph.D. in 1897.

After receiving his Ph.D. Perrin joined the faculty of the University of Paris in 1898, and became professor of physical chemistry in 1910. He remained in that position until the German occupation of France in 1940.

Perrin's work in 1895 on the nature of cathode rays helped with the discovery of the electron by J. J. Thomson (1856-1940) in 1897. Well in advance of the experimental work of Ernst Rutherford (1871-1937), Perrin suggested that the atom resembles a miniature solar system. He computed Avogadro's number, and explained solar energy as resulting from the thermonuclear reactions of hydrogen.

Perrin's most influential work was on Brownian motion—the continual agitation of small suspended particles in colloidal solutions. Beginning about 1908, using the newly developed ultramicroscope, Perrin carefully observed the manner of sedimentation of these particles and provided experimental confirmation of the theoretical work of Albert Einstein (1879-1955) and Marian Smoluchowski (1872-1817) on Brownian motion.

Perrin's observations also allowed him to estimate the size of water molecules and atoms as well as their quantity in a given volume. This was the first time the size of atoms and molecules could be reliably calculated from actual visual observations.

Perrin's work helped raise atoms from the status of useful hypothetical objects to observable entities whose reality could no longer be denied. For this achievement he was honored with the Nobel Prize for physics in 1926.

During World War I, like many top-class scientists, Perrin worked on underwater acoustic detection of submarines.

Perrin escaped to the United States at the time of the defeat of France in 1940. He became one of the founders of the French University of New York.

Perrin died in New York City on April 17, 1942. His remains were subsequently interred in the Panthéon in Paris.

1. Jean Baptiste Perrin, Timeline of Nobel Prize Winners, Physics, http://peace.nobel.brainparad.com jean_baptiste_perrin.
2. Jean Baptiste perrin, Wikipedia, http://en.wikipedia.org/wiki/Jean_Baptiste_Perrin.
3. Perrin, Jean-Baptiste, Microsoft Encarta Encyclopedia Online, http://ca.encarta.msn.com/ encyclopedia_761583303/perrin_jean-baptiste, 2007.

Guglielmo Marconi (1874-1937)

Invented the first practical radio-signaling system

Marchese Guglielmo Marconi, the son of an Italion land owner, was born on April 25, 1874 near Bologna, Italy. He was educated at the University of Bologna.

Marconi began experimenting with Hertzian (radio) waves in 1894. A few years earlier, Heinrich Hertz (1857-1894) had produced sparks in one circuit and detected them via radio waves in another circuit a few meters away. By 1895, Marconi had developed an apparatus with which he succeeded in sending signals to a point several kilometers away by means of a directional antenna.

Marconi tried unsuccessfully to interest the Italian Ministry of Posts and Telegraphs. However, in 1896 his cousin arranged an introduction to Nyilliam Preece, Engineer-in-Chief of the British Post Office. Encouraging demonstrations in London and on Salisbury Plain followed, and in 1897 Marconi obtained a patent, and established the Wireless Telegraph and Signal Company Limited, which opened the world's first "radio" factory at Chelmsford, England in 1898.

On May 13, 1897, Marconi sent the first ever wireless communication over water. It transversed the Bristol Channel from Lavernock Point (South Wales) to Flat Holm Island, a distance of 14 kilometres. And in 1899 Marconi established communication across the English Channel between England and France. In 1901, he communicated signals across the Atlantic Ocean between Poldhu, in Cornwall, England, and St. John's, in Newfoundland.

Marconi's system was soon adopted by the British and Italian navies, and by 1907 had been so much improved that transatlantic wireless telegraph service was established for public use.

Maarconi shared the 1909 Nobel Prize in physics with Karl Ferdinand Braun (1850-1918) in recognition of their contributions to the development of wireless telegraphy.

During World War I, Marconi was in charge of the Italian wireless service, and developed shortwave transmission as a means of secret communication. In the remaining years of his life he experimented with shortwaves and microwaves.

Marconi's work laid the foundation for the invention of the radio. He died on July 20, 1937.

1. Guglielmo Marconi, Marconiusa.com, www.marconiusa.org/marconi.
2. Guglielmo Marconi, Microsoft Encarta Encyclopedia Online, http://uk.encarta.msn.com/ encyclopedia_761556697/marconi_guglielmo_marchese, 2007.
3. Guglielmo Marconi, Wikipedia, http://en.wikipedia.org/wiki/Guglielmo_Marconi.

Johannes Stark (1874-1957)

Discovered the nature of electron spectral lines, contributing to the understanding of the structure of the atom

Johannes Stark was born in Schickenhof, Bavaria on April 15, 1874. He attended the University of Munich, where he studied physics, mathematics, chemistry and crystallography, and earned his Ph.D. in physics at the University in 1897. His doctoral dissertation was on Newton's electrochronic rings in a certain type of dim media.

In 1905, Stark used the Doppler effect to explain a shift in the frequencies of light waves radiated by swiftly moving ions. The shift indicated that the electrons were moving at different speeds.

In 1908, Stark became professor at the RWTH Aachen University. His abrasive personality caused conflicts that often led to career moves.

In 1913, Stark maintained a concentrated electrical field in a ray tube, and placed a photographic plate in the path of light rays that passed through a mixture of hydrogen and helium. When developed, the photographic plate showed that the spectral line produced by the hydrogen ions had split into several lines. As electrons moved from one energy level to another, the frequency of the light they gave off had changed, creating the split spectral line. Later research showed that Stark's work supported the quantum theory.

In 1919, Stark received the Nobel Prize in physics for his discovery of the nature of electron spectral lines. This discovery contributed to the understanding of the structure of the atom, and came to be known as the **Stark effect**.

Stark's abrasive personality caused conflicts that often led to career moves. During the Nazi regime, he used his authority to fight against modern theoretical physics, which he called "Jewish physics," and under Nazi law to remove Jewish professors from universities.

Stark withdrew from academia in 1922 and used his Nobel Prize money to finance a porcelain factory in Germany, which later failed.

After failure of his business, Stark's Nazi connections helped him win powerful positions, first as president of the State Physical-Technical Institute in 1933 and then as president of the federally funded German Research Association in 1934.

In 1947, a German court classified Stark as a leading Nazi, and sentenced him to four years in a labor camp. He died on June 21, 1957.

1.   Johannes Stark, Answers.com, www.answers.com/topic/johannes-stark?cat=technology.
2.   Johannes Stark, Microsoft Encarta Encyclopedia Online, http://encarta.msn.com/encyclopedia_761583359/johannes_stark, 2007.
3.   Johannes Stark, Wikipedia, http://en.wikipedia.org/wiki/Johannes_Stark.

Charles Barkla (1877-1944)

Discovered the characteristic X-rays of elements

Charles Glover Barkla was born in Widnes in the northwest of England on June 27, 1877. He studied at the Liverpool Institute and Liverpool University. After taking his master's degree in 1899 at Liverpool, Barkla went to Trinity College, Cambridge, but because of his passion for singing, he transferred to King's College to sing in the choir. At King's College he started his important research on X-rays.

Barkla received his Ph.D. from the University College in Liverpool in 1904. He then taught at Liverpool until 1909 when he became Wheatstone Professor at King's College, London. From 1913 onward he was professor of natural philosophy at Edinburgh University.

Barkla's scientific work concerned the properties of X-rays—in particular, the way in which X-rays are scattered by various materials. He showed in 1903 that the scattering of X-rays by gases depends on the molecular weight of the gas.

In 1904, Barkla observed the polarization of X-rays—a result that indicated that X-rays are a form of electromagnetic radiation. Further confirmation of this was obtained in 1907 when he performed experiments on the direction of scattering of a beam of X-rays to resolve a controversy with William Henry Bragg (1862-1942), who argued that X-rays were particles.

Barkla also demonstrated X-ray fluorescence, in which primary X-rays are absorbed and the excited atoms then emit characteristic secondary X-rays. The frequencies of the characteristic X-rays depend on the atomic number of the element.

Barkla found that secondary radiation from elements with the heaviest atoms and molecules had two components. One was the X-rays that scattered unchanged. A second, more penetrating type of radiation was produced by the element itself and had characteristics specific to the element. In further investigations, Barkla showed that two kinds of the second type of radiation were produced by heavy elements. He named the more penetrating of these types K radiation and the less penetrating L radiation.

For his discovery of the characteristic X-rays of elements, Barkla was awarded the 1917 Nobel Prize in phsics.

Barkla died on October 23, 1944.

1. Charles G. Barkla, Microsoft Encarta Encyclopedia Online, http://encarta.msn.com/encyclopedia_761583100/barkla_charles_glover, 2007.
2. Charles Glover Barkla, InfoPlease, www.answers.com/barkla?cat=technology.
3. Charles Glover Barkla, Wikipedia, http://en.wikipedia.org/wiki/Charles_Glover_Barkla.

Francis Aston (1877-1945)

Invented the mass spectrometer

Francis William Aston was born in Harborne, Birmingham, England on September 1, 1877. He studied science at Malvern College for two years before entering Mason College, Birmingham, in 1893.

For financial reasons, Aston left college and began working as a fermentation chemist at a brewery in Wolverhampton, where he stayed for three years. In 1903, he returned to the college, which had become the University of Birmingham.

In 1909, J.J. Thomson (1856-1940) invited Aston to move to the Cavendish Laboratory at Cambridge and work as his assistant. Aston accepted the post, which left him more time for research.

Thomson had developed a method of measuring atomic weights by using combinations of magnetic and electric fields to produce curves on a photographic plate. Aston helped refine those experiments.

Aston's work was interrupted by the First World War when he worked as a chemist at Farnborough in the Royal Aircraft Factory, investigating the canvas used to cover airplanes.

After the war, at Cavendish Laboratory, Aston constructed his first mass spectrograph. The apparatus clearly showed the two types of neon, and also showed isotopes for chlorine and many other elements. Aston continued to improve his mass spectrograph, and succeeded in analyzing most of the chemical elements.

Aston's first spectrograph had established the Whole Number Rule—the masses of all elements except hydrogen are a whole number of atomic mass units. The improved accuracy of his second spectrograph measured the differences from this rule, which gave details of the forces holding atoms together. Aston's third and final mass spectrograph had an accuracy of 1 in 100,000.

Aston was awarded the 1922 Nobel Prize in Chemistry for his discovery, by means of his mass spectrograph, of isotopes, in a large number of non-radioactive elements, and for his enunciation of the Whole-number Rule."

In his private life, Aston played the piano, violin, and cello. He never married and lived in Trinity College, Cambridge for 35 years until he died on November 20, 1945.

1. Francis W. Aston, NobelPrize.org, Biography, http://nobelprize.org/nobel_prizes/chemistry/laureates/1922/aston-bio.
2. Francis William Aston, Cambridge Physicists, www-outreach.phy.cam.ac.uk/camphy/physicists/physicists_aston.
3. Francis William Aston, Wikipedia, http://en.wikipedia.org/wiki/Francis_William_Aston.

H. Thomas Milhorn, MD, PhD

Lise Meitner (1878-1968)

Part of a team that discovered nuclear fission

Lise Meitner was born into a Jewish family in Vienna, Austria on November 17, 1878. Her father was one of the first Jewish lawyers in Austria. Originally named Elise, she shortened her name to Lise. She receiving her doctorate from the University of Vienna in 1906.

In 1918, Meitner and German chemist Otto Hahn (1879-1968) discovered the first long-lived isotope of the element protactinium. That year, Meitner was given her own physics section at the Kaiser Wilhelm Institute of Chemistry. In 1923, she discovered the cause of the Auger effect, the emission from surfaces of electrons with signature energies. The effect is named for Pierre Auger (1899-1993), a French scientist who independently discovered the effect in 1925.

From 1926 to 1933, Meitner was a professor of physics at the University of Berlin. In 1938, she fled Nazi-controlled Austria to Holland, using a diamond ring to bribe the border guards. Failing to find a position in the Netherlands, she moved on to Sweden, where she joined the atomic research staff at the University of Stockholm.

Meitner was the first to realize that the nucleus of an atom could be split into smaller parts—Uranium-235 nuclei could be split to form barium and krypton, accompanied by the ejection of several neutrons and a large amount of energy. However, it was politically impossible for the exiled Meitner to publish jointly with Hahn. He published the findings in 1939, and Meitner published the physical explanation two months later with her nephew, physicist Otto Frisch (1904-1979), and named the process "nuclear fission."

Meitner recognized the possibility for a chain reaction of enormous explosive potential. She refused an offer to work on the Los Alamos project, declaring that whe would have nothing to do with a bomb.

In 1944, Hahn was awarded the Nobel Prize for physics for his research into fission, but Meitner was ignored, partly because Hahn had downplayed her role. As a result, Meitner is often mentioned as one of the most glaring examples of scientific achievement that was overlooked by the Nobel committee. In 1997, the chemical element with the atomic number 109 was given the official name meitnerium (Mt) in her honor.

Meitner became a Swedish citizen in 1949, but moved to Britain in 1960, and died in Cambridge on October 27, 1968.

1. Lise Meitner (1878 - 1968), AtomicArchive.com, www.atomicarchive.com/Bios/Meitner.
2. Lise Meitner, Microsoft Encarta Encyclopedia Online, http://encarta.msn.com/
    encyclopedia_761561025/lise_meitner, 2007.
3. Lise Meitner, Wikipedia, http://en.wikipedia.org/wiki/Lise_Meitner.

Albert Einstein (1879-1955)

Explained the photoelectric effect and developed the theory of relativity

Albert Einstein was born in Ulm, Germany on March 14, 1879. He spent his youth in Munich, where his family owned a small shop that manufactured electrical machinery.

After secondary school, Einstein entered the Swiss Federal Institute of Technology in Zürich, graduating in 1900. He often skipped class to play his violin, using the notes of others to pass. His professors did not think highly of him and would not recommend him for a university position. In 1902, he secured a position as an examiner in the Swiss patent office in Bern.

In 1905, Einstein received his doctorate from the University of Zürich for a theoretical dissertation on the dimensions of molecules. He also published three theoretical papers of central importance to the development of 20th-century physics. In the first, he made significant predictions about the motion of particles (Brownian motion) that are randomly distributed in a fluid.

The second paper, on the photoelectric effect, contained a revolutionary hypothesis concerning the nature of light. Einstein not only proposed that under certain circumstances light can be considered as consisting of particles, but he also hypothesized that the energy carried by any light particle, called a photon, is proportional to the frequency of the radiation.

Einstein's third major paper contained what became known as the special theory of relativity. This contained two postulates: (1) Physical laws are the same in all inertial reference systems, and (2) the principle of the constancy of the speed of light.

Einstein published his general theory of relativity in 1916. In this theory, the interactions of bodies are explained as the influence of bodies on the geometry of space-time, having the three dimensions from Euclidean space and time as the fourth dimension. Using this, Einstein predicted the bending of starlight in the vicinity of a massive body, such as the Sun. Einstein's equation, $E = mc^2$, is perhaps the most famous equation in the world.

In 1921, Einstein received the Nobel Prize in physics for his services to Theoretical Physics, and especially for his explanation of the photoelectric effect. He died in fhs United States in 1955.

1. Albert Einstein, MacTutor, www-groups.dcs.st-and.ac.uk/~history/Printonly/Einstein.
2. Albert Einstein, Microsoft Encarta Encyclopedia online, http://encarta.msn.com/ encyclopedia_761562147/albert_einstein, 2007.
3. Albert Einstein, Wikipedia, http://en.wikipedia.org/wiki/Albert_Einstein.

Max von Laue (1879-1960)

Dscovered the diffraction of X-rays by crystals

Max Theodore Felix von Laue was born in Pfaffendorf (near Koblenz), Germany on October 9, 1979. Starting in 1899, he studied mathematics, physics, and chemistry. Then, in 1902 he went to the Friedrich-Wilhelms-University of Berlin, where he studied under Max Planck (1858-1947), who had given birth to the quantum theory revolution in 1900. Laue completed his doctorate in 1903. His dissertation was on interference phenomena in plane-parallel plates.

From 1906 to 1909 Laue worked as an asstant to Planck in Berlin, he pursued the application of entropy to radiation fields and the thermodynamic significance of the coherence of light waves. From 1909 to 1912, he worked under Arnold Sommerfeld (1868-1951) at the Institute for Theoretical Physics at Ludwig Maximilians University of Munich. There, Laue successfully diffracted X-rays with crystals, and wrote the first volume of his book on relativity.

From 1914 to 1919, Laue was at the Johann Wolfgang Goethe University of Frankfurt am Main. In 1916, he was engaged in vacuum tube development, at the Bayerische Julius-Maximilians-Universität Würzburg, for use in military telephony and wireless communications. In 1919, Laue was called to the Humboldt University of Berlin, a position he held until 1943. In 1921, he published the second volume of his book on relativity, and in 1932 showed that the threshold of the applied magnetic field which destroys superconductivity varies with the shape of the body.

Laue received the 1914 Nobel Prize in physics for his discovery of X-ray crystallography, which provided the means to determine the arrangement of atoms in some substances.

On April 23, 1945, French troops entered Hechingen. This was followed by the seizing of equipment involved in the German nuclear energy effort. Laue was taken to Huntington, England and interned at Farm Hall with other German scientists. He was returned to Germany early in 1946, where he was instrumental in re-establishing and organizing German science after World War II.

While driving to his laboratory, von Laue's car was struck by a motor cyclist. The cyclist was killed and Laue's car overturned. Laue died from his injuries 16 days later on April 24, 1960 in Berlin.

1. Max Theodor Felix Von Laue (1879 - 1960), RTStudents.com, www.rtstudents.com/radiology/max-von-laue.
2. Max Theodor Felix Von Laue, Microsoft Encarta Encyclopedia Online, http://ca.encarta.msn.com/ encyclopedia_761583392/Laue_Max_Theodor_Felix_von, 2007.
3. Max von Laue, Wikipedia, http://en.wikipedia.org/wiki/Diffraction.

Owen Richardson (1879-1959)

Developed a theory of the emission of charged atomic particles from a heated surface

Owen Willans Richardson was born in Dewsbury, Yorkshire, England on April 26, 1879. He graduated with a B.S. in physics, chemistry, and botany from Trinity College, Cambridge in 1900, and earned a D.Sc. degree in physics from University College, London in 1904.

Richardson's earliest research dealt with the properties of the electron. As early as 1901 he began the studies for which he was best known—thermionics, a term that Richardson himself coined.

In 1906, Richardson moved to the United States to accept a position as head of the physics department at Princeton University, where he taught for seven years.

In 1913, Richardson returned to England just before the start of World War I to accept the Wheatstone Chair of Physics at King's College at the University of London. During the war, he worked to improve the technology of military-communications systems.

In 1924, Richardson was appointed Yarrow Research Professor of the Royal Society, an honor that freed him from teaching duties to devote his full attention to research.

When Richardson began his thermionic research, many scientists believed that electron emissions were chemical in nature. Richardson, however, believed that negatively and positively charged radiation emanating directly from heated metal was physical in nature. Using platinum, and later tungsten, he showed that heating metal results in a loss of electrons from the metal's surface. He formulated the equation that expresses that phenomenon, now called **Richardson's law**.

Richarson received the Nobel Prize for physics in 1928 for his work on thermionic phenomenon, which formed the basis for the development of telephone, radio, television, and X-ray technology.

Richardson also did research on the photoelectric effect, gyrometrics (the study of the magnetic properties of a rotating electrical particle), the emission of electrons by chemical reactions, soft X-rays (electromagnetic radiation with a wavelength in the range of 10 to 0.01 nanometers), and the spectrum of hydrogen.

Richardson was knighted in 1939, retired in 1944, and died on February 15, 1959.

1. Owens willans Richardson, RookRags, www.bookrags.com/Owen_Willans_Richardson.
2. Owens Willans Richardson, Wikipedia, http://en.wikipedia.org/wiki/Owen_Willans_Richardson.
3. Sir Owen Willans Richardson, Microsoft Encarta Encyclopedia Online, http://encarta.msn.com/ encyclopedia_761583323/richardson_sir_owen_willans, 2007.

Paul Ehrenfest (1880-1933)

Applied quantum mechanics to rotating bodies and helped develop the statistical theory of nonequilibrium thermodynamics

Paul Ehrenfest was born to Jewish parents in Vienna Austria on January 18, 1880. His parents owned and operated a grocery store. As a child, Paul's health was poor. He was sickly, had dizzy spells, and suffered frequent nosebleeds.

Ehrenfest majored in chemistry at the Technische Hochschule, but he also took courses at the University of Vienna. In 1901, he transferred to Göttingen. He obtained his Ph.D. in Vienna in 1904. His dissertation was entitled *The motion of rigid bodies in fluids and the mechanics of Hertz.*

Ehrenfest spent 1908 to 1912 in St. Petersburg, Russia, having married a Russian woman. He failed to find a position.

In 1912, Ehrenfest was offered and accepted a position as professor of physics at the University of Leiden, where he remained the rest of his career.

Ehrenfest's most important research, the theory of adiabatic invariants, took place from 1912 to 1933. It is a concept from classical mechanics that on one hand refines certain methods of the provisional mechanics of the atom and on the other hand makes a link between atom mechanics and statistical mechanics.

Ehrenfest made major contributions to quantum physics, including the theory of phase transitions and the **Ehrenfest theorem**. His name is also given to the **Ehrenfest paradox**, an apparent paradox in relativity, and to **Ehrenfest Time**, the time characterizing the departure of quantum dynamics for observables from classical dynamics.

Ehrenfest was also interested in developing mathematical theories in economics. This interest was stimulated by his notion that there should be an analogy between thermodynamics and economic processes.

From the correspondence with his close friends it appears that Ehrenfest suffered from severe depression. By August 1932, Einstein was so worried that he wrote to the Board of the University of Leiden, expressing deep concern and suggesting ways in which Ehrenfest's workload could be reduced.

In Amsterdam, on September 25, 1933, Ehrenfest lost a battle with depression. Having made arrangements for the care of his other children, he first shot his younger son Wassik, who had Down syndrome, then killed himself.

1. O'Connor, J. J. and E F Robertson, Paul Ehrenfest, MacTutor, www-groups.dcs.st-and.ac.uk/~history/Biographies/Ehrenfest.
2. Paul Ehrenfest, Wikipedia, http://en.wikipedia.org/wiki/Paul_Ehrenfest.

Clinton Davisson (1881–1958)

Proved the wave nature of moving electrons by means of diffraction by crystals

Clinton Joseph Davisson was born in Bloomington, Illinois on October 22, 1881. While teaching at Princeton, he did doctoral thesis research with Owen Richardson (1879-1959), receiving his Ph.D. in physics from Princeton in 1911. His thesis was entitled *On The Thermal Emission of Positive Ions From Alkaline Earth Salts.* Davisson was then appointed assistant professor at the Carnegie Institute of Technology.

In 1917, during the First World War, Davisson took a leave from the Carnegie Institute to do war-related research with the Engineering Department of the Western Electric Company (later to become Bell Telephone Laboratories).

At the end of the war, Davisson accepted a permanent position at Western Electric to do basic research. He remained at Western Electric until 1946, at which time he accepted a research professor appointment at the University of Virginia. He remained at the University of Virginia until his retirement in 1954.

At Western Electric, Davisson pursued his interest in thermionics, the emission of electrically charged particles from conducting materials, such as metals, which were heated to high temperatures. His experiments in thermionics revised the classic theory on conduction and the thermal energy of electrons.

In 1927, Davisson and Lester Germer (1897-1871) performed an experiment showing that electrons were diffracted at the surface of a crystal of nickel. This Davisson-Germer experiment confirmed the hypothysis of Louis de Broglie (1892-1987) that particles of matter have a wave-like nature, which is a central tenet of quantum mechanics. Their observation allowed the first measurement of a wavelength "λ" for electrons, which agreed well with de Broglie's equation $\lambda = h / p$, where "h" is Planck's constant and "p" is the electron's momentum.

In 1937, Davisson was awarded the Nobel Prize in physics for his discovery of electron diffraction. He shared the Nobel Prize with George Paget Thomson (1892-1975), who independently discovered electron diffraction at about the same time as Davisson.

Davisson died on February 1, 1958.

1.   Clinton Davisson, Biography, NobelPrize.org,
     http://nobelprize.org/nobel_prizes/physics/laureates/1937/davisson-bio.
2.   Clinton Davisson, Wikipedia, http://en.wikipedia.org/wiki/Clinton_Davisson
3.   Clinton J. Davisson, Microsoft Encarta Encyclopedia Online, http://encarta.msn.com/
     encyclopedia_761583156/clinton_j_davisson, 2007.

Theodor von Kármán (1881-1963)

Contributed to fluid mechanics, turbulence theory, and supersonic flight

Theodore von Kármán was born into a Jewish family at Budapest, Austria-Hungary on May 11, 1881 After graduating from high school, Kármán had to undertake compulsory military service for one year. He served as an artillery cadet in the Austro-Hungarian army.

In 1903, von Kármán was appointed an assistant in hydraulics at the Palatine Joseph Polytechnic, a position which he held for three years. He also acted as a consultant for a German locomotive manufacturer.

Von Kármán studied engineering at Royal Technical University, graduating in 1902. He received his doctorate in 1908 from the University of Göttingen. Then he taught at Göttingen for four years. In 1912, he accepted a position as director of the Aeronautical Institute at RWTH Aachen. His time there was interrupted by service in the Austro-Hungarian army from 1915 to 1918, during which time he designed an early helicopter.

In 1930, von Kármán went to the United States as professor and director of the Guggenheim Aeronautics Laboratory at the California Institute of Technology. He made theoretical contributions to aerodynamics, hydrodynamics, and mathematical analysis in thermodynamics. He also developed the first theory of supersonic drag, designed supersonic wind tunnels, and initiated research that led to the first breaking of the sound barrier by an aircraft. He later became a naturalized citizen of the United States.

In 1944, von Kármán and others founded the Jet Propulsion Laboratory, which is now a Federally funded research and development center managed and operated by the California Institute of Technology. In 1946 he became the first chairman of the Scientific Advisory Group which studied aeronautical technologies for the United States Army Air Forces.

Von Kármán was personally responsible for many key advances in aerodynamics, notably his work on supersonic and hypersonic airflow characterization. He used mathematical tools to study fluid flow, and interpretated those results to guide practical designs. He was instrumental in the design of swept-back wings that are used in modern jet aircraft. He never married. He died while on a visit to Aachen, Germany on May 6, 1963.

1. O'Connor, J. J. and E F Robertson, Theodore von Kármán, MacTutor, www-groups.dcs.st-and.ac.uk/ ~history/Biographies/Karman.
2. Theodore von Kármán, Microsoft Encarta Encyclopedia Online, http://encarta.msn.com/ encyclopedia_761561340/K%C3%A1rm%C3%A1n_Theodore_von, 2007.
3. Theodore von Kármán, Wikipedia, http://en.wikipedia.org/wiki/ heodore_von_K%C3%A1rm%C3%A1n.

Hans Geiger (1882-1945)

Invented the Geiger counter

$\mathbf{J}$ohannes "Hans" Wilhelm Geiger was born at Neustadt-an-der-Haardt, Germany on September 30, 1882. His father was professor of Indology at the University of Erlangen.

In 1902, Geiger began studying physics and mathematics at the University of Erlangen, and was awarded a doctorate in 1906. In 1907, he began working with Ernest Rutherford (1871-1937) at the University of Manchester in England, where he created the **Geiger counter**.

The detection component of the Geiger counter is a metal tube filled with gas under low pressure. The tube and a copper cathode are under a high voltage. When an alpha particle emitted from the nucleus of a radioactive atom passes through the tube, it ionizes a gas molecule, leaving the molecule with an electric charge. The ion is then attracted to the cathode, and as it collides with other gas molecules it produces more ions. This brief cascade of ions produces a momentary electrical current. The Geiger counter records each cascade electronically, indicating each cascade with a click. Rapidly repeating clicks indicate an increased level of radioactivity.

In 1911, Geiger and John Nuttall (1890-1958) discovered the **Geiger-Nuttall law**, which relates the decay constant of a radioactive isotope with the energy of the alpha particles emitted. Geiger and Nuttall performed experiments that led to Rutherford's atomic model.

In 1912, Geiger became leader of the Physical-Technical Reichsanstalt in Berlin, and in 1925 he became a professor in Kiel. In 1928, he and his student Walther Müller (1905-1979) created an improved version of the Geiger counter, the **Geiger-Müller counter**.

From 1925 to 1929, Geiger was on the faculty of the University of Kiel. From 1929 to 1936, he was a professor at the University of Tübingen. And from 1936 to 1945, he was a professor at Technische Hochschule, Berlin.

During World War II, Geiger served as an artillery officer in the German army. His loyalty to the Nazi Party led him to betray his Jewish colleagues, many of whom had helped him in his research before he became a member of the Party. He died in Potsdam, Germany on September 24, 1945, a few months after World War II ended.

1. Geiger-Nuttall Law, Wikipedia, http://en.wikipedia.org/wiki/Geiger-Nuttall_law.
2. Hans Geiger, Microsoft Encarta Encyclopedia Online, http://encarta.msn.com/encyclopedia_761589241/Hans_Geiger, 2007.
3. Hans Geiger, NNDB, www.nndb.com/people/123/000099823.
4. Hans Geiger, Wikipedia, http://en.wikipedia.org/wiki/Hans_Geiger.

James Franck (1882-1964)
Provided experimental verification of the quantum theory

James Franck was born in Hamburg, Germany on August 26, 1882. He completed his Ph.D. in 1906, and received his venia legendi (permission for lecturing) for physics in 1911, both at the University of Berlin. He then taught at the University of Berlin until 1918.

During World War I, Franck served in the German Army. Following the war, he became head of the physics division of the Kaiser Wilhelm Gesellschaft for Physical Chemistry. In 1920, he became ordinarius professor of experimental physics and Director of the Second Institute for Experimental Physics at the University of Göttingen. While at the university, he worked on quantum physics with Max Born (1882-1970), who was Director of the Institute of Theoretical Physics.

In collaboration with Gustav Hertz (1877-1975), Franck conducted experiments on the effects produced by bombarding atoms in mercury vapor with electrons and tracing the energy changes that resulted from the collisions. They found that electrons with insufficient velocity simply bounced off the mercury atoms, but that an electron with a higher velocity lost precisely 4.9 eV of energy to an atom. If the electron had more than 4.9 eV of energy, the mercury atom still absorbed only that amount. The Franck-Hertz experiment gave proof to the theory of Niels Bohr (1885-1962) that an atom can absorb internal energy only in precise and definite amounts, or quanta. For the research, which provided experimental verification of the quantum theory, Franck shared the 1925 Nobel Prize in physics with Hertz.

In 1933, after the Nazis came to power, Franck left his post in Germany and continued his research in the United States, first at Johns Hopkins University, and then, after a year in Denmark, in Chicago, where he became involved in the Manhattan Project during World War II. Franck became a leader of those scientists in the Manhattan Project who sought to stop the bomb's use against Japan, suggesting instead that the bomb be exploded in an unpopulated area to demonstrate its power to the Japanese government.

Franck also made important contributions to the study of photosynthesis. He died in Göttingen, Germany on May 21, 1964.

1. James Franck, Jewish Virtual Library, www.jewishvirtuallibrary.org/jsource/biography/James_Franck.
2. James Franck, Microsoft Encarta Encyclopedia Online, http://encarta.msn.com/encyclopedia_761578882/franck_james, 2007.
3. James Franck, Timeline of Nobel Prize Winners, http://peace.nobel.brainparad.com/james_franck..
4. James Franck, Wikipedia, http://en.wikipedia.org/wiki/James_Franck.

Max Born (1882-1970)
Did fundamental research in quantum mechanics

Max Born, the son of an anatomist and embryologist, was born in Breslau (now Wrocław, Poland), on December 11, 1882. He studied at the University of Breslau, followed by Heidelberg University and the University of Zurich. He then received his Ph.D. at the University of Göttingen. In 1909, Born joined the facutly at Göttingen. From 1915 to 1919, except for a period in the German army, he was extraordinarius professor of theoretical physics at the University of Berlin, where he formed a life-long friendship with Albert Einstein (1979-1955). In 1919, he became ordinarius professor at the University of Frankfurt am Main.

Born's early work involved crystals, particularly the vibrations of atoms in crystal lattices. The **Born–Haber cycle** is a theoretical cycle of reactions and changes by which it is possible to calculate the lattice energy of ionic crystals.

In 1921, Born was appointed professor of theoretical physics at Göttingen. In 1925, he and Pascual Jordan (1902-1980) developed the matrix mechanics introduced by Werner Heisenberg (1901-1976). Up until that time, matrices had seldom been used by physicists. They were considered to belong to the realm of pure mathematics. Born also showed how to interpret the theoretical results of Louis de Broglie (1892-1987) and the experiment results of Clinton J. Davisson (1881-1958), which showed that particles have wavelike behavior. Born's interpretation was that the particles exist, but are guided by a wave. At any point, the square of the amplitude indicates the probability of finding a particle there.

Born escaped to England in 1933, a refugee from Germany, and acquired British citizenship in 1939. During his first three years in England, he conducted research at the University of Cambridge. He was Tait Professor of Natural Philosophy at the University of Edinburgh from 1936 to 1953.

In 1954, born shared the Nobel Prize for physics with Walther Bothe (1891-1957) for his fundamental research in quantum mechanics, especially for his statistical interpretation of the wavefunction. Bothe's share was for his coincidence method. Born died in Göttingen, Germany on January 5, 1970. He is buried in Göttingen, where his tombstone displays his fundamental equation of matrix mechanics $pq - qp = h/2\pi i$.

1. Born, Max, Answers.com, www.answers.com/max+born&r=67.
2. Max Born, Microsoft Encarta Encyclopedia Online, http://uk.encarta.msn.com/ encyclopedia_761555160/born_max, 2007.
3. Max Born, Wikipedia, http://en.wikipedia.org/wiki/Max_Born. http://nobelprize.org /nobel_prizes/physics/laureates/1954/born-bio.

Percy Bridgman (1882-1961)

Invented an apparatus to produce extremely high pressures and made discoveries in high-pressure physics

Percy Williams Bridgman was a pioneer in investigating the effects of enormous pressures on the behavior of matter—solids, liquids, and gases. He was born on April 21, 1882 in Cambridge, Massachusetts. He entered Harvard University as an undergraduate and received his Ph.D. from there in 1908. Bridgman joined the Harvard physics department as a research fellow in 1908, became an instructor in 1910, and a full professor in 1919. He remained there until his retirement.

In 1905, Bridgeman began investigating the properties of matter under extremely high pressure. He developed a device that allowed him to create pressures eventually exceeding 10 gigapascals. This apparatus led to many new findings, including the effect of pressure on electrical resistance, and on liquid and solid states. Bridgeman also studied electrical conduction in metals and properties of crystals. He developed the **Bridgman seal,** which seals a high pressure volume by the use of an unsupported area to create a higher pressure between two pistons. A viscous material, such as copper or soap stone, is used in the generated higher pressure area to seal the intended pressure area.

The most striking effect of the enormous pressures Bridgeman created was the change in the melting point of many substances. Bridgman also found different crystalline forms of matter, which are stable under very high pressure but unstable under low pressure.

Bridgman also developed a set of thermodynamic equations. The **Bridgman's thermodynamic equations** are a basic set of equations, derived using a method of generating a large number of thermodynamic identities involving a number of thermodynamic quantities.

Bridgman was awarded the 1946 Nobel Prize in physics for the invention of an apparatus to produce extremely high pressures and for the discoveries he made high pressure physics.

Bridgman committed suicide by gunshot on August 20, 1961 after living with metastatic cancer for some time.

In 1955, the General Electric Company announced the production of synthetic diamonds, which their scientists, working on methods and information derived from Bridgman's work, had produced from ordinary carbon.

1. Percy W. Brigdman, NobelPrize.org, http://nobelprize.org/nobel_prizes/physics/laureates/1946/bridgman-bio.
2. Percy Williams Bridgman, BookRags, http://www.bookrags.com/Percy_Williams_Bridgman.
3. Percy Williams Bridgman, Wikipedia, http://en.wikipedia.org/wiki/Percy_Williams_Bridgman.

Walther Meissner (1882-1974)

Co-discovered the fact that a superconductor expels a magnetic field

Fritz Walther Meissner was born in Berlin, Germany on December 16, 1882. He studied mechanical engineering at the Technische Hochschule Berlin-Charlottenburg from 1901 to 1904, followed by two years of mathematics and physics at the University of Berlin.

In 1907, Meissner received his doctorate in physics from the University of Berlin, having completed his dissertation under the direction of Max Planck (1958-1947).

After receiving his doctoral degree, Meissner entered the Physikalisch-Technische Bundesanstalt, where he established the world's third largest Helium-liquifier.

Superconductivity had been discovered by Dutch physicist Heike Kamerlingh Onnes (1853-1926) in 1911. Since that time, however, little had been learned about the properties of superconductors. In 1933, Meissner and Robert Ochsenfeld (1901-1993) measured the flux distribution outside of tin and lead specimens, which were cooled below their transition temperature in the presence of a magnetic field. They discovered that below the superconducting transition temperature the specimens became perfectly diamagnetic, cancelling all flux inside. The experiment demonstrated for the first time that superconductors were more than just perfect conductors. Their discovery has become known as the **Meissner effect.**

In 1934, Meissner became chairman of technical physics at the Technical University of Munich.

After World War II, Meissner became president of the Bavarian Academy of Sciences and Humanities, and founded the Academy's Commission for Low Temperature Research. Laboratories were located in Herrsching am Ammersee until 1965, when they were moved to Garching.

In 1952, Meissner officially retired, but continued carrying out research for many years. He died in Munich, Germany on November 16, 1974. On the hundred-year anniversary of his birth, in 1982, the Institute for Low Temperature Research in Munich was renamed in Meissner's honor—The Walther Meissner Institute of Low-Temperature Research. The Institute carries out long-term projects which are too ambitious for the lifetime or capacity of any single researcher or which require the collaboration of specialists in various disciplines.

1. Meissner Effect, Wikipedia, http://en.wikipedia.org/wiki/Meissner_effect.
2. Walther Meissner (1882-1974), Magnet Lab, www.magnet.fsu.edu/education/tutorials/pioneers/ meissner.
3. Walther Meissner, Wikipedia, http://en.wikipedia.org/wiki/Walther_Meissner.

H. Thomas Milhorn, MD, PhD

Victor Hess (1883-1964)

Discovered the source of cosmic rays

Victor Franz Hess was born on June 24, 1883 in Waldstein Castle, near Peggau in Steiermark, Austria. His father was a royal forester in Prince Öttingen-Wallerstein's service. In 1901, Hess entered the University of Graz as an undergraduate student, and continued postgraduate studies in physics until he received his Ph.D. in 1910. He then became an Assistant at the Institute of Radium Research, Viennese Academy of Sciences, where he stayed until 1920.

For many years, scientists had been puzzled by the levels of ionizing radiation measured in the atmosphere. The assumption at the time was that the radiation originated from ground minerals, and therefore would decrease with the distance from the Earth. To study this phenomenon, Hess first increased the precision of the measuring equipment, and then during 1911 to 1912 personally took the equipment aloft in a balloon on 10 different occasions. He systematically measured the radiation at altitudes up to 5.3 km. His work showed that the level of radiation decreased up to an altitude of about one kilometer, but above that height the level increased to above the level on Earth. He concluded that radiation was penetrating the atmosphere from outer space. Hess's finding was confirmed in 1925 by Robert Millikan (1868-1953), who named the radiation cosmic rays. Hess's discovery opened the door to many new discoveries in nuclear physics.

Hess took leave of absence in 1921 and traveled to the United States, where he worked at the U.S. Radium Corporation in New Jersey and as Consulting Physicist for the U.S. Bureau of Mines in Washington D.C. In 1923, he returned to the University of Graz where in 1925 he was appointed the Ordinary Professor of Experimental Physics. Then, in 1931 he became Professor and Director Institute of Radiology at The University of Innsbruck.

Hess was awarded a share of the Nobel Prize in Physics in 1936 for his discovery of cosmic radiation. The other share of the Prize went to Carl Anderson (1905-1936) for his discovery of the positron.

To escape Nazi persecution, Hess relocated to the United States with his Jewish wife in 1938, and took a position at Fordham University as Professor of Physics. In 1944, he became a naturalized United States citizen. Hess retired from Fordham University in 1956. He died on December 17, 1964 in Mount Vernon, New York.

1.   Biographical Information about Victor A. Hess, www.mpi-hd.mpg.de/hfm/HESS/public/hessbio.
2.   Victor F. Hess-Biography, http://scienzapertutti.lnf.infn.it/biografie/hess-bio_fra.
3.   Victor Francis Hess, Wikipedia, http://en.wikipedia.org/wiki/Victor_Francis_Hess.

Niels Bohr (1885-1962)

Investigated the structure of atoms and of the radiation emanating from them

Niels Henrik David Bohr was born in Copenhagen, Denmark on October 7, 1885. His father was professor of physiology at the University of Copenhagen. Bohr received his doctorate from the University of Copenhagen in 1911. He went on to conduct experiments under J. J. Thomson (1856-1940) at Trinity College, Cambridge and Ernest Rutherford (1871-1937) at the University of Manchester.

Rutherford had shown that the atom consisted of a positively charged nucleus, with negatively charged electrons in orbit around it. Bohr expanded upon this theory by proposing that electrons travel only in certain successively larger orbits. He suggested that the outer orbits could hold more electrons than the inner ones, and that these outer orbits determine the atom's chemical properties. Bohr also suggested that when an electron jumps from an outer orbit to an inner one, it emits a discrete amount of electromagnetic radiation. Later, other physicists expanded this theory into quantum mechanics.

In 1916, Niels Bohr became a professor at the University of Copenhagen and director of the newly constructed Institute of Theoretical Physics in 1920. In 1922, he was awarded the Nobel Prize in physics for his investigation of the structure of atoms and of the radiation emanating from them. His son, Aage Bohr, would win the Nobel Prize in 1975.

Bohr also conceived the principle of complementarity—that items could be separately analyzed as having several contradictory properties. For example, physicists currently consider light to be both a wave and a stream of particles—two apparently mutually exclusive properties. In 1943, shortly before he was to be arrested by the German police, Bohr escaped to Sweden, then traveled to London, and eventually to the United States where he worked at the top-secret Los Alamos laboratory in New Mexico on the Manhattan Project to develop an atomic bomb.

After the war, Bohr returned to Copenhagen, advocating the peaceful use of nuclear energy. He died in Copenhagen on November 18, 1962. He is buried in the Assistens Kirkegård in the Nørrebro section of Copenhagen. The chemical element with atomic number 107 named **bohrium** (Bh) in honor of Niels Bohr.

1. Niels Bohr, Wikipedia, http://en.wikipedia.org/wiki/Niels_Bohr.
2. O'Connor and E F Robertson, Niels Henrik David Bohr, www-groups.dcs.st-and.ac.uk/~history/Printonly/Bohr_Niels., 2003.
3. Niels Bohr, Microsoft Encarta Encyclopedia Online, http://encarta.msn.com/encyclopedia_761576813/niels_bohr, 2007.
4. Niels Bohr, Physicist, LucidCafe, www.lucidcafe.com/library/95oct/nbohr.

Arthur Dempster (1886-1950)
Discovered the isotope uranium-235

Arthur Jeffrey Dempster was born in Toronto, Canada on August 14, 1886. He received his bachelor's degree in 1909 from the University of Toronto and his master's degrees in 1910 from the same institution. He then traveled to study in Germany, but left at the outset of World War I for the United States, where he received his Ph.D. in physics at the University of Chicago in 1916. He then joined the faculty there. And in 1927 he was made professor of physics.

In 1918, Dempster developed the first modern mass spectrometer, which was over 100 times more accurate than previous versions, and established the basic theory and design of mass spectrometers that is used to this day.

Dempster's research over his career centered about the mass spectrometer and its applications. In 1935, he showed that uranium did not consist solely of the isotope uranium–238. Seven out of every thousand uranium atoms were uranium–235. It was this isotope that was later predicted by Niels Bohr (1885-1962) to be capable of sustaining a chain reaction that could release large amounts of atomic fission energy. This knowledge subsequently allowed the development of the atom bomb, as well as nuclear power.

Regarded as the principal authority on positive rays, Dempster made extensive studies that brought about the discovery that protons go through helium without being appreciably deflected. He discovered that the protons of a hydrogen atom have wave characteristics and that they vibrate at a million times the frequency of light waves.

Dempster left the University of Chicago during World War II to work on the Manhattan Project that developed the world's first nuclear weapons, but returned to Chicago when his work was competed.

From 1943 to 1946, Dempster was chief physicist of the University of Chicago's Metallurgical Laboratory, which was founded to study the materials necessary for the manufacture of atomic bombs.

In 1946, Dempster took a position as a division director at the Argonne National Laboratory in Argonne, Illinois.

Dempster died of a heart attack while on vacation in Stuart, Florida on March 11, 1950.

1. Arthur Jeffrey Dempster, Answers.com, www.answers.com/topic/arthur-jeffrey-dempster?cat=technology.
2. Arthur Jeffrey Dempster, Wikipedia, http://en.wikipedia.org/wiki/Arthur_Jeffrey_Dempster.
3. Dr. A. J. Dempster, Physicist, 63, Dead, On This Day, New York Times, March 11, 2007.

Manne Siegbahn (1886-1978)

Worked with X- ray spectroscopy and developed instruments for precise measurement of X-ray wavelengths

Karl Manne Georg Siegbahn was born in Örebro, Sweden on December 3, 1886. His father was a stationmaster of the State Railways. Siegbahn obtained his Ph.D. at the Lund University in 1911. His thesis work involved magnetic field measurements.

Siegbahn became a lecturer at Lund in 1911, and from 1922 to 1937 was a professor of physics at the University of Uppsala.

From 1937 until his retirement in 1964 Siegbahn was Research Professor of Experimental Physics at the Royal Swedish Academy of Sciences in Stockholm and the first director of its Nobel Institute of Experimental Physics.

Siegbahn's early work was in electricity and magnetism, but by 1914 he had turned his attention to X-ray spectroscopy. X-ray spectroscopy is based on the fact that each element, when bombarded by fast-moving electrons, emits X rays of a characteristic wavelength and frequency. He improved the X-ray spectrometer in various ways so it detected and measured X-rays with more precision, and allowed him to discover previously unknown series of X-rays, including the M series.

Siegbahn's work contributed to the understanding of the atom and its structure, and supported the prevailing model that electrons were arranged in spherical shells around the nucleus of an atom. His research yielded information about virtually all the elements from sodium to uranium, and made possible the analysis of unknown substances.

Siegbahn's 1923 publication, *Spectroscopy of X Rays*, became a standard reference, his measurements of X-ray wavelengths were relied upon for their precision, and other physicists adopted his precise instrumentation.

Siegbahn's research ultimately resulted in many current applications of X-ray spectroscopy in such diverse fields as nuclear physics, chemistry, astrophysics, and medicine.

Siegbahn received the 1924 Nobel Prize for physics for his discoveries and research in the field of X-ray spectroscopy. His son, Kai Manne Börje Siegbahn (1918-2007), would share the 1981 Nobel Prize for physics.

Siegbahn died in Stockholm, Sweden on September 26, 1978.

1. Karl Manne Georg Siegbahn, Microsoft Encarta Encyclopedia Online, http://encarta.msn.com/ encyclopedia_761583352/siegbahn_karl_manne_georg, 2007.
2. Karl Manne Georg Siegbahn, Timeline of Nobel Prize Winners, http://peace.nobel.brainparad.com/ karl_manne_georg_siegbahn.
3. Manne Siegbahn-biography, www.angelfire.com/folk/hhh0/siegbahn-bio.

H. Thomas Milhorn, MD, PhD

Erwin Schrödinger (1887-1961)

Developed an equation that describes the space and time dependence of
quantum mechanical systems

Erwin Rudolf Josef Alexander Schrödinger was born in Erdberg, Vienna
on August 12, 1887. He received his doctorate in theoretical physics from
the University of Vienna in 1910.

Between 1914 and 1918, Schrödinger participated in war work as a
commissioned officer in the Austrian fortress artillery. In 1921, he became
professor of physics at the University of Zürich in Switzerland, where he
did his most important work. In January 1926, Schrödinger published in
the Annalen der Physik the paper *Quantisation as an Eigenvalue Problem*
in which he described what is now known as the time-independent
Schrödinger equation:

$$ih\frac{\partial \Psi}{\partial t} = -\frac{\hbar^2}{2m}\frac{\partial^2 \Psi}{\partial x^2} + V(x)\Psi(x,t) \equiv \tilde{H}\Psi(x,t)$$

The equation gave the correct energy eigenvalues for the hydrogen-like
atom. Schrödinger submitted a second paper four weeks later that solved
the quantum harmonic oscillator, the rigid rotor, and the diatomic
molecule. A third paper in May showed the equivalence of his approach to
that of Werner Heisenberg (1901-1976), and gave a treatment of the Stark
effect, which is the shifting and splitting of spectral lines of atoms and
molecules due to the presence of an electric field. A fourth paper showed
how to treat problems in which the system changes with time, as in
scattering problems.

In 1927, Schrödinger joined Max Planck (1858-1947) at the Friedrich
Wilhelm University in Berlin. In 1933, Schrödinger became a Fellow of
Magdalen College at the University of Oxford. Soon after his arrival, he
shared the Nobel Prize in physics with Paul Dirac (1902-1984) for their
discovery of new productive forms of atomic theory. Schrödinger's
position at Oxford did not work out because of his unconventional
personal life, living with both his wife and his mistress. He accepted a
position at the University of Graz in Austria in 1936.

After 17 years in Ireland, in 1956 Schrödinger returned to Austria to a
chair at the University of Vienna. He stayed there until he died of
tuberculosis on January 4, 1961.

1. Erwin Schrödinger, Microsoft Encarta Encyclopedia Online, http://encarta.msn.com/
   encyclopedia_761555158/schrodinger, 2007.
2. Erwin Schrödinger, Wikipedia, http://en.wikipedia.org/wiki/Erwin_Schr%C3%B6dinger.
3. O'Connor, J. J. and E. F. Robertson, Erwin Rudolf Josef Alexander Schrödinger, MacTutor,
   www-groups.dcs.st-and.ac.uk/~history/Biographies/Schrodinger.

Gustav Hertz (1887-1975)

Proved that atoms can absorb internal energy only in discrete amounts

Gustav Ludwig Hertz was a German experimental physicist and a nephew of Heinrich Rudolf Hertz (1857-1894), who carried out a number of important studies on electromagnetic waves, and for whom the unit of frequency (Hz) is now named. Gustav, the son of a lawyer, was born in Hamburg, Germany on July 22, 1887. Between 1906 and 1911, he studied mathematics and physics, but eventually decided to concentrate on a career in experimental physics. He received his doctorate at the Humboldt University of Berlin in 1911.

From 1912 to 1914, Hertz worked at the University of Berlin. It was during this time that he and James Franck (1882-1964) performed experiments known as the Franck-Hertz experiments. They studied the effect of the impact of electrons on atoms of mercury vapor. They found that electrons with insufficient velocity simply bounced off the mercury atoms, but that an electron with a higher velocity lost precisely 4.9 eV of energy to an atom. If the electron had more than 4.9 eV of energy, the mercury atom still absorbed only that amount. This research was an important confirmation of the Bohr model of the atom, with electrons orbiting the nucleus with specific, discrete energies. Proving that atoms can absorb internal energy only in definite amounts, they thus demonstrated the quantum theory of the German physicist Max Planck (1858-1947).

In 1914, Hertz was drafted into the German Army. After being severely injured in combat, he was released in 1915. Following the war, from 1920 to 1925, he worked in the physics laboratory at the Philips Incandescent Lamp Factory at Eindhoven. Hertz and Franck were awarded the 1925 Nobel Prize in physics for this work. That same year, Hertz was made Director of the Physics Institute of the University of Halle.

In 1928, Franck moved to the Charlottenburg Technological University in Berlin as Director of the Physics Institute. There, he developed a method of separating the isotopes of neon. As a result of rising anti-Semitism, Hertz resigned from this position in 1935.

In 1945, Hertz went to the USSR to continue his work in atomic research. In 1954, he returned to East Germany where he was appointed Professor and Director of the Physics Institute in Leipzig. Hertz retired in 1967. He died in Berlin, East Germany on October 30, 1975.

1. Gustav Hertz, Jewish Virtual Library, www.jewishvirtuallibrary.org/jsource/biography/hertz.
2. Gustav Hertz, Microsoft Encarta Encyclopedia Online, http://encarta.msn.com/ encyclopedia_761578076/gustav_hertz, 2007.
3. Gustav Hertz, Wikipedia, http://en.wikipedia.org/wiki/Gustav_Ludwig_Hertz.

Chandrasekhara Raman (1888-1970)

Discovered that an extremely small fraction of scattered light has a
frequency different from the incident photons

Chandrasekhara Venkata Raman was born on November 7, 1888 in
Tiruchirapalli, Tamil Nadu, India. His father was a lecturer in mathematics
and physics. Raman entered Presidency College, Madras in 1902, and in
1907 received his M.A. degree. He then joined the Indian Finance
Department as an Assistant Accountant General in Calcutta. After office
hours, he began going to the Association for the Cultivation of Science
laboratory and doing experiments.

In 1917, Raman resigned from his government service and took up the
newly created Palit Professorship in Physics at the University of Calcutta.
There, he discovered the **Raman effect,** which is described as follows:
When light is scattered from an atom or molecule, most photons are
elastically scattered. The scattered photons have the same frequency and
wavelength as the incident photons. However, an extremely small fraction
of the scattered light is scattered by an excitation, with the scattered
photons having a frequency different from, and usually lower than, the
frequency of the incident photons. In a gas, the Raman effect can occur
with a change in vibrational, rotational, or electronic energy of a molecule.
The Ramah spectrascope is based on this phenomenon.

Raman was knighted in 1929, and awarded the 1930 Nobel Prize in
physics for his work on the scattering of light and for the discovery of the
Raman effect.

Raman also worked on the acoustics of musical instruments. He
worked out the theory of transverse vibration of bowed strings on the basis
of superposition velocities. He was also the first to investigate the
harmonic nature of the sound of the Indian drums.

In 1934, Raman became the director of the newly established Indian
Institute of Science in Bangalore. In 1947, he was appointed the first
National Professor by the new government of Independent India.

Raman retired from the Indian Institute of Science in 1948, and a year
later established the Raman Research Institute in Bangalore Karnataka,
serving as its director. Raman died on November 21, 1970 in Bangalore,
Karnataka. His nephew, Subramanyan Chandrasekhar (1910-1925), would
also win a Nobel prize in physics in 1983.

1.   C. V. Ramane, Wikipedia, http://en.wikipedia.org/wiki/C._V._Raman.
2.   Chandrasekhara Venkata Raman, 2000, lokpriya.comwww.lokpriya.com/personalities/ scientists/
     cvraman.
3.   Chandrasekhara Venkata Raman, Microsoft Encarta Encyclopedia Online,
     http://encarta.msn.com/ encyclopedia_761561584/raman_sir_chandrasekhara_venkata, 2007.
4.   Ramn Scattering, Wikipedia, http://en.wikipedia.org/wiki/Raman_scattering.

Fritz Zernike (1888-1966)

Invented the phase-contrast microscope

Fritz Zernike, the son of two mathematics teachers, was born in Amsterdam, the Netherlands on July 16, 1888. He studied chemistry, mathematics, and physics at the University of Amsterdam. In 1913, he became assistant to Jacobus Cornelius Kapteyn (1851-1922) at the astronomical laboratory of Groningen University.

In 1914, Zernike was responsible jointly with Leonard Salomon Örnstein (1880-1941) for the derivation of the **Örnstein-Zernike relation** in critical-point theory, which in statistical mechanics is an integral equation that describes how the correlation between two molecules can be calculated. In 1915, Zernike obtained a position in theoretical physics at Groningen University.

Conventional optical microscopes were inadequate for viewing the detail in living specimens, particularly if the specimens were transparent. The problem was that in a microscope image, there are variations in the phase of light, determined by the path that light travels, that the eye cannot detect. Zernike discovered that the effects due to changes in the optical path can be transformed into changes in light intensity, which the eye can detect. To accomplish this transformation, he invented the phase-contrast microscope, which uses a diffraction plate, or phase plate, inserted between the two components of the objective lens. Manipulation of the direct and diffracted light creates an optical-path difference that results in greater light intensity, thus revealing detail indiscernible with a conventional microscope.

Zernike also contributed to the efficient description of the imaging defects, or aberrations, of optical imaging systems (microscopes and telescopes). His orthogonal circle polynomials provided the optics community with a tool to separate the various aberrations and to solve the long-standing problem of the optimum balancing of the various aberrations of an optical instrument. Since the 1960's, Zernike's circle polynomials have been widely used in optical design, optical metrology, and image analysis.

Zernike was awarded the Nobel prize for physics in 1953 for his invention of the phase contrast microscope. He died in Amersfoort, Netherlands, March 10, 1966.

1. Fritz Zernike, Biography, NobelPrize.org, http://nobelprize.org/nobel_prizes/physics/ laureates/1953/zernike-bio.
2. Fritz Zernike, Microsoft Encarta Encyclopedia, 2003.
3. Fritz Zernike, Wikipedia, http://en.wikipedia.org/wiki/Frits_Zernike.
4. The Phase Contrast Microscope, Microscopes, http://nobelprize.org/educational_games/ physics/microscopes/phase/index.

Otto Stern (1888-1969)

Developed a method of studying the magnetic moments of atoms, atomic nuclei, and protons

Otto Stern was born in Sohrau (Żory) in the German Empire's Kingdom of Prussia (now in Poland) on February 17, 1888. In 1892, he moved with his parents to Breslau. He completed a doctoral degree in physical chemistry at the University of Breslau in 1912.

After receiving his doctorate, Stern joined Albert Einstein (1879-1955) at the University of Prague, and later followed him to the University of Zurich, where he became Privatdocent of Physical Chemistry at the Eidgenössische Technische Hochschule in 1913. Stern's earliest work involved statistical thermodynamics and quantum theory.

In 1914, Stern went to the University of Frankfort, and remained there until 1921, except for a period of military service.

From 1921 to 1922 Stern was Associate Professor of Theoretical Physics at the University of Rostock.

In 1923, Stern became Professor of Physical Chemistry and Director of the laboratory at the University of Hamburg. He and Walther Gerlach (1889-1979) developed the molecular beam method, which proved to be a powerful tool for investigating the properties of molecules, atoms and atomic nuclei. By shooting a beam of silver atoms through a non-uniform magnetic field onto a glass plate, they found that the beam split into two distinct beams, instead of broadening into a continuous band. This experiment verified the space quantization theory, which stated that atoms can align themselves in a magnetic field only in a few directions, instead of in any direction, as classical physics had suggested.

In 1933 Stern measured the magnetic moment of the proton by using a molecular beam, and found that it was actually about 2-1/2 times the theoretical value. That same year, he moved to the United States, where he was appointed Research Professor of Physics at the Carnegie Institute of Technology in Pittsburgh. He remained there until 1945.

In 1943, Stern received the Nobel Prize in physics for his contribution to the development of the molecular beam method and his discovery of the magnetic moment of the proton. He died on August 17, 1969.

1. Otto Stern, Biography, NobelPrize.org, http://nobelprize.org/nobel_prizes/physics/laureates/1943/stern-bio.
2. Otto Stern, Microsoft Encarta Encyclopedia Online, http://encarta.msn.com/encyclopedia_761563620/stern_otto, 2007.
3. Otto Stern, Timeline of Nobel Prize Winners, http://peace.nobel.brainparad.com/otto_stern.
4. Otto Stern, Wikipedia, http://en.wikipedia.org/wiki/Otto_Stern.

Paul Ewald (1888-1985)
Developed X-ray diffraction methods

Paul Peter Ewald was born on January 23, 1888 in Berlin, Germany. He studied physics, chemistry, and mathematics at Gonville and Caius College in Cambridge during the winter of 1905. Then in 1906 and 1907 he continued his studies at the University of Göttingen, where his interest primarily was mathematics. In 1907, he continued his mathematical studies at the Ludwig Maximilians University of Munich under Arnold Sommerfeld (1868-1951) at his Institute for Theoretical Physics. Ewald was granted his doctorate in 1912.

During World War I, Ewald served in the German military as a medical technician. At the conclusion of the war, he returned to Ludwig Maximilians University as an assistant to Sommerfeld.

In 1921, Ewald was appointed to the faculty at Stuttgart Technische Hochschule, and in 1930 he became director of the university's Institute for Theoretical Physics.

The main thrust of Ewald's work was X-ray crystallography. The **Ewald construction** and the **Ewald sphere** are named after him. The Ewald construction is a method of interpreting the results of Max von Laue (1870-1960) in terms of reciprocal lattices. It uses a simple geometric construction that demonstrates the relationship in simple terms.

The Ewald sphere is a geometric construct used in X-ray crystallography which demonstrates the relationship between the wavelength of the incident and diffracted X-ray beams, the diffraction angle for a given reflection, and the reciprocal lattice of the crystal. Ewald's sphere can be used to find the maximum resolution available for a given X-ray wavelength and the unit cell dimensions. It is often simplified to the two-dimensional **Ewald's circle.**

Toward the end of World War II, Ewald was concerned that peace would result in the establishment of multiple, competing national journals of crystallography. So, in 1944, he proposed the establishment of an International Union of Crystallography that would have sole responsibility for publishing crystallographic research.

Ewald moved to the United States in 1949 and took a position at the Polytechnic Institute of Brooklyn as a professor and head of the Physics Department. He retired as head of the department in 1957 and from teaching in 1959. Ewald died on August 22, 1985 in Ithaca, New York.

1. Ewald's Sphere, Wikipedia, http://en.wikipedia.org/wiki/Ewald_sphere.
2. Paul Peter Ewald, Answers.com, www.answers.com/topic/paul-peter-ewald.
3. Paul Peter Ewald, NNDB, www.nndb.com/people/207/000099907.

William L. Bragg (1890-1971)

Analyzed crystal structure by X-rays

William Lawrence Bragg was born in North Adelaide, South Australia on March 30, 1890. His father, William Henry Bragg (1862-1942), was Professor of Mathematics and Physics at the University of Adelaide. At age five, Lawrence fell from his tricycle and broke his arm. His father had recently read about Röntgen's experiments in Europe and used the newly discovered X-rays to examine the broken arm. This is the first recorded surgical use of X-rays in Australia.

Bragg went to the University of Adelaide at age 15 to study mathematics, chemistry, and physics, graduating in 1908. In the same year his father accepted a job at the University of Leeds, and brought the family back to England. Lawrence entered Trinity College, Cambridge in 1909, and, majoring in physics, graduated in 1911.

Bragg is most famous for **Bragg's law of diffraction,** which deals with the diffraction of X-rays by crystals. Bragg's law makes it possible to calculate the positions of the atoms within a crystal from the way in which an X-ray beam is diffracted by the crystal lattice. He made this discovery in 1912, during his first year as a research student in Cambridge.

At age 25, Bragg and his father shared the 1915 Nobel Prize for physics for analysis of crystal structure by means of X-rays.

After World War II, Bragg held positions at Trinity College and then the University of Manchester.

In 1937, Bragg moved to the National Physical Laboratory as director, but soon accepted an invitation to Cambridge as the Cavendish Professor of Experimental Physics. He was knighted in 1941.

Bragg stayed at Cambridge until 1953, when he moved to the Royal Institution, London, as director of the Davy-Faraday Research Laboratory, a position once held by his father.

In 1948, Bragg became interested in the structure of proteins, and played a major part in the 1953 discovery of the structure of DNA in that he provided support to Francis Crick (1916-2004) and James Watson (1928-). The X-ray method that Bragg had developed forty years before was at the heart of the discovery.

In 1953, Bragg accepted the job of Resident Professor at the Royal Institution in London. He died on July 1, 1971 at the age of 81.

1. Sir Lawrence Bragg, Microsoft Encarta Encyclopedia Online, http://ca.encarta.msn.com/ encyclopedia_761583123/Bragg_Sir_(William)_Lawrence, 2007.
2. Sir William Lawrence Bragg, Timeline of Nobel Prize Winners, Physics, http://peace.nobel.brainparad.com/william_lawrence_bragg.
3. William Lawrence Bragg, Wikipedia, http://en.wikipedia.org/wiki/William_Lawrence_Bragg.

James Chadwick (1891-1974)

Discovered the neutron

James Chadwick was born in Bollington, Cheshire, England on October 20, 1891. He graduated from Manchester University in 1911 and remained to work with Ernest Rutherford (1871-1937).

Chadwick was studying in Germany at the start of World War I, and was interned in Ruhleben Prisoner of War Camp just outside Berlin. During his internment, he set up a laboratory in the stables where he worked on the ionization of phosphorus and the photo-chemical reaction of carbon monoxide and chlorine. He spent most of the war years in Ruhleben.

Chadwick returned to England in 1919 to carry out research at Cambridge University. In 1923, he became the assistant director of research at the Cavendish Laboratory.

In 1932, Chadwick made a fundamental discovery in nuclear science; he discovered the particle in the nucleus of an atom that subsequently became known as the neutron. This accounted for the unknown mass in the nucleus. Because of its neutrality, this particle did not have to overcome any electric barrier and was capable of penetrating and splitting the nuclei of even the heaviest elements.

Chadwick's discovery made it possible to create elements heavier than uranium in the laboratory. It particularly led to the study of nuclear reactions caused by slowed neutrons, and led to the discovery of nuclear fission, which triggered the development of the atomic bomb.

In 1935, Chadwick became professor at the University of Liverpool. That same year he received the Nobel Prize for physics for his discovery of the neutron.

Chadwick spent much of his time from 1943 to 1945 in the United States, principally at the Los Alamos Scientific Laboratory in New Mexico. He was one of the first in Britain to stress the possibility of the development of an atomic bomb, and was the chief scientist associated with the British atomic bomb effort. He was knighted in 1945.

In 1959, Chadwick became a Fellow of Gonville and Caius College, University of Cambridge. He died at Cambridge on July 23, 1974.

1. Cain, Jeannette, Chadwick, James: 1891-1974, Light-Science.com, www.light-science.com/chadwick.
2. Chadwick, Sir James, Microsoft Encarta Encyclopedia Online, http://au.encarta.msn.com/encyclopedia_761558353/chadwick_sir_james, 20037.
3. James Chadwick, Wikipedia, http://en.wikipedia.org/wiki/James_Chadwick.

Walther Bothe (1891-1957)

Invented the coincidence circuit and made several discoveries with it

Walther Wilhelm Georg Bothe was born in Oranienburg, Germany on January 8, 1891. He studied physics from 1908 until 1912 at the University of Berlin under Max Planck (1858-1947), earning his doctorate in 1914. His thesis work involved the study of the molecular theory of refraction, reflection, scattering, and absorption of light rays. During World War I he was taken prisoner by the Russians and spent a year in captivity in Siberia.

After the war, in 1920, Bothe began working at the radiation laboratory at Physikalisch-Technische Reichsanstalt in Berlin. In 1924, he discovered that if a single particle is detected by two or more Geiger counters, the detection will be practically coincident in time. Using this observation, he constructed a circuit allowing several counters in coincidence to determine the angular momentum of a particle.

In 1925, this new process allowed Bothe and Hans Geiger (1882-1945) to study the coincidences between the scattered X-ray and the recoiling electron, leading them to discover small-scale conservation of energy and momentum. Bothe also used the coincidence method to discover penetrating radiation coming from the upper atmosphere. His data indicated that the radiation was not composed exclusively of gamma rays, but was also composed of high energy particles, now known as mesons.

In 1927, Bothe began applying the coincidence method to the transmutation of light elements by bombardment with alpha particles. And in the 1930s, he found that the radiation emitted by beryllium when it is bombarded with alpha particles was a new form of penetrating high energy radiation, which was later shown to be neutrons.

In 1934, Bothe became Director of the Institute of Physics at the Max Planck Institute for Medical Research. In 1938, Wolfgang Gentner (1906-1980) and Bothe published a paper on the energy dependence of the nuclear photo-effect, which was the first decisive evidence that the absorption spectra of nuclei are accumulative and continuous.

For his discovery of the method of coincidence and the discoveries subsequently made by it, Bothe was awarded, jointly with Max Born (1882-1970), the Nobel Prize for physics for 1954. Bothe died on February 8, 1957.

1. Walther Bothe, Biography, NobelPrize.org, http://nobelprize.org/nobel_prizes/physics/laureates/ 1954/bothe-bio.
2. Walther Bothe, Wikipedia, http://en.wikipedia.org/wiki/Walther_Bothe.
3. Walther Wilhelm Georg Bothe, Microsoft Encarta Encyclopedia Online, http://ca.encarta.msn.com/ encyclopedia_761583122/Bothe_Walther_Wilhelm_Georg, 2007.

Arthur Compton (1892-1962)

Discovered that X-ray wavelengths increase due to scattering of the
radiant energy by free electrons

Arthur Holly Compton was born in Wooster, Ohio on September 10,
1892. His father was dean of the University of Wooster, which Arthur
attended. In 1913, he went to Princeton, where his obtained his Ph.D. On
graduation in 1916, he was named instructor in physics at the University
of Minnesota. The experiments begun there eventually led him to state that
magnetization of a material depends not on the orbits of the electrons in it,
but on the electron's own elementary characteristics; he was the first to
suggest the existence of quantized electron spin.

In 1922, Compton found that X-ray wavelengths increase due to
scattering of the radiant energy by free electrons, and that scattered quanta
have less energy than the quanta of the original ray. This discovery,
known as the **Compton effect** or **Compton scattering** demonstrated the
dual nature of electromagnetic radiation, and earned Compton a share of
the Nobel Prize for physics in 1927 with Charles Wilson (1869-1959).
Wilson's share of the prize was for his method of making the paths of
electrically charged particles visible by condensation of vapor

In 1923 Compton became professor of physics at the University of
Chicago. In 1941, he was placed in charge of the government committee
charged with investigating the properties and manufacture of uranium. In
1942, he appointed Robert Oppenheimer (1904-1964) as the Committee's
top theorist. When the Committee's work was taken over by the Army in
the summer of 1942, it became the Manhattan Project.

Immediately after the Japanese attack on Pearl Harbor on December 7,
1941, Compton convinced the government to consolidate plutonium
research at the University of Chicago and for an ambitious schedule that
called for producing the first atomic bomb in January 1945. The objectives
at Chicago were to produce chain-reacting piles of uranium to convert to
plutonium, find ways to separate the plutonium from the uranium, and to
design a bomb. In December 1942, a team of scientists, directed by Enrico
Fermi (1901-1954), achieved a sustained chain reaction in the world's first
nuclear reactor. In 1946, Compton became Washington University's ninth
Chancellor. He died on March 15, 1962, and is buried in the Wooster
Cemetery in Wooster, Ohio.

1. Arthur Compton, Microsoft Encarta Encyclopedia Online, http://encarta.msn.com/
    encyclopedia_761570704/arthur_compton, 2007.
2. Arthur Holly compton, 1892-1962, Selected Papers Of Great American Physicists, http://aip.org/
    history/gap/Compton/Compton.
3. Arthur Holly Compton, Wikipedia, http://en.wikipedia.org/wiki/Arthur_Compton.

George Thomson (1892-1975)

Independently proved the wave properties of electrons

George Paget Thomson, the son of physicist and Nobel laureate Joseph John (J. J.) Thomson (1856-1940), was born in Cambridge, England on May 3, 1892. Thomson studied mathematics and physics at Trinity College, Cambridge until the outbreak of World War I in 1914, when he joined the Queen's Regiment of Infantry.

After a brief service in France, Thomson became a fellow at Cambridge, and then taught physics at Cambridge from 1919 to 1922 and at the University of Aberdeen in Scotland from 1922 to 1930.

In 1927, Thomson performed an experiment in which he passed electrons through a thin gold foil onto a photographic plate behind the foil. The plate revealed a diffraction pattern, a series of concentric circles with alternate darker and lighter rings. Thomson thus demonstrated that electrons could be diffracted like a wave, a discovery proving the principle of wave-particle duality, which had first been suggested by Louis de Broglie (1892-1987) in the 1920s.

In 1930, Thomson was appointed professor at the Imperial College of Science and Technology in London, where he remained until 1952. During his time there he became interested in nuclear physics, and when the fission of uranium by neutrons was discovered at the beginning of 1939, he foresaw military and peacetime possibilities. He persuaded the British Air Ministry to procure a ton of uranium oxide for experiments. These experiments were incomplete at the outbreak of the war, and Thomson was made Chairman of the British Committee to investigate the possibilities of an atomic bomb. This committee reported in 1941 that a bomb was possible.

For his work with electron diffraction, Thomson was jointly awarded the Nobel Prize for physics in 1937 for his work in discovering the wave-like properties of the electron. The prize was shared with Clinton Joseph Davisson (1881-1958), who had made the same discovery independently

Thomson was also noted for his work in aerodynamics and nuclear energy. He was knighted in 1943.

In 1952, Thomson became Master of Corpus Christi College, Cambridge. He died on September 10, 1975.

1. George Paget Thomson, Biography, NobelPrize.org, http://nobelprize.org/nobel_prizes/physics/laureates/1937/thomson-bio.
2. George Paget Thomson, Wikipedia, http://en.wikipedia.org/wiki/George_Paget_Thomson.
3. George Thomson, Microsoft Encarta Encyclopedia Online, http://encarta.msn.com/encyclopedia_761575224/thomson_sir_george_paget, 2007.

Louis de Broglie (1892-1987)

Created the field of wave mechanics

Louis Victor de Broglie was born in Dieppe, Seine-Maritime, France on August 15, 1892. Upon the death of his older brother in 1960, he became the 7th duc de Broglie. He received his first degree in history. Afterward, he turned his attention to mathematics and physics, obtaining this degree in 1913.

With the outbreak of Word War I in 1914, de Broglie was conscripted for military service, and posted to the wireless section of the army, where he remained until the end of the war in 1918.

De Broglie's 1922 doctoral thesis, *Research on Quantum Theory*, introduced his theory of electron waves, which included the wave-particle duality theory of matter based on the work of Albert Einstein (1879-1955) and Max Planck (1858-1947). This research culminated in the **de Broglie hypothesis,** stating that any moving particle has an associated wave. De Broglie thus created a new field in physics, wave mechanics, uniting the physics of light and matter. The existence of these matter waves was confirmed experimentally in 1927.

In 1928, de Broglie became professor in the faculty of sciences, University of Paris.

For his discovery of the wave nature of electrons, de Broglie was awarded the Nobel Prize in physics in 1929.

Between 1930 and 1950, de Broglie's work was chiefly devoted to the study of the various extensions of wave mechanics. He made major contributions to the fostering of international scientific co-operation.

De Broglie was named permanent secretary of the Academy of Sciences in 1942, and adviser to the French Atomic Energy Commission in 1945.

In his later work, de Broglie developed a causal explanation of wave mechanics, in opposition to the wholly probabilistic models of quantum mechanics. Today, this explanation is known as the **de Broglie-Bohm theory**, since it was refined by David Bohm (1917-1992) in the 1950s.

Among the applications of de Broglie's work has been the development of electron microscopes to get much better image resolution than optical ones because of shorter wavelengths of electrons compared with photons. De Broglie never married. He died on March 19, 1987.

1. Broglie, Louis Victor, duc de, Infoplease, www.infoplease.com/ce6/people/A0809038.
2. Louis de Broglie, Biography, NobelPrize.org,
    http://nobelprize.org/nobel_prizes/physics/laureates/1929/broglie-bio.
3. Louis de Broglie, Microsoft Encarta Encyclopedia Online, http://encarta.msn.com/
    encyclopedia_761571357/louis_de_broglie, 2007.
4. Louis de Broglie, Wikipedia, http://en.wikipedia.org/wiki/Louis_de_Broglie.

Robert Watson-Watt (1892-1973)

Developed radar

Robert Alexander Watson-Watt was born in Brechin in Angus, Scotland on April 13, 1892. He was a descendant of James Watt (1736-1819), the inventor of the practical steam engine. Watson-Watt graduated with a B.Sc. in engineering from University College, Dundee (part of the University of St. Andrews) in 1912, and accepted an assistantship there.

In 1915, Watson-Watt joined the Meteorological Office, which was interested in his ideas on the use of radio for the detection of thunderstorms. Lightning gives off a radio signal as it ionizes the air, and he planned to detect this signal to warn pilots of approaching thunderstorms. He was initially stationed at the Wireless Station of the Air Ministry Meteorological Office in Aldershot, England. His early experiments were successful in detecting the lightening radio signal, and he quickly proved to be able to do so at long ranges. He located the signal with a directional antenna, which could be manually turned to maximize the signal, thus pointing to the storm. To see the fleeting signal, he used a cathode-ray oscilloscope with a long-lasting phosphor. Such a system represented a significant part of a complete radar system, and was in use as early as 1923.

Then in 1924 Watson-Watt moved to Ditton Park, to the west of London. The National Physical Laboratory (NPL) already had a research station there, and in 1927 they were combined as the Radio Research Station, with Watson-Watt in charge. After a further re-organization in 1933, Watson-Watt became Superintendent of the Radio Department of NPL in Teddington.

In 1935, Watson-Watt demonstrated his first aircraft detection system (radar). It consisted of two receiving antennas located about ten kilometers away from one of the British Broadcasting Company's shortwave broadcast antennas at Daventry. Signals traveling directly from the station were filtered out, and a Heyford bomber flown around the site was detected. It has been said that there would have been no success in the Battle of Britain without radar, and consequently Britain would not have survived.

Watson-Watt was knighted in 1942. He died on December 5, 1973 and is buried with his wife in the church yard at Pitlochry, England.

1.   Robert Alexander Watson-Watt, Wikipedia, http://en.wikipedia.org/wiki/Robert_Watson-Watt.
2.   Robert Watson Watt, Radar, Inventer of the week, Lemelson-MIT Program, http://web.mit.edu/invent/iow/watsonwatt.
3.   Sir Robert Watson-Watt, Microsoft Encarta Encyclopedia Online, http://encarta.msn.com/encyclopedia_761569742/Watson-Watt_Sir_Robert_Alexander, 2007.

Satyendra Bose (1894-1974)

Developed a statistical method of handling bosons

Satyendra Nath Bose was born in Kolkata (Calcutta), India on January 1, 1894. His father worked in the Engineering Department of the East India Railway. Bose spoke several languages, and could also play the Esraj, a musical instrument similar to a violin.

Bose was awarded a B.Sc. in 1913 and an M.Sc. in 1915 in applied mathematics from Presidency College in Calcutta. From 1916 to 1921 he was a lecturer in the physics department of the University of Calcutta. In 1921, he joined the department of Physics of the then recently founded Dacca University, again as a lecturer.

In 1924, Bose wrote a paper deriving Planck's quantum radiation law without any reference to classical physics. After initial setbacks to his efforts to publish it, he sent the article to Albert Einstein (1879-1955) in Germany, who recognized the importance of the paper. Einstein translated it into German and submitted it on Bose's behalf to Zeitschrift für Physik. As a result of this recognition, Bose was able to leave India for the first time and spent two years in Europe, during which time he worked with Louis de Broglie (1892-1987), Marie Curie (1867-1934), and Einstein.

Einstein further extended Bose's work on the behavior of photons, which are particles of zero spin, resulting in what is now known as **Bose-Einstein statistics**. At low temperatures, a large number of bosons may occupy the same energy state, resulting in what is known as a **Bose condensation**. This behavior applies to any group of particles of integral spin, or bosons as they are now called.

Bose returned to Dhaka in 1926 and became a professor. He did not have a doctorate, and so ordinarily he would not be qualified for the post, but Einstein had recommended him. Bose later became head of the department.

Bose's work ranged from X-ray crystallography to unified field theories. He also published an equation of state for real gases with Megh Nad Saha (1893-1956).

Bose continued teaching at Dhaka University until 1945. At that time he returned to Calcutta and taught at Calcutta University until he became Vice Chancellor of Vishwabharati University in 1958. He died on February 4, 1974. He is honored as the namesake of the **boson**.

1. J. J. O'Conner and E. F. Robertson, Satyendranath Bose, www-groups.dcs.st-and.ac.uk/~history/Biographies/Bose.
2. Satyendra Nath Bose, Microsoft Encarta Encyclopedia Onlilne, http://encarta.msn.com/encyclopedia_761580024/Bose_Satyendra_Nath, 2007.
3. Satyendra Nath Bose, Wikipedia, http://en.wikipedia.org/wiki/Satyendra_Nath_Bose.

H. Thomas Milhorn, MD, PhD

Pyotr Kapitsa (1894-1984)

Developed inventions and made discoveries in the area of low-temperature physics

Pyotr Leonidovich Kapitsa, the son of a military engineer, was born in Kronstadt, Russia on July 9, 1894. He graduated from the Petrograd Polytechnical Institute in 1918, and then taught electrical engineering there for two years.

In 1920, Kapitsa worked in the Cavendish Laboratory in Cambridge with Ernest Rutherford (1871-1937) for over 10 years.

During the 1920s, Kapitsa developed techniques for creating ultrastrong magnetic fields by injecting high currents into specially constructed air-core electromagnets for brief periods of time.

In 1928, Kapitsa discovered the linear dependence of resistivity on magnetic field for various metals placed in very strong magnetic fields.

Kapitsa was director of the Mond Laboratory in Cambridge from 1930 to 1934. In 1934, he developed a new and original apparatus for producing significant quantities of liquid helium, based on the adiabatic principle. That same year, he was on a visit to the Soviet Union to participate in a scientific conference when he was not permitted to leave the country.

In Russia, Kapitsa studied the properties of liquid helium. This led to the 1937 discovery of the superfluidity of helium. In 1939, he developed a new method for liquefaction of air with a low pressure cycle using a special high-efficiency expansion turbine. Consequently, during World War II he was assigned to head the Department of Oxygen Industry and attached to the USSR Council of Ministers, where he adapted his low pressure expansion techniques for industrial purposes. **Kapitsa resistance** is a resistance to the flow of heat across the interface between liquid helium and a solid that produces a temperature discontinuity.

In the years after the war, Kapitsa invented high power microwave generators, and discovered a new kind of continuous high pressure plasma discharge, with electron temperatures over a million K.

Belatedly, in 1978 Kapitsa was awarded the Nobel Prize in physics for the work in low-temperature physics he did around 1937. He shared the Prize with Arno Penzias (1933-) and Robert Wilson (1936-), who won for unrelated work in astronomy.

Kapitsa died on April 8, 1984.

1. Piotr Leonidovich Kapitsa, Microsoft Encarta Encyclopedia Online, http://uk.encarta.msn.com/encyclopedia_761564281/kapitza_peter_leonidovich, 2007.
2. Pyotr Kapitsa, Biography, NobelPrize.org, http://nobelprize.org/nobel_prizes/physics/laureates/1978/kapitsa-bio.
3. Pyotr Kapitsa, Wikipedia, http://en.wikipedia.org/wiki/Pyotr_Leonidovich_Kapitsa.

Igor Tamm (1895-1971)

Discovered and explained the electromagnetic radiation emitted when a charged particle passes through an insulator at a speed greater than the speed of light in that medium

Igor Yevgenyevich Tamm, the son of an engineer, was born in Vladivostok, Russian Empire (now Russia) on July 8, 1895. In 1913 and 1914, he studied at the University of Edinburgh. Then he moved to Moscow State University, and graduated from there in 1918. Tamm was awarded the degree of Doctor of Physico-Mathematical Sciences.

In 1934, Pavel Čerenkov (1904-1990) discovered a blue glow is emitted by charged particles traveling at very high speeds through water. This became known as the Čerenkov effect or Čerenkov radiation. Tamm and Il'ja Frank (1908-1990) provided the theoretical explanation of this effect, which results when electromagnetic radiation is emitted when a charged particle passes through an insulator at a speed greater than the speed of light in that medium.

The team also helped construct the **Čerenkov detector**, which assisted them in observing the Čerenkov effect with other high-energy particles.

Between 1924 and 1930, Tamm developed the quantum theory of acoustical vibrations and the scattering of light in solid bodies, as well as the theory of interactions of light with electrons.

In 1933, Tamm theorized on the existence of surface states (**Tamm's levels**) of electrons in semiconductors.

In 1945, Tamm developed an approximation method for many-body physics. This is a field of physics that attempts to navigate the gulf of complexity separating the quantum behaviors of atoms, molecules, and subatomic particles. Because Sidney Dancoff (1914-1951) developed it independently in 1950, it is now called the **Tamm-Dancoff approximation**.

Tamm was awarded the Nobel Prize for physics in 1958, jointly with Čerenkov and Frank, for discovering and explaining the Čerenkov effect. This work resulted in the development of new methods for detecting and measuring the velocity of high-speed nuclear particles, and became of great importance for research in nuclear physics.

Tamm died in Moscow, Soviet Union (now Russia) on April 12, 1971.

1. Igor Y. Tamm, Biography, NobelPrize.org, http://nobelprize.org/nobel_prizes/physics/laureates/1958/tamm-bio.
2. Igor Yevgenyevich Tamm, Microsoft Encarta Encyclopedia Online, http://encarta.msn.com/encyclopedia_761572163/tamm_igor_yevgenyevich, 2007.
3. Igor Yevgenyevich Tamm, Wikipedia, http://en.wikipedia.org/wiki/Igor_Tamm.

John Cockcroft (1897-1967)

Used accelerated particles to study the atomic nucleus

John Cockcroft, the eldest son of a cotton mill owner, was born in Todmorden, England on May 27, 1897. He studied mathematics at Victoria University of Manchester from 1914 to 1915, and was a signaler in the Royal Artillery from 1915 to 1918.

After the war, Cockcroft studied electrotechnical engineering at Manchester College of Technology from 1919 until 1920. He received a mathematics degree from St. John's College, Cambridge in 1924, and then began research work under Ernest Rutherford (1871-1937) at the Cavendish Laboratory at Cambridge and collaborated with Russian physicist Pyotr Kapitsa (1894-1984) on the production of intense magnetic fields and low temperatures.

In 1928, Cockcroft, with Ernest Walton (1903-1995), began to work on the acceleration of protons. They designed a linear accelerator, and in 1932 they bombarded lithium with high energy protons, and succeeded in transmuting it into helium and other chemical elements. This was the first time an atomic nucleus of one element had been changed to a different nucleus.

At the outbreak of the Second World War, Cockcroft took the post of Assistant Director of Scientific Research in the Ministry of Supply, working on radar.

In 1944, Cockcroft moved to Canada. There, he took charge of the Canadian Atomic Energy project and became Director of the Montreal Laboratory and Chalk River Laboratories.

In 1946, Cockcroft returned to England to set up the Atomic Energy Research Establishment at Harwell, charged with developing Britain's atomic power program. His 12 years there saw the production of the British atomic bomb, an advance in the peaceful use of atomic energy, and construction of the nuclear energy power station at Calder Hall. Cockcroft was knighted in 1948.

In 1951, Cockcroft and Walton, were awarded the Nobel Prize in physics for their work on the use of accelerated particles to study the atomic nucleus.

In 1959, Cockcroft became the first Master of Churchill College, Cambridge. He served as chancellor of the Australian National University from 1961 to 1965. Cockcroft died on September 18, 1967.

1. John Cockcroft, Biography, NobelPrize, http://nobelprize.org/nobel_prizes/physics/laureates/1951/ cockcroft-bio.
2. John Cockcroft, BookRags, www.bookrags.com/John_Cockcroft.
3. John Cockcroft, Wikipedia, http://en.wikipedia.org/wiki/John_Cockcroft.

Patrick Blackett (1897-1974)

Developed the counter-controlled cloud chamber

Patrick Maynard Stuart Blackett, the son of a stockbroker, was born at Kensington, London on November 18 1897. He attended the Royal Naval College, Osborne on the Isle of Wight. Then, he went to Dartmouth in Hanover, New Hampshire. In 1914, he became a midshipman, and served in the British Navy as a sub-lieutenant during the First World War.

In 1919, impressed by the Cavendish Laboratory at Cambridge, Blackett decided to leave the navy and study mathematics and physics.

After graduating from Magdalen College, Cambridge in 1921, Blackett spent 10 years working at the Cavendish Laboratory as an experimental physicist with Ernest Rutherford (1871-1937), and in 1923 became a fellow of Kings College, Cambridge. In 1924, using a cloud chamber, he obtained the first photographs of the transmutation of nitrogen into an oxygen isotope.

In 1932, working with Giuseppe Occhialini (1907-1993), Blackett devised a system of geiger counters which only took photographs when a cosmic ray particle traversed the cloud chamber. Using this counter-controlled cloud chamber, they found 500 tracks of high energy cosmic ray particles.

In 1933, Blackett discovered 14 tracks, which confirmed the existence of the positron, and revealed the opposing spiral traces of positron-electron pair production. This work and that on annihilation radiation (collision between a positron and an electron in which both are transformed into gamma radiation) made him one of the first experts on anti-matter. That same year he moved to Birkbeck College, University of London as Professor of Physics. Then in 1937 he went to the University of Manchester where he was elected to the Langworthy Professorship and created a major international research laboratory.

During World War II Blackett was chief adviser on operational research to the British navy. In 1948, he was awarded the Nobel Prize for physics for his investigation of cosmic rays using his invention of the counter-controlled cloud chamber.

Blackett was appointed Head of the Physics Department of Imperial College London in 1953, and retired in July, 1963. He died on July 13, 1974.

1. Patrick Blackett, Baron Blackett, Wikipedia, http://en.wikipedia.org/ wiki/ Patrick_Blackett,_Baron_Blackett.
2. Patrick Blackett, Microsoft Encarta Encyclopedia Online, http://encarta.msn.com/ encyclopedia_761571669/blackett, 2007.
3. Patrick M. S. Balckett, Biography, http://nobelprize.org/nobel_prizes/physics/laureates/ 1948/blackett-bio.

Isidor Rabi (1898-1988)

Developed a resonance method for recording the magnetic properties of atomic nuclei

Isidor Isaac Rabi was born on July 29, 1898 in Rymanów, Austria (now in Poland). His family immigrated to the United States the following year and settled in New York. Rabi's father made ends meet by working at a variety of jobs, such as delivering ice and laboring in factories.

Rabi received a B.S. in Chemistry from Cornell University in 1919 and a Ph.D. from Columbia University in 1927. His dissertation research was on the magnetic properties of crystals.

A fellowship allowed Rabi to spend the next two years in Europe working with Niels Bohr (1885-1962), Werner Heisenberg (1901-1976), Wolfgang Pauli (1900-1958), and Otto Stern (1888-1969). He then joined the Columbia University faculty, where he remained until his retirement.

In 1930, Rabi began research on the magnetic properties of atomic nuclei in an effort to ascertain the nature of the force binding the protons in the nuclei. He adapted and built upon the molecular beam method of Otto Stern (1888-1969). Rabi's research led to the invention of a method to measure the hyperfine transition levels of atoms by using known radio frequencies to cause transitions between these energy levels.

In 1940, Rabi was granted leave from Columbia to work as Associate Director of the Radiation Laboratory at the Massachusetts Institute of Technology on the development of radar.

During World War II, at the request to of Robert Oppenheimer (1904-1964), Rabi became a visiting consultant at Los Alamos, working on the atom bomb. After the war, Rabi continued his research, which contributed to the inventions of the LASER and the atomic clock.

He was awarded the 1944 Nobel Prize in physics for developing a resonance method for recording the magnetic properties of atomic nuclei.

Rabi chaired Columbia's physics department from 1945 to 1949. When Columbia created the rank of University Professor in 1964, Rabi was the first to be appointed to it. He retired from teaching in 1967, but remained active in the department until his death on January 11, 1988 after a prolonged illness. The **Rabi cycle**, which is the cyclic behavior of a two-state quantum system in the presence of an oscillatory driving field, is named in honor of Rabi.

1.   Isidor Isaac Rabi (1898-1988, Magnetic Lab, www.magnet.fsu.edu/education/tutorials/ pioneers/rabi.
2.   Isidor Isaac Rabi, Microsoft Encarta Encyclopedia Online, http://encarta.msn.com/ encyclopedia_761552382/rabi_isidor_isaac, 2007.
3.   Isidor Isaac Rabi, Wikipedia, http://en.wikipedia.org/wiki/Isidor_Isaac_Rabi.

Leó Szilárd (1898-1964)

Suggested the possibility of a nuclear chain reaction

Leó Szilárd, the son of a civil engineer, was born in Budapest, Austri-Hungary on February 11, 1898. He became a student in engineering at Budapest Technical University in 1916, but had to join the Austro-Hungarian Army in 1917. At the end of the war, in 1919, he resumed his studies at Budapest Technical University. However, he soon decided to leave Hungary because of the rising antisemitism under the Horthy regime.

In 1923, Szilárd received his doctorate from the Humboldt University of Berlin. He was appointed an assistant to Max von Laue (1979-1960) at the University of Berlin's Institute for Theoretical Physics in 1924. In 1927, he became an instructor in physics at the University of Berlin.

When Adolf Hitler came to power in Germany, Szilard went to England, where he began work in nuclear physics. In 1938, he moved to the United States as guest researcher at Columbia University.

Szilárd drafted a confidential letter to Franklin D. Roosevelt explaining the possibility of nuclear weapons and warning of Nazi work on such weapons. He recommended the development of a program which could lead to a nuclear bomb. He convinced Albert Einstein (1979-1955) and Paul Wigner (1902-1955), also Jewish physicists who had fled Hitler's Germany, to sign the letter. The letter led directly to the establishment of research into nuclear fission by the U.S. government and ultimately to the creation of the Manhattan Project.

Szilárd moved to the University of Chicago, where he and Enrico Fermi (1901-1954) constructed the first reactor. It used uranium as the fuel and graphite to moderate the reaction. They achieved the first self-sustaining nuclear chain reaction in 1942.

In 1947, Szilárd switched fields of study because of his disinchantment with nuclear weapons. He moved from physics to molecular biology. In 1960, he was diagnosed with bladder cancer, which he overcame.

In 1961, Szilárd published a book of short stories *The Voice of the Dolphins* in which he wrestled with the moral and ethical issues raised by the Cold War and his own role in the development of atomic weapons.

In 1962, Szilárd founded the Council for Abolishing War. He died of a heart attack in his sleep in La Jolla, California on May 30, 1964.

1. Leó Szilárd www.spartacus.schoolnet.co.uk/SCszilard.
2. Leó Szilárd, Wikipedia, http://en.wikipedia.org/wiki/Le%C3%B3_Szil%C3%A1rd.
3. Szilárd, Leó, Microsoft Encarta Encyclopedia Online, http://au.encarta.msn.com/encyclopedia_761571588/Szilard_Leo, 2007.

H. Thomas Milhorn, MD, PhD

Vladimir Fock (1898-1974)

Made fundamental contributions to quantum theory

Vladimir Aleksandrovich Fock was born in St. Petersburg, Russia on December 22, 1898. In 1914, while he was still in high school, World War I broke out, and St. Petersburg was renamed Petrograd. In 1916, upon graduation, he volunteered for the army. He was given a short course on artillery and then sent to fight as an artillery officer on the front.

In 1918, Fock entered Petrograd University, majoring in mathematics and physics. By that time Russia was in the middle of a civil war, following the Russian Revolution. In 1919, a new State Optical Institute was opened in Petrograd, and a special group of students was formed to make sure the brightest students got a good education. Fock was a member of the group.

Before graduating from Petrograd University in 1922, Fock had already published two papers, one on quantum mechanics and the other on mathematical physics. After graduating, he continued postgraduate studies at Petrograd University, and became a professor there in 1932.

From 1919 to 1923 and 1928 to 1941 Fock collaborated with the State Institute of Optics, from 1924 to 1936 with the Leningrad Institute of Physics and Technology, and from 1934 to 1941 and 1944 to 1953 with the Lebedev Physical Institute.

Fock's primary scientific contribution was in the development of quantum physics. He also contributed significantly to the fields of mechanics, theoretical optics, theory of gravitation, and physics of continuous medium.

In 1926, Fock generalized the Klein-Gordon equation. He gave his name to **Fock space** (an algebraic system used in quantum mechanics to describe quantum states with a variable or unknown number of particles), **Fock representation,** and **Fock state** (in quantum mechanics, any state of the Fock space with a well-defined number of particles in each state),

In 1939, he developed the **Hartree-Fock** method (an approximate method for the determination of the ground-state wavefunction and ground-state energy of a quantum many-body system).

The Hartree-Fock method finds its typical application in the solution of the electronic Schrödinger equation of atoms, molecules, and solids, but it has also found widespread use in nuclear physics. Fock died in St. Petersburg, Russia on December 27, 1974.

1. O'Connor, J. J. and E F Robertson, Vladimir Aleksandrovich Fock, MacTutor, www-groups.dcs.st-and.ac.uk/~history/Printonly/Fock, 2006.
2. Vladimir Aleksandrovich Fock www.mathsoc.spb.ru/pantheon/fock.
3. Vladimir Fock, Wikipedia, http://en.wikipedia.org/wiki/Vladimir_Aleksandrovich_Fock.

John van Vleck (1899-1980)

Did theoretical investigations of the electronic structure of magnetic and disordered systems

John Hasbrouck van Vleck, the son of mathematician, was born in Middletown, Connecticut on March 13, 1899. When he was seven years old his father accepted a professorship at the University of Wisconsin.

Van Vleck received his B.S. from the University of Wisconsin in 1920 and his M.S. and Ph.D. degrees from Harvard University in 1921 and 1922, respectively.

After stints on the faculty at Harvard, the University of Minnesota, and the University of Wisconsin-Madison, van Vleck returned to Harvard. There, he developed fundamental theories of the quantum mechanics of magnetism and the bonding in metal complexes.

During World War II, van Vleck studied the possibility of nuclear weapons at the University of California at Berkeley, and worked on improving radar technology at Harvard's Radio Research Laboratory. He participated in the Manhattan Project by serving on the Los Alamos Review committee. The committee's important contribution was a reduction in the size of the firing gun for the Little Boy bomb. This eliminated additional design-weight and sped up production of the bomb for its eventual release over Hiroshima.

After the war, van Vleck served as the chairperson of the physics department and the dean of engineering and applied physics at Harvard.

In the year 1961-1962, van Vleck was George Eastman Visiting Professor at the University of Oxford.

Van Vleck applied quantum mechanics to ferromagnetism (the type of magnetism in ordinary magnets) and paramagnetism (the type of magnetism displayed by substances that have slight sensitivity to being magnetized). He also developed a theory that allowed scientists to calculate the energy levels of an atom or ion in a crystal, important to the development of LASERs. In addition, he worked on magnetic resonance, which has become an important tool in medicine.

The 1977 Nobel Prize for physics was awarded to Van Vleck, Philip Anderson (1923-) and Nevill Mott (1905-1986) for their fundamental theoretical investigations of the electronic structure of magnetic and disordered systemss. Van Vleck died on October 27, 1980.

1.   John H. Van Vleck, Autobiography, NobelPrize.org, http://nobelprize.org/nobel_prizes/physics/laureates/1977/vleck-autobio.
2.   John Hasbrouck Van Vleck , Wikipedia, http://en.wikipedia.org/wiki/Professor.
3.   John Hasbrouck Van Vleck, Microsoft Encarta Encyclopedia Online, http://encarta.msn.com/encyclopedia_761583386/van_vleck_john_hasbrouck, 2007.

20th Century Physics
(1900-Present)

Dennis Gabor (1900-1979)

Invented the hologram method

Dennis Gabor, the son of a director of a mining company, was born in Budapest, Hungary on June 5, 1900. He studied at the Technical University of Budapest and at the Charlottenburg Technical University in Berlin, acquiring a doctorate in electrical engineering. His doctoral work involved developing one of the first high speed cathode ray oscillographs and making the first iron-shrouded magnetic electron lens.

During Gabor's career, he concentrated his research on electron and plasma physics, electron microscopy, and physical optics. He analyzed the properties of high voltage electric transmission lines by using cathode-beam oscillographs, which led to his interest in electron optics. Studying the fundamental processes of the oscillograph, he was led to other electron-beam devices, such as electron microscopes and TV tubes.

Beginning in 1927, Gabor made his first successful inventions—the high pressure quartz mercury lamp with superheated vapour and the molybdenum tape seal, since used in millions of street lamps.

In 1933, when Hitler came to power, Gabor fled Germany, and was invited to Britain to work at the development department of the British Thomson-Houston company in Rugby, Warwickshire.

Gabor's interest in electron optics led him to the invention of holography, a lensless system of three-dimensional photography. The basic idea was that for perfect optical imaging, the total of all the information has to be used; not only the amplitude, as in usual optical imaging, but also the phase. In this manner, a complete holo-spatial picture can be obtained.

At the time Gabor developed holography, coherent light sources were not available, so the theory had to wait more than a decade until its first practical applications were realized.

The invention in 1962 of the LAZER, the first coherent light source, was followed by the first hologram in 1963, after which holography became commercially available.

In 1948, Gabor moved from Rugby to Imperial College London, and in 1958 became professor of Applied Physics. He remained there until his retirement in 1967.

For his invention of holography, Gabor was awarded the 1971 Nobel Prize in physics. Gabor died on February 9, 1979.

1. Dennis Gabor, Autobiography, NobelPrize.org, http://nobelprize.org/nobel_prizes/physics/ laureates/ 1971/ gabor-autobio.
2. Dennis Gabor, Microsoft Encarta Encyclopedia Online, http://encarta.msn.com/encyclopedia_761562367/dennis_gabor, 2007.
3. Dennis Gabor, Wikipedia, http://en.wikipedia.org/wiki/Dennis_Gabor.

H. Thomas Milhorn, MD, PhD

Fritz London (1900-1954)

Explored how quantum theory could work and be observed on the
macroscopic scale

Fritz Wolfgang London was born in Breslau, Germany (now Wroclow,
Poland) on March 7, 1900. His father taught mathematics in Bonn. London
attended the universities of Bonn, Frankfurt, and Munich, and received his
Ph.D. in philosophy from Munich in 1921. After teaching at several
secondary schools in Germany, he decided to study physics. He returned to
Munich to study under Arnold Sommerfeld (1868-1951).

London began to explore the ways in which quantum principles could
be used to explain visible phenomena. His research into the hydrogen
molecule in the late 1920s, along with German physicist Walter Heitler
(1904-1981), advanced the existing knowledge of chemical bonding, and
marked the beginning of modern quantum chemistry. London wrote about
the mechanism of chemical reactions and about the quantum mechanical
interpretation of the Van der Waals forces.

In 1932, London began to investigate superconductivity. He and his
brother, Heinz London (1907-1970), found that an extremely thin outer
layer in the superconducting material contains the electrical current.

In 1933, London was forced to resign his post at the University of
Berlin by the Nazi regime. He left Germany for England, where he worked
at Oxford University's Clarendon Laboratory. He then spent two years at
the Institut Henri Poincaré in Paris before taking a job as professor of
theoretical chemistry at Duke University in North Carolina.

London predicted that in its fluid phase the light helium isotope
Heium-3 should show a different type of degeneracy from that of Helium-
4. This prediction was ultimately confirmed experimentally.

London's prediction of the magnetic moment of a rotating
superconductor, the **London moment,** was also confirmed. And his
predictions on the properties of systems with Bose-Einstein and Fermi-
Dirac statistics have been extensively used in the investigations of ultra-
cold atoms and molecules in the nineties and beyond.

London remained at Duke until his death in Durham, North Carolina
on March 30, 1954.

London's predictions in the field of low temperature physics have
deeply influenced the development of the fields of superconductors,
quantum fluids, and quantum solids.

1. Fritz London (1900-1954), Duke Physics, www.phy.duke.edu/about/FritzLondon.
2. Fritz London, Wikipedia, http://en.wikipedia.org/wiki/Fritz_London.
3. The World of Chemistry on Fritz London, BookRags, www.bookrags.com/biography/fritz-london-woc.

George Uhlenbeck (1900-1988)

Co-discovered that the electron has an intrinsic spin

George Eugene Uhlenbeck was born in the Batavia, Java (now Jakarta, Indonesia) on December 6, 1900. When he was six years old the family moved permanently to The Hague, Holland. After studying chemical engineering for one semester at the Institute of Technology in Delft, he switched to physics and mathematics at the University of Leiden.

After graduating, he began postgraduate studies at Leiden under the supervision of Paul Ehrenfest (1889-1933). To support himself financially, he took a part-time job as a teacher in a girls' school in Leiden. He received his master's degree in 1923.

After a brief stay in Italy, Uhlenbeck returned to the University of Leiden as Ehrenfest's assistant. Soon after taking up the appointment, working with graduate student Samuel Goudsmit (1902-1978), Uhlenbeck discovered electron spin. Uhlenbeck then spent two months in Copenhagen writing his doctoral dissertation—systematizing statistical notions and expanding on the electron spin ideas.

After receiving his doctoral degree in 1927, Uhlenbeck accepted a position at the University of Michigan in Ann Arbor. He returned to the Netherlands to take up a chair in Utrecht in 1935. He took leave from Utrecht in 1938 to spend a year as visiting professor at Columbia University in New York.

During the Second World War, from 1943 until 1945, Uhlenbeck worked at Massachusetts Institute of Technology as a member of the team pursuing the development of radar. Then, after the war ended he returned to Ann Arbor where, after spending 1948-49 at Princeton, he was named Henry Cahart professor at Michigan in 1954.

In 1960, Uhlenbeck moved to the Rockefeller Institute in New York where he remained on the staff until he retired in 1971.

As well as fundamental work on quantum mechanics, Uhlenbeck worked on atomic structure and the kinetic theory of matter. He extended the equations of Ludwig Boltzmann (1844-1906) to dense gasses and wrote two important papers on Brownian motion. However, the main topic on which he worked throughout his career was statistical physics. The aim of this topic was to understand the relationship between physics at the atomic level and that at the macroscopic level.

Uhlenbeck died in Boulder, Colorado on October 31, 1988.

1. George Uhlenbeck, NNDB, www.nndb.com/people/204/000099904.
2. O'Connor, J. J. and E. F. Robertson, George Eugene Uhlenbeck, MacTutor, www-groups.dcs.st-and.ac.uk/~history/Biographies/Uhlenbeck.

H. Thomas Milhorn, MD, PhD

Wolfgang Pauli (1900-1958)
Discovered the exclusion principle

Wolfgang Ernst Pauli was born in Vienna, Austria on April 25, 1900. He attended the Ludwig-Maximilians University of Munich, where he worked under Arnold Sommerfeld (1868-1951). He received his doctorate in 1921 for a thesis on the quantum theory of ionized molecular hydrogen.

Pauli spent 1923 to 1928 as a lecturer at the University of Hamburg. In 1924, he proposed a new quantum degree of freedom (or quantum number) with two possible values to resolve inconsistencies between observed molecular spectra and the developing theory of quantum mechanics.

In 1925, Pauli formulated the **Pauli exclusion principle**, which states that no two electrons can exist in the same quantum state, identified by four quantum numbers, including his new two-valued degree of freedom. This new degree of freedom was later identified as electron spin.

In 1926, shortly after Werner Heisenberg (1901-1976) published the matrix theory of quantum mechanics, Pauli used it to derive the observed spectrum of the hydrogen atom. In 1927, he introduced the 2x2 Pauli matrices as a basis of spin operators, thus solving the nonrelativistic theory of spin. In 1928, Pauli was appointed Professor of Theoretical Physics at the ETH Zurich in Switzerland. In 1930, he proposed the existence of a neutral particle with a small mass (no greater than one percent the mass of a proton) to explain the continuous spectrum of beta decay. In 1934, Enrico Fermi (1901-1954) incorporated Pauli's particle, which he called a neutrino, into his theory of beta decay. The existence of the neutrino was subsequentlyt confirmed experimentally.

After the outbreak of the Second World War, Pauli moved to the United States, where he was Professor of Theoretical Physics at Princeton University. There, he proved the spin-statistics theorem, a critical result of quantum field theory, which states that particles with half-integer spin are fermions, while particles with integer spin are bosons.

In 1945, Pauli received the Nobel Prize in physics for his discovery in 1925 of the Pauli exclusion principle.

In 1949, after returning to Zurich, Pauli and Felix Villars (1921-2002) published a paper on **Pauli-Villars regularization**, which provides an important prescription for removing infinities from quantum field theories. Pauli died of pancreatic cancer on December 15, 1958.

1. O'Connor, J. J. and E F Robertson, MacTutor, www-groups.dcs.st-and.ac.uk/~history/Printonly /Pauli, October 2003.
2. Wolfgang Pauli, Microsoft Encarta Encyclopedia Online, http://encarta.msn.com/ encyclopedia_761553051/wolfgang_pauli, 2007.
3. Wolfgang Pauli, Wikipedia, http://en.wikipedia.org/wiki/Wolfgang_Pauli.

Enrico Fermi (1901-1954)
Demonstrated the first controlled atomic fission reaction

Enrico Fermi was born in Rome, Italy on September 29, 1901. He earned his doctorate in physics from the University of Pisa in 1922. From 1922 to 1924, he studied with Max Born (1882-1970) in Göttingen, Germany. Then, in 1924, he returned to Italy to teach mathematics at the University of Florence.

In 1926, Fermi devised a method for calculating the behavior of a system composed of particles that obeyed the exclusion principle of Wolfgang Pauli (1900-1958). This principle states that no two particles can have identical quantum numbers. The method that Fermi developed became known as **Fermi statistics**, and the particles that obey the Pauli exclusion principle became known as **fermions**. Paul Dirac (1902-1984) independently developed an equivalent theory using a different approach.

In 1927, Fermi became professor of theoretical physics at the University of Rome. And in 1933 he published a theory that explained beta decay (the transformation of a neutron into a proton, an electron, and a neutrino). Fermi's explanation of beta decay introduced a fundamental force called the weak nuclear force.

In 1934, Fermi discovered that shooting neutrons through paraffin wax at a sample of uranium atoms slowed the neutrons down and increased the intensity of the resulting radioactivity. Fermi was awarded the 1938 Nobel Prize in physics for his work with neutrons and radioactivity. After accepting the Nobel Prize in Sweden, Fermi immigrated to the United States rather than return to Italy and Mussolini's dictatorship. He became a professor at Columbia University in New York in 1939, and in 1941 moved to the University of Chicago.

In 1939, a group of physicists warned U.S. President Franklin D. Roosevelt that Germany might be working on an atomic bomb. In 1942, the Manhattan Project officially began. By the end of the year, Fermi had designed and presided over the first controlled fission reaction.

In 1944, Fermi became a United States citizen. In August of 1945 the United States dropped atomic bombs on two cities in Japan, Hiroshima and Nagasaki. And in 1946 Fermi returned to Chicago, where he continued to work on radioactivity until his death on November 28, 1954.

1. Enrico Fermi, Biography, NobelPrize.org, http://nobelprize.org/nobel_prizes/physics /laureates/ 1938/fermi-bio.
2. Enrico Fermi, Microsoft Encarta Encyclopedia Online, http://encarta.msn.com/encyclopedia_761578253/enrico_fermi, 2007.
3. Enrico Fermi, Wikipedia, http://en.wikipedia.org/wiki/Enrico_Fermi.

Ernest Lawrence (1901-1958)

Invented, developed, and utilized the cyclotron

Ernest Orlando Lawrence was born in Canton, South Dakota on August 8, 1901. His father was a supertendant of schools. Lawrence received his Ph.D. in physics at Yale University in 1925. He remained at Yale, becoming an assistant professor in 1927.

Lawrence's early work was on ionization phenomena and the measurement of ionization potentials of metal vapors.

In 1928, Lawrence was appointed associate professor of physics at the University of California at Berkeley, and two years later became a full professor. The following year he founded the university radiation laboratory in Berkeley, and became its director in 1936.

Lawrence is best known for the invention, development, and utilization of the cyclotron at Berkley. His work on the cyclotron began in 1929. A cyclotron is a machine used to accelerate charged particles to high energies. It consists of two back-to-back D-shaped cavities sandwiched between two electromagnets. A radioactive source is placed in the center of the cyclotron, and the electromagnets are turned on. The radioactive source emits charged particles, which circle around inside the D-shaped cavities. The two D-shaped cavities are hooked to a radio wave generator, which gives one cavity a positive charge and the other cavity a negative charge. After a moment, the radio wave generator reverses the charges on the cavities. The charges are continually switched back and forth, thus accelerating the particle.

In 1939, Lawrence was awarded the Nobel Prize in physics for this work on the cyclotron.

Larger and more powerful versions of the cyclotron were later built by Lawrence. In 1941, the instrument was used to generate artificially the cosmic particles called mesons, and later the studies were extended to antiparticles.

During World War II, Lawrence made vital contributions to the development of the atomic bomb. After the war, he played a part in the attempt to obtain international agreement on the suspension of atomic-bomb testing, being a member of the United States delegation at the 1958 Geneva Conference on this subject.

Lawrence died on August 27, 1958 at Palo Alto, California.

1. Cyclotron, Glossary, Jefferson Lab, http://education.jlab.org/glossary/cyclotron.
2. Ernest Lawrence, Biography, NobelPrize.com, nobelprize.org/nobel_prizes/physics/laureates/ 1939/lawrence-bio.
3. Ernest Lawrence, Wikipedia, http://en.wikipedia.org/wiki/Ernest_Lawrence.

Robert Van de Graaff (1901-1967)

Invented the Van de Graaf Generator

Robert Jemison Van de Graaff was born in Tuscaloosa, Alabama on December 20, 1901. He received his B.S. in 1922 and master's degree in engineering in 1923, both at the University of Alabama.

Van de Graaff worked briefly as an engineer at the Alabama Power Company before studying at the Sorbonne in Paris (1924-1925), where his interests turned to atomic physics.

The following year, Van de Graaff went to Oxford University as a Rhodes Scholar to undertake research into the mobility of gaseous ions. He generated the ions by means of a Wimshurst machine, a portable high-voltage generator invented in the late 19th century. Van de Graaff concluded that a much more efficient means would be to store the generated ions inside a large hollow sphere. He went on to receive a B.S. in physics at Oxford in 1926 and his Ph.D. in 1928.

In 1929, Van de Graaff returned to the United States to join the Palmer Physics Laboratory at Princeton University as a National Research Fellow. There, he developed his first electrostatic generator, which produced 80,000 volts. By 1933, he had constructed a much larger generator, capable of generating seven million volts.

The Van de Graaff generator uses a motorized insulating belt (usually made of rubber) to conduct electrical charges from a high voltage source on one end of the belt to the inside of a metal sphere on the other end. Electricl charge residing on the outside of the sphere builds up to produce an electrical potential much higher than that of the primary high voltage source.

From 1931 to 1934 Van de Graaff was a research associate at the Massachusetts Institute of Technology (MIT). There, in 1931 he constructed his first large machine in an unused aircraft hangar. He became an associate professor at MIT in 1934, and stayed there until 1960.

During World War II, Van de Graaff was director of the High Voltage Radiographic Project. After the war, he co-founded the High Voltage Engineering Corporation.

During the 1950s, Van de Graaff invented the insulating-core transformer (producing high-voltage direct current). He also developed tandem generator technology. He died in Boston, Massachusetts on January 16, 1967.

1. Robert J. Van de Graaff, Wikipedia, http://en.wikipedia.org/wiki/Robert_J._Van_de_Graaff.
2. Robert Jemison Van de Graaff, Institute of Chemistry, http://chem.ch.huji.ac.il/history/graaff.
3. Van de Graaff, Robert Jemison, Microsoft Encarta Encyclopedia Online,
    http://uk.encarta.msn.com/encyclopedia_121503383/Van_de_Graaff_Robert_Jemison, 2007.

Werner Heisenberg (1901-1976)

Stated that the simultaneous determination of two paired quantities of a particle has an unavoidable uncertainty.

Werner Karl Heisenberg was born in Würzburg, Germany on December 5, 1901. His family moved to Munich in 1910, where his father was professor of Greek language at the University of Munich. In 1920, Heisenberg entered the University of Munich, where he finished his undergraduate and graduate work in three years. In 1923, he presented his doctoral dissertation on turbulence in streams of fluid.

In 1925, Heisenberg invented matrix mechanics, the first formalization of quantum mechanics, which he further developed with the help of Max Born (1882-1970) and Pascual Jordan (1902-1980). He taught at the University of Leipzig from 1927 to 1941.

Heisenberg is best known for his **uncertainty principle**, which he developed in 1927. It states that the simultaneous determination of two paired quantities, for example the position and momentum of a particle, has an unavoidable uncertainty. The more precisely one value is known, the greater the range of possibilities that exist for the other.

In the late 1920s and early 1930s, Heisenberg collaborated with Wolfgang Pauli (1900-1958) and Paul Dirac (1902-1984) in developing an early version of quantum electrodynamics, which is a relativistic quantum field theory of electrodynamics.

In 1932, Heisenberg was awarded the Nobel Prize in physics for his role in the creation of quantum mechanics. Quantum mechanics describes matter in terms of both particles and waves.

After the discovery of the neutron by James Chadwick (1891-1974) in 1932, Heisenberg wrote a three part paper which described the modern picture of the nucleus of an atom. He proposed the proton-neutron model of the nucleus and used it to explain the nuclear spin of isotopes. He discussed the binding energies and the stability of the various nuclear components. These papers opened the way for others to apply quantum theory to the atomic nucleus.

From 1942 to1945 Heisenberg taught at the University of Berlin. During World War II he headed Germany's failed effort in nuclear fission research. In 1958, he became director of the Max Planck Institute for Physics and Astrophysics. He died on February 1, 1976.

1. O'Connor, J. J. and E. F. Robertson, Werner Heisenberg, MacTutor, www-groups.dcs.st-and.ac.uk/~history/Biographies/Heisenberg.
2. Werner Heisenberg, Microsoft Encarta Encyclopedia Online, http://encarta.msn.com/encyclopedia_761575759/heisenberg, 2007.
3. Werner Heisenberg, Wikipedia, http://en.wikipedia.org/wiki/Werner_Heisenberg.

Alfred Kastler (1902-1984)

Developed methods that used light to manipulate and study the energy levels of electrons in atoms

Alfred Kastler was born on May 3, 1902 in the town of Guebwiller in Alsace, which belonged to Germany at the time, but is now part of France. He earned a B.S. in 1926 at the École Normale Supérieure in Paris. After graduation, he taught high school in several towns before accepting a position in 1931 as a research assistant at the University of Bordeaux, where he received his Ph.D. in physics in 1936.

Kastler was a lecturer at Clermont-Ferrand University from 1936 to 1938. He spent 1938 to 1941 as a professor of physics at the University of Bordeaux. In 1941, he returned to the École Normale Supérieure as assistant professor of physics, and eventually became the head of the department.

Kastler developed two methods for using light to energize atoms so that they could be studied—double resonance and optical pumping. Double resonance, developed in the late 1940s, first applies a beam of light to a group of atoms to excite them to a particular energy level. Radio waves are applied to the atoms after they have been excited and before they return to their unexcited state. This causes the characteristics of the atoms' energy levels to change slightly. The light emitted in this state is slightly different than that emitted by the atoms when they returned to their unexcited states when only the beam of light had been applied. By finding the exact frequency of the radio waves that changes the characteristics of an energy level, Kastler was able to precisely map each energy level.

Optical pumping, developed in 1950, is a way to get a group of atoms into a uniform energy level. Polarized light is focused on a group of atoms in their unexcited state. Some atoms jump to a higher energy level and some do not. When the energized atoms return to the unexcited state, and the light is focused on them again, none jump to a higher energy level, indicating that the energy states of all the atoms are the same. The technique was later applied in the development of the LASER, the magnetometer, and improvements to the atomic clock.

Kastler retired in 1968, and then served as director of research at the French National Center for Scientific Research from 1968 to 1972. He died on January 7, 1984.

1. Alfred Kastler, Biography, http://nobelprize.org/nobel_prizes/physics/laureates/1966/kastler-bio. NobelPrize.org,
2. Alfred Kastler, Wikipedia, http://en.wikipedia.org/wiki/Alfred_Kastler.
3. AlfredKastler, Microsoft Encarta Encyclopedia Online, http://ca.encarta.msn.com/encyclopedia_761583228/alfred_kastler, 2007.

Eugene Wigner (1902-1905)

Contributed to the theory of the atomic nucleus and elementary particles

Eugene Paul Wigner was born in Budapest, Austria-Hungry into a Jewish Middle class family on November 17, 1902. His father was the director of a leather-tanning factory. Beginning in 1921, Wigner studied chemical engineering at the Technische Hochschule in Berlin. He subsequently obtained his doctorate in engineering in 1925 at the same institution. His dissertation contained the first theory of the rates of association and dissociation of molecules.

Having completed his doctorate, Wigner returned to Budapest to join his father's tannery firm as planned. However, he was not happy there, and took a position at Kaiser Wilhelm Institute in Berlin.

Later, at Göttingen, Wigner laid the foundation for the theory of symmetries in quantum mechanics, and in 1927 introduced what is now known as the **Wigner D-matrix**. A year later he returned to Berlin.

From 1930 to 1933 Wigner spent part of the year at Princeton and part at Berlin. His Berlin post vanished under the Nazi rules passed in 1933, so he became full-time at Princeton.

Wigner went to Wisconsin in 1936 and remained there a year before returning to Princeton as professor of mathematical physics.

In the late 1930s, Wigner extended his research into atomic nuclei, and developed a general theory of nuclear reactions. In 1937, he became a naturalized citizen of the United States.

Wigner was one of scientists who informed President Franklin D. Roosevelt in 1939 of the possible military use of atomic energy. He subsequently worked on the Manhattan Project at the University of Chicago during World War II from 1942 to 1945, and in 1946 became Director of Research and Development at Clinton Laboratories (now Oak Ridge National Laboratories). He led the group that designed the first very high-powered nuclear reactors, which were built at Hanford, Washington for the production of the isotope plutonium-239.

Wigner shared the 1963 Nobel Prize for physics with Maria Goeppert-Mayer (1906-1972) and Hans Jensen (1907-1973) for their work in elucidating the structure of the atomic nucleus.

Wigner retired from Princeton University on January 1, 1995, and died on January 1, 1995.

1. Eugene Wigner, Biography, NobelPrize.org, http://nobelprize.org/nobel_prizes/physics/laureates/ 1963/wigner-bio.
2. Eugene Wigner, Wikipedia, http://en.wikipedia.org/wiki/Eugene_Wigner.
3. Wigner, Eugene Paul, Microsoft Encarta Encyclopedia Online, http://au.encarta.msn.com/ encyclopedia_761555154/wigner_eugene_paul, 2007.

Paul Dirac (1902-1984)

Discovered new, productive forms of atomic theory

Paul Adrien Maurice Dirac was born in Bristol, England on August 8, 1902. His father was a Swiss immigrant who taught French, and his mother was a librarian. Dirac studied electrical engineering at the Merchant Venturer's Technical College, completing his degree in 1921. He then completed a B.A. in applied mathematics at the University of Bristol in 1923.

Dirac noticed an analogy between the old Poisson brackets of classical mechanics and the recently-proposed quantization rules in the matrix formulation of quantum mechanics by Werner Heisenberg (1901-1976). This observation allowed Dirac to obtain the quantization rules in a more illuminating manner. For this work, published in 1926, he received a Ph.D. from Cambridge. During that same year he founded quantum electrodynamics, a relativistic quantum field theory of electrodynamics.

In 1928, Dirac took an important step towards bringing quantum physics into conformity with the special theory of relativity of Albert Einstein (1879-1955) by devising an equation, now called the **Dirac equation,** that could describe the behavior of electrons at any speed up to the speed of light. This equation provided an explanation of one of the electron's intrinsic properties—spin.

Dirac proposed in 1931 that there should exist a class of particles with the same mass and spin as the electron, but with the opposite electrical charge. The existence of these antiparticles was later verified by Carl Anderson (1905-1991) in 1932. For his prediction, Dirac became recognized as the discoverer of antimatter.

Dirac's *Principles of Quantum Mechanics*, published in 1930, quickly became one of the standard textbooks on the subject, and is still used today

Dirac was the Lucasian Professor of Mathematics at Cambridge from 1932 to 1969. During World War II, he conducted important theoretical and experimental research on uranium enrichment by gas centrifuge.

For the discovery of new productive forms of atomic theory, Dirac shared the Nobel Prize in physics in 1933 with Erwin Schrödinger (1887-1961).

Dirac spent the last 10 years of his life as professor of physics at Florida State University. He died on October 20, 1984.

1. Paul Adrein Maurice Dirac, Featured Physicist, The Physical World, www.physicalworld.org/ restless_universe/html/ru_dira.
2. Paul Dirac, Microsoft Encarta Encyclopedia Online, http://encarta.msn.com/ encyclopedia_761579313/paul_dirac, 2007.
3. Paul Dirac, Wikipedia, http://en.wikipedia.org/wiki/Paul_Dirac.

Walter Brattain (1902-1987)

Co-invented the transistor

Walter Houser Brattain was born to American parents in Amoy, China on February 19, 1902. He was raised in the state of Washington on a cattle ranch owned by his parents. He earned his B.S. in physics and mathematics at Whitman College in Walla Walla, Washington in 1924, his M.A. from the University of Oregon in 1926, and his Ph.D. in physics at the University of Minnesota in 1929. His thesis was on electron impact in mercury vapor. In 1928 and 1929 Brittain worked at the National Bureau of Standards in Washington, D.C., and in 1929 was hired by Bell Telephone Laboratories.

Before World War II, the chief area of Brattain's research was the surface properties of solids. His early work was concerned with thermionic emission and adsorbed layers on tungsten. He continued on into the field of rectification and photo-effects at semiconductor surfaces, beginning with a study of rectification at the surface of cuprous oxide. This work was followed by similar studies of silicon.

During World War II, Brattain devoted his time to developing methods of submarine detection under a contract with the National Defense Research Council at Columbia University.

Following the war, Brattain returned to Bell Laboratories and soon joined the semiconductor division of the newly-organized Solid State Department of the Laboratories. William Shockley was the director of the semiconductor division, and early in 1946 he had initiated an investigation of semiconductors intended to produce a practical solid state amplifier. Brattain continued in the same line of research as before the war, this time investigating the surface properties of both silicon and germanium.

Brattains' chief contributions to solid state physics were the discovery of the photo-effect at the free surface of a semiconductor, the invention of the point-contact transistor jointly with John Bardeen (1908-1991), and work leading to a better understanding of the surface properties of semiconductors. Together with John Bardeen and William Shockley (1910-1989) Brattain invented the transistor. First announced in 1948, the transistor was perfected by 1952 for commercial use in portable radios, hearing aids, and other devices. The three men shared the 1956 Nobel Prize in physics for their invention. Brattain died on October 13, 1987.

1. Walter H. Brattain, Biography, NobelPrize.org, http://nobelprize.org/nobel_prizes/physics/laureates/1956/brattain-bio.
2. Walter Houser Brattain, Microsoft Encarta Encyclopedia Online, http://au.encarta.msn.com/encyclopedia_781531789/brittain_vera_mary, 2007.
3. Walter Houser Brattain, Wikipedia, http://en.wikipedia.org/wiki/Walter_Houser_Brattain.

Cecil Powell (1903-1969)

Developed a photographic method of studying nuclear processes and discovered the pion

Cecil Frank Powell, the son of a gunsmith, was born in Tonbridge, Kent, England on December 5, 1903. He graduated from Sidney Sussex College of the University of Cambridge in 1925 with a B.S. in natural science, and earned his Ph.D. in physics in 1927 at the Cavendish Laboratory at Cambridge. In 1928, he became a research assistant at the University of Bristol.

Powell became interested in subatomic-particle-detection devices. The drawback of the cloud chamber at the time was that it required a resting period each time it was used. Powell began experimenting with photographic emulsions, but found that none of those available were of adequate quality to record evidence of the particles. He finally persuaded a photographic film company to create a new emulsion that would work. To obtain data on cosmic rays, Powell sent the new emulsion aloft in hydrogen balloons to high altitudes to capture traces of cosmic rays. These expeditions resulted in Powell's discovery of a new particle, the pion (also called a pi meson). The pion was thought to be the hypothetical particle proposed in 1935 by Yukawa Hideki (1907-1981) of Japan in his theory of the nuclear binding force between protons and neutrons.

In 1936, Powell visited the West Indies as seismologist of an expedition investigating volcanic activity. He returned to Bristol in the following year, and in 1948 he was established as Melville Wills Professor of Physics at the University of Bristol.

Powell was awarded the 1950 Nobel Prize in physics for his photographic methods and for discovering the pion.

Increasingly concerned about social problems attached to scientific and technological advances, Powell served from 1961 to 1963 as the chairperson of the Scientific Policy Committee of the European Organization for Nuclear Research. In 1963, he was appointed vice chancellor of the University of Bristol, as well as director of the university's H. H. Wills Physics Laboratory.

Powell died on August 9, 1969, a few months after retiring from the University of Bristol. At the time, he was walking in the foothills of the Alps near the Valsassina region of Italy.

1. Cecil Frank Powell, Microsoft Encarta Encyclopedia Online, http://encarta.msn.com/ encyclopedia_761583311/powell_cecil_frank, 2007.
2. Cecil Frank Powell, Wikipedia, http://en.wikipedia.org/wiki/Cecil_Frank_Powell.
3. Cecil Powell, Biography, NobelPrize.org, http://nobelprize.org/nobel_prizes/physics/ laureates/1950/powell-bio.

Ernest Walton (1903-1995)

Transmutated atomic nuclei by artificially accelerated atomic particles

Ernest Thomas Sinton Walton, the son of a Methodist minister, was born in Dungarvan, County Waterford, Ireland on October 6, 1903. He entered Trinity College, Dublin in 1922, and was awarded a B.S. in 1926 and a master's in 1927. He was then accepted as a research student at Cambridge University under the supervision of Ernest Rutherford (1871-1937), director of Cavendish Laboratory. He was awarded his Ph.D. in 1931.

During the early 1930s, Walton and John Cockcroft (1897-1967) built an apparatus that split the nuclei of lithium atoms by bombarding them with a stream of protons in a high-voltage tube. The atom splitting produced helium atoms in place of the lithium. This was experimental verification of theories about atomic structure that had been proposed earlier by several other physicists. The linear particle accelerator they built is now called the **Cockcroft-Walton generator**. It helped usher in an era of particle-accelerator-based experimental nuclear physics.

Walton later succeeded in accelerating protons to such a high velocity that they were able to penetrate atomic nuclei of light elements and start nuclear reactions.

Walton was especially interested in the relations created during this process between the energies of the protons before they hit the nuclei and those of the created nuclear particles. These experiments demonstrated that atomic nuclei contain enormous energies. The research also provided the first experimental confirmation of the equations of Albert Einstein (1879-1955) for the equivalence of mass and energy. During World War II, Walton refused to participate in the Manhattan Project to build the atomic bomb. After the war, he was President of the Irish branch of the Pugwash organization of scientists working against nuclear weapons.

Walton and Cockcroft were awarded the 1951 Nobel Prize in physics for their atomic research with particle accelerators. Walton was the first Irishman to win a Nobel Prize in science.

On his retirement from Trinity College, Dublin in 1974, Walton returned to Northern Ireland, and for the remaining twenty years of his life he lived in Belfast, where he died on June 25, 1995.

Physicists in laboratories in many countries continue to use his method for accelerating charged particles to produce nuclear reactions.

1. Ernest T. S. Walton, Biography, NobelPrize.org, http://nobelprize.org/nobel_prizes/physics/laureates/1951/walton-bio.
2. Ernest T. S. Walton, Microsoft Encarta Encyclopedia Online, http://uk.encarta.msn.com/encyclopedia_781536080/Walton_Ernest_Thomas_Sinton, 2007.
3. Ernest Walton, Wikipedia, http://en.wikipedia.org/wiki/Ernest_Walton.

John von Neumann (1903-1957)

Formulated a fully quantum mechanical generalization of statistical mechanics

John von Neumann was born János von Neumann in Budapest, Hungary on December 28, 1903. His father was a lawyer who worked in a bank. János received his Ph.D. in mathematics, with minors in experimental physics and chemistry, from the University of Budapest in 1926 at the age of 22. He lectured at Berlin from 1926 to 1929 and at Hamburg from 1929 to 1930.

In 1930, von Neumann, his mother, and his brothers immigrated to the United States, where Von Neumann joined the faculty of the Institute for Advanced Study at Princeton University. On arrival, he changed his name to John, and in 1937 he became a naturalized citizen of the United States.

After obtaining U.S. citizenship, von Neumann took an interest in applied mathematics, and then developed an expertise in explosions—phenomena which are difficult to model mathematically. This led him to a large number of military consultancies, primarily for the Navy, which in turn led to his involvement in the Manhattan Project. This involvement included frequent trips by train to the project's secret research facilities in Los Alamos, New Mexico. In 1955, he became a member of the U.S. Atomic Energy Commission.

Von Neumann is noted for his fundamental contributions to the theory of quantum mechanics, particularly the concept of "rings of operators" (now known as **Neumann algebras**), and also for his pioneering work in applied mathematics, mainly in statistics and numerical analysis. He is also known for the design of high-speed electronic computers.

Along with Edward Teller (1908-2003) and Stanislaw Ulam (1909-1984), von Neumann worked out key steps in the nuclear physics involved in thermonuclear reactions and the hydrogen bomb.

Von Neumann was diagnosed with cancer in 1957, possibly caused by exposure to radiation during his witnessing of atomic bomb tests. He died in great pain within a year of the initial diagnosis. The cancer had spread to the brain, impairing his mental abilities. While at Walter Reed Hospital in Washington, D.C. he was kept under military security for fear he might reveal military secrets while heavily medicated. Von Neumann was buried at Princeton Cemetery in Princeton, New Jersey.

1. John von Neumann, Microsoft Encarta Encyclopedia, http://encarta.msn.com/encyclopedia_761579159/Neumann_John_von, 2007.
2. John von Neumann, Wikipedia, http://en.wikipedia.org/wiki/John_von_Neumann.
3. O'Conner, J. J. and E. F. Robertson, John von Neumann, MacTutor, www-history.mcs.st-and.ac.uk/history/Biographies/Von_Neumann.

Louis Néel (1904-2000)

Made discoveries concerning antiferromagnetism and ferrimagnetism

Louis Eugène Félix Néel was born in Lyon, France on November 22, 1904. He studied at the Ecole Normal Supérieure in Paris from 1924 to1928, and was appointed lecturer in 1928.

In 1932, Néel obtained his doctorate in sciences at the University of Strasbourg, and was appointed Professor at the Faculty of Science in 1937.

Néel discovered in the early 1930s the phenomenon of antiferromagnetism. With antiferromagnetism, the spins of electrons align in a regular pattern, with neighboring spins pointing in opposite directions. Generally, antiferromagnetic materials exhibit antiferromagnetism at a low temperature, and become disordered above a certain temperature. This transition temperature is now called the **Néel temperature**. Above the Néel temperature, the material is typically paramagnetic (magnetism which occurs only in the presence of an externally applied magnetic field).

During the Second World War, Néel worked on the defense of ships of the French fleet against German magnetic mines, and invented an effective new method of protection—neutralization.

After the war, in 1946, Néel went to Grenoble and established the Laboratoire d'Electrostatique et de Physique du Métal, which became one of the external laboratories of the Centre National de la Recherche Scientifique.

In 1947, Néel developed a quantitative theory of ferrimagnetic fields. He demonstrated the magnetic memory of beds of rock, which helped explain the physics of terrestrial magnetism. Ferrimagnetism is a phenomenon in ferrites where there can be incomplete cancellation of antiferromagnetic arranged spins, giving a net magnetic moment

In 1954, Néel became Director of the Institut Polytechnique de Grenoble and the Ecole Française de Papeterie.

In 1970, Néel was awarded, jointly with Hannes Alfven (1908-1995), the Nobel Prize for physics for his pioneering studies of the magnetic properties of solids. Alfven's share of the award was for magnetohydrodynamics, with applications in plasma physics.

In 1970, Néel was appointed President of the Institut National Polytechnique. That same year he

Néel died on November 17, 2000.

1. Antiferromagnetism, Wikipedia, http://en.wikipedia.org/wiki/Antiferromagnetism.
2. Louis Eugène Félix Néel, Wikipedia, http://en.wikipedia.org/wiki/
    Louis_Eug%C3%A8ne_F%C3%A9lix_N%C3%A9el.
3. Néel, Louis-Eugène-Félix, Microsoft Encarta Encyclopedia Online, http://encarta.msn.com/
    encyclopedia_761583284/n%C3%A9el_louis_eug%C3%A8ne_f%C3%A9lix, 2007.

Pavel Čerenkov (1904-1990)

Discovered and explained the electromagnetic radiation emitted when a charged particle passes through an insulator at a speed greater than the speed of light in that medium

Pavel Alekseevič Čerenkov was born in 1904 to peasant parents in the small village of Nižniaja Čigla in Voronež Oblast, Russia. He graduated from the Department of Physics and Mathematics of Voronež State University in 1928.

In 1930, Čerenkov took a post as a senior researcher in the Lebedev Institute of Physics.

In 1940, Čerenkov was awarded the degree of Doctor of Physico-Mathematical Sciences.

In 1953, he was confirmed as Professor of Experimental Physics.

In 1959, Čerenkov became the head of the institute's photo-meson processes laboratory.

Čerenkov became interested in the emission of blue light from a bottle of water subjected to radioactive bombardment. This phenomenon, associated with charged atomic particles moving at velocities higher than the speed of light in the local medium, proved to be of great importance in subsequent experimental work in nuclear physics, and for the study of cosmic rays. It became known as the **Čerenkov effect** or **Čerenkov radiation**.

The **Čerenkov detector,** which makes use of the Čerenkov effect, became a standard piece of equipment in atomic research for observing the existence and velocity of high-speed particles. The device was installed in Sputnik III.

Čerenkov also shared in the development and construction of electron accelerators and in the investigations of photo-nuclear and photo-meson reactions.

In 1958, Čerenkov, Il′ja Frank (1908-1990) and Igor Tamm (1895-1971) were awarded the Nobel Prize in physics for their discovery and explanation of the Čerenkov effect.

Čerenkov died in Moscow on January 6, 1990, and was buried in Novodevič'e Cemetery.

1. Pavel A. Cherenkov, Biography, NobelPrize.org, http://nobelprize.org/nobel_prizes/ physics/ laureates/1958/cerenkov-bio.
2. Pavel Alekseevič Čerenkovm, Wikipedia, http://en.wikipedia.org/wiki/ Pavel_Alekseyevich_%C4%8Cerenkov.
3. Pavel Alekseyevich Cherenkov, Microsoft Encarta Encyclopedia Online, http://it.encarta.msn.com/encyclopedia_761557279/%C4%8Cerenkov_Pavel_Alekseevi%C4%8 D, 2007.

H. Thomas Milhorn, MD, PhD

Robert Oppenheimer (1904-1967)
Headed the Manhattan Project to develop the nuclear fission bomb

Julius Robert Oppenheimer, the son of a wealthy textile importer, was born in New York City on April 22, 1904. He attended Harvard University, where he majored in chemistry, but also studied Greek, architecture, classics, art, and literature.

Oppenheimer was accepted for postgraduate work at the Cavendish Laboratory in Cambridge; however, his clumsiness in the laboratory made it apparent that his forte was theoretical, not experimental physics, so he left in 1926 for the University of Göttingen to study under Max Born (1882-1970). He received his Ph.D. in 1927 at the age of 22, and then did postdoctoral work at Harvard and the California Institute of Technology.

After being diagnosed with a mild case of tuberculosis, he and his brother Frank spent some weeks at a ranch in New Mexico. Afterward, he accepted a faculty position in physics at the University of California, Berkeley. He was the chief founder of the American school of theoretical physics at Berkeley. At the Institute for Advanced Study, he became senior professor of theoretical physics.

Oppenheimer's notable achievements in physics include the **Born-Oppenheimer approximation**, as well as work on electron-positron theory; the **Oppenheimer-Phillips process**; quantum tunneling; relativistic quantum mechanics; quantum field theory; black holes; and cosmic rays.

Oppenheimer is best known for his role as the director of the Manhattan Project, the World War II effort to develop the first nuclear weapons at the Los Alamos laboratory in New Mexico. For that work, he is known as the father of the atomic bomb.

After the war, Oppenheimer was a chief advisor to the newly created United States Atomic Energy Commission, and used that position to lobby for international control of atomic energy.

In 1947, Oppenheimer became director of the Institute for Advanced Studies in Princeton, New Jersey. Because of his outspoken political opinions, he was stripped of his security clearance in a much-publicized and politicized hearing in 1954.

Oppenheimer died on February 18, 1967. He is remembered as a founding father of the American school of physics.

1. J. Robert Oppenheimer (1904 -1967), AtomicArchive.com, www.atomicarchive.com/Bios/ Oppenheimer.
2. J. Robert Oppenheimer, Microsoft Encarta Encyclopedia Online, http://encarta.msn.com/ encyclopedia_761575357/Oppenheimer, 2007.
3. Robert Oppenheimer, Wikipedia, http://en.wikipedia.org/wiki/Robert_Oppenheimer.

Carl Anderson (1905-1991)

Discovered the positron

Carl David Anderson, the son of Jewish immigrants, was born in New York City on September, 5, 1905. He studied physics and engineering at the California Institute of Technology (Caltech), receiving his Ph.D. in 1930. For his doctoral thesis he studied the space distribution of photoelectrons ejected from various gases by X-rays.

From 1930 to 1933, Anderson was a research fellow at Caltech. He became an assistant professor of physics there in 1933 and a full professor in 1939.

At Caltech, under the supervision of Robert Millikan (1868-1953), Anderson began investigations into cosmic rays, and in 1932 encountered unexpected particle tracks in his cloud chamber photographs. He interpreted these tracks as having been created by a particle with the same mass as the electron, but with opposite electrical charge. Thus, he validated the theoretical prediction of Paul Dirac (1902-1984) of the existence of the positron. He did this by by shooting gamma rays produced by thorium carbide into other materials, resulting in creation of positron-electron pairs.

Anderson shared the 1936 Nobel Prize in physics with Victor Hess (1883-1964) for his discovery of the positron. Hess's portion of the Prize was for his discovery of cosmic radiation.

Also in 1936, Anderson and his first graduate student, Seth Neddermeyer (1907-1988), discovered the muon (historically known as a mu-meson), a subatomic particle 207 times more massive than the electron (The muon has a negative charge and a spin of 1/2. It has a mean lifetime of 2.2 microseconds.

Like all fundamental particles, the muon has an antimatter partner of opposite charge but equal mass and spin—the antimuon, also called a positive muon.) The existance of the muon had been predicted in 1935 by Yukawa Hideki (1907-1981).

Anderson spent his entire career at Caltech. During World War II, he conducted research in rocketry for the Navy.

Anderson died on January 11, 1991, and is buried in the Forest Lawn, Hollywood Hills Cemetery in Los Angeles.

1. Carl D. Anderson, Biography, NobelPrize.org, http://nobelprize.org/nobel_prizes/physics/laureates/1936/anderson-bio.
2. Carl David Anderson, Microsoft Encarta Encyclopedia Online, http://encarta.msn.com/encyclopedia_761569228/anderson_carl_david, 2007.
3. Carl David Anderson, Wikipedia, http://en.wikipedia.org/wiki/Carl_David_Anderson.

Emilio Segrè (1905-1989)

Co-discovered the antiproton

Emilio Gino Segrè, the son of an industrialist, was born in Tivoli, Italy on February 1, 1905. He enrolled in the University of Rome La Sapienza as an engineering student, but switched to physics in 1927. He earned his doctorate in 1928, having studied under Enrico Fermi (1901-1954).

Segrè was in the Italian Army from 1928 to 1929. Then, in 1939 he worked with Otto Stern (1888-1969) in Hamburg and Pieter Zeeman (1865-1943) in Amsterdam as a Rockefeller Foundation fellow. He was assistant professor of physics at the University of Rome from 1932 to 1936. From 1936 to 1938 he was Director of the Physics Laboratory at the University of Palermo.

Segrè obtained a molybdenum strip from the Berkeley Radiation Laboratory's cyclotron deflector in 1937. He found that it was emitting anomalous forms of radioactivity. After careful chemical and theoretical analysis, he discovered that some of the radiation was being produced by a previously unknown element, which he named technetium. It was the first artificially synthesized (not occurring in nature) chemical element.

While Segrè was on a summer visit to California in 1938, Mussolini's Fascist government passed anti-Semitic laws barring Jews from university positions. As a result, Segrè accepted a rather lowly job as a research assistant at the Berkley Radiation Laboratory, and later became a lecturer in the physics department.

While at Berkeley, he helped discover the element astatine and the isotope plutonium-239, which was later used to make Fat Man, the atom bomb dropped on Nagasaki. From 1943 to 1946, Segrè worked at the Los Alamos National Laboratory as a group leader for the Manhattan Project. He subsequently taught at Columbia University, University of Illinois, and University of Rio de Janeiro. On his return to Berkeley in 1946, he became a professor of physics.

In 1955, with Owen Chamberlain (1920-2006), using the betatron particle accelerator at Lawrence Berkeley National Laboratory, Segrè detected the antiproton. For this discovery, he and Chamberlain shared the 1959 Nobel Prize in physics.

In 1974, Segrè returned to the University of Rome as a professor of nuclear physics. He died of a heart attack on April 22, 1989.

1. Emilio G. Segrè, Wikipedia, http://en.wikipedia.org/wiki/Emilio_G._Segr%C3%A8.
2. Emilio Gino Segrè, Microsoft Encarta Encyclopedia Online, http://encarta.msn.com/encyclopedia_761567941/segr%C3%A8_emilio_gino, 2007.
3. Emilio Segrè, Biography, NobelPrize.org, http://nobelprize.org/nobel_prizes/physics/laureates/1959/segre-bio.

Ernst Ruska (1906-1988)

Designed the first electron microscope

Ernst August Friedrich Ruska was born in Heidelberg, Germany on December 25, 1906. He earned an engineering degree from the University of Berlin in 1931, and a Ph.D. in 1933.

Since electron waves were shorter than ordinary light waves, Ruska deduced that they would allow for greater magnification. In 1932, he and German physicist Max Knoll (1897-1969), under whom Ruska obtained his doctorate, built the first electron microscope. Despite being crude, it was capable of magnifying objects 400 times.

In 1937, Ruska accepted a research position at Siemens-Reiniger-Werke. He was responsible for the development of television receivers and transmitters, as well as photoelectric cells with secondary amplification. Convinced of the great practical importance of electron microscopy for pure and applied research, he continued the development of high-resolution electron microscopes with larger materials

At Berlin-Spandau in 1937, Ruska took part in seting up the Laboratory for Electron Optics, and developed the first customised electron microscopes (the Siemens Super Microscope). By the beginning of 1945, around 35 institutions were equipped with one.

In 1955, Ruska became director of the Institute for Electron Microscopy of the Fritz Haber Institute. Concurrently, he served as professor at the Technical University of Berlin from 1957 until his retirement in 1972.

For his work on the electron microscope, Ruska shared the 1986 Nobel Prize in physics with Gerd Karl Binnig (1947- of Germany and Heinrich Rohrer (1933-) of Switzerland, who designed the scanning tunneling microscope.

Although modern electron microscopes can magnify an object two million times, they are still based upon Ruska's prototype and his correlation between wavelength and magnification.

Researchers now use electron microscopes to examine microorganisms and cells, a variety of large molecules, medical biopsy samples, metals and crystalline structures, and the characteristics of various surfaces.

Ruska died on May 27, 1988.

1. Ernst Ruska, Autobiography, NobelPrize.org, http://nobelprize.org/nobel_prizes/ physics/laureates/1986/ruska-autobio.
2. Ernst Ruska, Microsoft Encarta Encyclopedia Online, http://encarta.msn.com/ encyclopedia_761583331/Ernst_Ruska, 2007.
3. Ernst Ruska, Wikipedia, http://en.wikipedia.org/wiki/Ernst_Ruska.

Felix Bloch (1905-1983)

Developed a new method for the precise measurements of the strength of
the magnetic field of the atomic nucleus

Felix Bloch was born in Zürich, Switzerland on October 23, 1905. Initially studying engineering at the Federal Institute of Technology in Zürich, he soon changed to physics and graduated in 1927. He continued his studies at the University of Leipzig, gaining his doctorate in 1928. His dissertation dealt with the quantum mechanics of electrons in crystals, and developing the theory of metallic conduction.

From 1928 to 1932, Bloch served as a researcher at a number of universities in Europe, and was a professor at the University of Leipzig from 1932 to 1933. He left Germany in 1933 upon Hitler's ascent to power, and worked at various institutions in Holland, Denmark, and Italy. He moved to the United States in 1934 after accepting an associate professorship of physics at Stanford University.

In 1945, Bloch lead a team that used a new method to measure the strength of the magnetic field of the nucleus using radio waves and nuclear magnetic resonance (NMR) to perform the measurements. With NMR, scientists can measure how much electromagnetic radiation of a specific frequency is absorbed by an atomic nucleus that is placed in a strong magnetic field. This method helps to reveal atomic and molecular structures. At the same time, Edward Purcell (1912-1997) and his research group at Harvard University made similar observations.

Bloch shared the Nobel Prize in physics with Purcell for their development of a new method for the precise measurements of the strength of the magnetic field of the atomic nucleus, called nuclear magnetic resonance. A number of important applications have come from NMR, including magnetic resonance imaging (MRI). MRI produces detailed internal images of the human body, which helps physicians diagnose disease and injuries.

Bloch worked on the Manhattan Project at Los Alamos, contributing to the effort to develop an atomic bomb and to improve radar technology. In 1954 and 1955, he served as the first director-general of the European Organization for Nuclear Research. After his return to Stanford University, he continued his investigations on nuclear magnetism, particularly in regard to the theory of relaxation. He died on September 10, 1983.

1.   Felix Bloch, Biography, NobelPrize.org, http://nobelprize.org/nobel_prizes/physics/laureates/ 1952/ bloch-bio.
2.   Felix Bloch, Microsoft Encarta Encyclypedia Online, http://au.encarta.msn.com/ encyclopedia_781533066/Bloch_Felix, 2007.
3.   Felix Bloch, Wikipedia, http://en.wikipedia.org/wiki/Felix_Bloch.

Nevel Mott (1905-1996)

Did theoretical investigations of the electronic structure of magnetic and disordered systems

Nevill Francis Mott was born in Leeds, England on September 30, 1905. Both his parents had been research students under J. J. Thomson (1856-1940) at the Cavendish Laboratory in Cambridge. Because of his father's job changes, the family moved first to Giggleswick in the West Riding of Yorkshire, Staffordshire, then to Chester, and finally to Liverpool, where his father was appointed Director of Education.

At 10 years of age Nevel began formal education at Clifton College in Bristol, then at St. John's College, Cambridge. He was to become a pioneer of solid-state physics and materials science.

In 1928, Mott was appointed to a lectureship at Manchester University. He returned to Cambridge in 1930 as a fellow and lecturer of Gonville and Caius College, where he worked on collision theory and nuclear problems.

In 1933, Mott moved to Bristol University as a professor of theoretical physics. Work there included a theory of transition metals, of rectification, hardness of alloys, and of the photographic latent image. He also worked on low-temperature oxidation and the metal-insulator transition.

In 1948, Mott became Professor of Physics and Director of the Henry Herbert Wills Physical Laboratory at Bristol. In 1954, he was appointed Professor of Physics at Cambridge, a post he held until 1971. There, he applied quantum theory to many problems in nuclear and solid-state physics. Additionally, he served as Master of Gonville and Caius College, Cambridge from 1959 to 1966. He was knighted in 1962.

Mott retired from Cambridge in 1965, but then conducted important research in the new field of non-crystalline semiconductors, in particular at Imperial College in London form 1971 to 1973.

Mott was awarded the 1977 Nobel Prize in physics, with Philip Anderson (1923-) and John Van Vleck (1899-1980) who had pursued independent research, for his theoretical investigations of the electronic structure of magnetic and disordered systems.

Mott died in Milton Keynes in Buckinghamshire on August 8, 1996.

1. Mott, Sir Nevill, Microsoft Encarta Encyclopedia Online, http://au.encarta.msn.com/ encyclopedia_781534827/mott_sir_nevill, 2007.
2. Nevill Francis Mott, Wikipedia, http://en.wikipedia.org/wiki/Nevill_Mott.
3. Sir Nevill F. Mott, Biography, NobelPrize.org, http://nobelprize.org/nobel_prizes/physics/ laureates/ 1977/mott-bio.

Hans Bethe (1906-2005)

Contributed to theories of stellar energy production and to the
development of nuclear weapons

Hans Albrecht Bethe was born on July 2, 1906 in Strasbourg, Alsace-Lorraine, which was part of Germany but now is in France. He obtained his doctorate in physics from the University of Munich, supervised by Arnold Sommerfeld (1868-1951). He then did postdoctoral stints in Cambridge and at the laboratory of Enrico Fermi (1901-1954) in Rome.

In 1929, Bethe made an important contribution to solid state physics and chemistry with his formulation of the basic concepts of crystal field theory. In 1930, he devised a formula for the energy loss of swift charged particles in matter, known as the **Bethe formula**, which continues to be important. He left Germany in 1933 for England when the Nazis came to power and he lost his job at the University of Tübingen.

In 1935, Bethe moved to the United States, where he joined the faculty at Cornell University. When the Second World War began, he collaborated with Edward Teller (1908-2003), then at George Washington University, on a theory of shock waves generated by the passage of a projectile through a gas, later of importance to missle reentry.

When Robert Oppenheimer (1904-1967) was put in charge of forming a secret weapons design laboratory in Los Alamos, New Mexico, he appointed Bethe as Director of the Theoretical Division. Bethe's work included calculating the critical mass of uranium-235 and the multiplication of nuclear fission in an exploding atomic bomb. Along with Richard Feynman (1918-1988), he developed a formula for calculating the explosive yield of the bomb. He also played a critical role in the development of the hydrogen bomb, although he had originally joined the project with the hope of proving it could not be made.

In 1947, Bethe was the first to explain the Lamb-shift in the hydrogen spectrum, and thus laid the foundation for the modern development of quantum electrodynamics.

Bethe's latter work included his contention that the source of the energy of the Sun is thermonuclear reactions in which hydrogen is converted into helium, a process known as stellar nucleosynthesis.

Bethe was awarded the 1967 Nobel Prize in physics for his contributions to the theory of nuclear reactions, especially his discoveries concerning nucleosynthesis. He died in Itica New York on March 6, 2005.

1. Hans Albrecht Bethe, Microsoft Encarta Encyclopedia Online, http://encarta.msn.com/encyclopedia_761572597/bethe_hans_albrecht, 2007.
2. Hans Bethe, Biography, NobelPrize.org, http://nobelprize.org/nobel_prizes/physics/laureates/1967/bethe-bio.

Maria Goeppert-Mayer (1906-1972)

Developed a model in which the atomic nucleus has a structure containing successive proton-neutron shells held together by complex forces

Maria Goeppert was born in Kattowitz, Upper Silesia (Germany) on June 28, 1906. Her family moved to Göttingen in 1910 when her father was appointed Professor of Pediatrics at the University of Göttingen. On her father's side, she was the seventh straight generation of university professors. Maria was educated in physics at the University of Göttingen.

In the spring of 1924 Maria enrolled at the University at Göttingen, with the intention of becoming a mathematician. But soon she found herself more attracted to physics. She received her doctorate in theoretical physics in 1930.

In 1931, Goeppert married physical chemist Joseph E. Mayer, and the couple moved to the United States, Mayer's home country. She adopted a hyphenated form of their names and anglicized the spelling. She became a U.S. citizen in 1933.

Goeppert-Mayer was not able to secure full-time work in her field until she was 53. Mayer performed most of her scientific work as a volunteer.

During her time at the University of Chicago and Argonne National Laboratories, Goeppert-Mayer developed a model in which the atomic nucleus has a structure containing successive proton-neutron shells held together by complex forces. She found evidence that atomic nuclei with 2, 8, 20, 28, 50, 82, or 126 nucleons (known as magic numbers) show unusual stability. She later demonstrated the theoretical basis for a shell model of atomic nuclei, with each magic number corresponding to a completed nuclear shell.

For this work, she and Hans Jensen (1907-1973), who had arrived at the same model independently, received a Nobel Prize in physics in 1963, together with Eugene Wigner (1902-1995) who had laid the foundation for their work.

Goeppert-Mayer was the second female laureate in physics after Marie Curie (1867-1934).

Goeppert-Mayer and Joseph joined the faculty of the University of California at San Diego in 1960. She died in San Diego on February 20, 1972 after a heart attack the previous year had left her comatose.

1. Maria Goeppert-Mayer, Biography, NobelPrize.org, http://nobelprize.org/nobel_prizes/physics/laureates/1963/mayer-autobio.
2. Maria Goeppert-Mayer, Microsoft Encarta Encyclopedia Online, http://encarta.msn.com/encyclopedia_761571961/goeppert-mayer_maria, 2007.
3. Maria Goeppert-Mayer, Wikipedia, http://en.wikipedia.org/wiki/Maria_Goeppert-Mayer.

Sin-Itiro Tomonaga (1906-1979)
Contributed to the development of quantum electrodynamics

Sin-Itiro Tomonaga was born in Tokyo on March 31, 1906. In 1913, his family moved to Kyoto when his father was appointed professor of philosophy at Kyoto Imperial University. Sin-Itro earned his B.S. in physics from Kyōto Imperial University in 1929. He was engaged in graduate work for three years at the same university.

Tomonaga was appointed a research associate at the Institute of Physical and Chemical Research in Tokyo. There, he started to work in a newly developed frontier of theoretical physics—quantum electrodynamics, which is a relativistic quantum field theory of electrodynamics.

From 1937 to 1939, Tomonaga studied nuclear physics and the quantum field theory with Werner Heisenberg (1901-1976) at the University of Leipzig.

In 1940, Tomonaga developed the intermediate coupling theory to clarify the structure of the meson cloud around the nucleon.

During World War II, Tomonaga did military research for the Japanese Navy, and worked on theories of quantum electrodynamics. British physicist Paul Dirac (1902-1984) had earlier applied quantum mechanics to an analysis of the electromagnetic field. His work predicted that particles, such as the electron, could have an infinite quantity of energy, which contradicted experimental observations. Tomonaga reworked Dirac's equations to eliminate the problematic infinite quantities, making Dirac's work consistent with observation, and permitting physicists to predict the properties of particles and radiation.

Julian Schwinger (1918-1994) and Richard Feynman (1918-1988), working independently, also modified Dirac's equations to eliminate the problematic infinities.

Tomonaga, Schwinger, and Feynman were jointly awarded the Nobel Prize in physics in 1965 for their work in modifying Dirac's equaitons.

Tomonaga additionally worked in the areas of quantum dynamics, the theory of neutrons, and electromagnetics.

Tomonaga became increasingly involved in scientific administration, serving as a member, and later president, of the Science Council of Japan. From 1957 on, he was active in movements against the deployment of nuclear weapons. Tomonaga died in Tokyo on July 8, 1979.

1. Shin'ichirō Tomonaga, Microsoft Encarta Encyclopedia Online, http://encarta.msn.com/ encyclopedia_761583379/tomonaga_shin%E2%80%99ichir%C5%8D, 2007.
2. Sin-Itiro Tomonaga, Biography, NobelPrize.org, http://nobelprize.org/nobel_prizes/ physics/ laureates/ 1965/tomonaga-bio.

Hans Jensen (1907-1973)

Developed a model in which the atomic nucleus has a structure containing successive proton-neutron shells held together by complex forces

Johannes Hans Daniel Jensen was born in Hamburg, Germany on June 25, 1907. He studied physics, mathematics, physical chemistry, and philosophy at the Universities of Hamburg and Freiburg, and obtained his Ph.D. in physics from Hamburg in 1932.

After graduating, Jensen became a scientific assistant at the Institute for Theoretical Physics of the University of Hamburg. In 1936, he obtained a D.Sc. from that university.

In 1941, Jensen became a Professor of Theoretical Physics at the Technische Hochschule in Hannover. In 1949, he was appointed professor at the University of Heidelberg, a position Jensen held until his retirement in 1969. For periods from 1951 to 1953 and again in 1961 he left Heidelberg to lecture in the United States.

In 1949, Jensen postulated that protons and neutrons in the nucleus of an atom are arranged in shells, like the atom's electrons. Maria Goeppert-Mayer (1906-1972) had arrived at the same conclusion independently. The two decided to work together to further explore the concept. They found evidence that atomic nuclei with 2, 8, 20, 28, 50, 82, or 126 nucleons (known as magic numbers) show unusual stability. Goeppert-Mayer later demonstrated the theoretical basis for a shell model of atomic nuclei, with each magic number corresponding to a completed nuclear shell.

Jensen was a visiting professor at the University of Wisconsin in 1951; the Institute of Advanced Study, Princeton in 1952; the University of California at Berkeley in 1952; the California Institute of Technology in 1953; Indiana University in 1953, the University of Minnesota in 1956; and the University of California at La Jolla in 1961.

In 1955, Jensen and Goeppert-Mayer coauthored the book *Elementary Theory of Nuclear Shell Structure,* which chronicled their discoveries.

Jensen shared the 1963 Nobel Prize for physics with Goeppert-Mayer for their proposal of the shell nuclear model and with Eugene Wigner (1902-1995) whose work provided the background needed by Goeppert-Mayer and Jensen.

In 1969, Jensen was made an honorary citizen of Fort Lauderdale, Florida. He died on February 11, 1973.

1. J. Hans D. Jensen, Biography, NobelPrize.org, http://nobelprize.org/nobel_prizes/physics/ laureates/ 1963/jensen-bio.
2. J. Hans Daniel Jensen, Microsoft Encart Encyclopedia Online, http://encarta.msn.com/ encyclopedia_761583223/j_hans_daniel_jensen, 2007.
3. J. Hans. D. Jensen, Wikipediai, http://en.wikipedia.org/wiki/J._Hans_D._Jensen.

H. Thomas Milhorn, MD, PhD

Hideki Yukawa (1907-1981)

Predicted the existence of a particle that was the carrier of the strong
nuclear force on the basis of theoretical work

Hideki Yukawa was born in Tokyo, Japan on January 23, 1907. Between 1932 and 1939 he was a lecturer at the Kyoto University and lecturer and Assistant Professor at the Osaka University.

Yukawa became a lecturer in physics at Kyōto University in 1932, and was made professor in 1939. He also taught from 1933 to 1939 at Ōsaka University.

In 1835, Yukawa published a paper entitled *On the Interaction of Elementary Particles* in which he proposed a new field theory of nuclear forces. The force between protons and neutrons was said to be mediated by a particle of about 200 electron masses. This force was of very short range, but strong enough to overcome the repulsive force of protons, and diminished rapidly with distance. Yukawa predicted that this force manifested itself by the transfer of particles between neutrons and protons.

Yukawa received the D.Sc. degree in 1938, and in 1939 became Professor of Theoretical Physics at Kyoto University.

In 1947, Cecil Powell (1903-1969) at Bristol University in the U.K. discovered the pion in cosmic radiation. It satisfied all of Yukawa's postulates. At the time, the pion was thought to be the particle that had been predicted by Yukawa as the carrier of the strong nuclear. However, there is some question about this; the pion (pi meson) may be just one of a family of strongly interacting mesons.

Yukawa also predicted K-capture, in which an electron in the lowest hydrogen energy level could be absorbed by the nucleus.

Yukawa received the 1949 Nobel Prize in physics for his prediction of the particle that was a carrier of the strong nuclear force. He was the first Japanese person to win the Nobel Prize.

Yukawa became a visiting professor to the Institute for Advanced Study at Princeton in 1948, and from 1949 to 1953 he was a visiting professor at Columbia University.

In 1950, Yukawa became professor emeritus at Ōsaka University, and in 1953 was named director of the Research Institute for Fundamental Physics at Kyōto University. He died on September 8, 1981.

1.   Hideki Yukawa, Biography, NobelPrize.org, http://nobelprize.org/nobel_prizes/physics/laureates/1949/yukawa-bio.
2.   Hideki Yukawa, Microsoft Encarta Encyclopedia Online, http://encarta.msn.com/encyclopedia_761575794/yukawa_hideki, 2007.
3.   Yukawa, Hideki (1907-1981, Eric Weisstein's World of Biography, Wolfram Research,)http://scienceworld.wolfram.com/biography/Yukawa.

Edward Teller (1908-2003)

Helped develop atomic and hydrogen bombs

Edward Teller was born into a Jewish family in Austria-Hungary on January 15, 1908. He left Hungary in 1926, partially due to limits on the number of students who could study at universities under Horthy's regime. He subsequently graduated in chemical engineering at the University of Karlsruhe in Germany, and received his Ph.D. in physics under Werner Heisenberg (1901-1976) at the University of Leipzig in 1930. Teller's Ph.D. dissertation dealt with one of the first accurate quantum mechanical treatments of the hydrogen molecular ion.

After two years at the University of Göttingen, Teller left Germany in 1933 because of the rise of Hitler. He went briefly to England and then to Copenhagen in 1934, where he worked under Niels Bohr (1885-1962).

In 1935, Teller became a professor of physics at George Washington University, where in 1941 he became interested in the use of nuclear energy, both fusion and fission. He became a naturalized citizen of the United States that same year.

In 1942, Teller was invited to participate in summer planning seminar at the University of California, Berkeley for the origins of the Manhattan Project. At the seminar, he diverted discussion from the fission weapon to the possibility of a fusion weapon—what he called the "Super" weapon. This would later become known as the hydrogen bomb. In 1943, Teller became part of the Theoretical Physics Division at the secret Los Alamos laboratory during the war, and continued to push his ideas for a fusion weapon, even though it had been put on a low priority. Despite making some valuable contributions, He irriated fellow scientists by refusing to engage in the calculations for the implosion of the fission bomb.

Teller returned to Los Alamos in 1950 to work on the hydrogen-bomb project. And in 1951, he developed the first workable design for a megaton-range hydrogen bomb.

The rift between Teller and many of his colleagues widened in 1954 when he testified against Robert Oppenheimer (1904-1967), the former head of Los Alamos and member of the Atomic Energy Commission, at Oppenheimer's security clearance hearing.

Teller suffered a stroke and died at his home on the Berkeley campus on September 9, 2003. He is remembered a the father of the hydrogen bomb.

1. Edward Teller, Atomic Archive.com, www.atomicarchive.com/Bios/Teller.
2. Edward Teller, Microsoft Encarta Encyclopedia Online, http://au.encarta.msn.com/
3. enclopedia_761557475/Teller_Edward, 2007.
4. Edward Teller, Wikipedia, http://en.wikipedia.org/wiki/Edward_Teller.

Il'ja Frank (1908-1990)

Discovered and explained the electromagnetic radiation emitted when a charged particle passes through an insulator at a speed greater than the speed of light in that medium

Il'ja Mikhailovich Frank, the son of a professor of mathematics, was born in Leningrad, Russia on October 23, 1908. He graduated from Moscow State University in 1930 with a degree in physics.

Frank worked at the State Optical Institute from 1931 to 1934. He accepted a position at the Lebedev Institute of Physics of the Academy of Sciences in Moscow in 1934.

In 1934, Pavel Čerenkov (1904-1990) discovered a blue glow is emitted by charged particles traveling at very high speeds through water. This became known as the **Čerenkov effect** or **Čerenkov radiation**. Frank and Igor Tamm (1895-1971) provided the theoretical explanation of this effect, which results when electromagnetic radiation emitted when a charged particle passes through an insulator at a speed greater than the speed of light in that medium.

The team also helped construct the **Čerenkov detector**, which assisted them in observing the Čerenkov effect with other high-energy particles.

Frank expanded his research on Čerenkov effect by studying how the phenomenon was affected by the optical properties of different media.

In 1941, Frank was promoted to officer in charge of the Atomic Nucleus Laboratory at the Lebedev Institute of Physics.

Frank joined the faculty at Moscow State University in 1944. In 1957, also became the Director of the Neutron Laboratory of the Joint Institute of Nuclear Investigations.

Frank was awarded the Nobel Prize for physics in 1958 jointly with Čerenkov and Tamm for discovering and explaining the Čerenkov effect. This work resulted in the development of new methods for detecting and measuring the velocity of high-speed nuclear particles and became of great importance for research in nuclear physics.

Frank's other work included collaboration with Cherenkov and Tamm in research on electron radiation. He also studied of gamma rays and neutron beams.

Frank died on June 22, 1990.

1. Il'ja M. Frank Biography, NobelPrize.org, http://nobelprize.org/nobel_prizes/physics/laureates/1958/frank-bio.
2. Illya M. Frank, Microsoft Encarta Encyclopedia Online, http://ca.encarta.msn.com/encyclopedia_761583178/Frank_Ilya_Mikhailovich, 2007.
3. Ilya Mikhailovich Frank Wikipedia, Wikipedia, http://en.wikipedia.org/wiki/Ilya_Frank.

John Bardeen (1908-1991)

Co-invented the transister and developed a fundamental theory of
conventional superconductivity

The only person to win two Nobel Prizes in physics, John Bardeen, was born in Madison, Wisconsin on May 23, 1908. His father was the first Dean of the Medical School of the University of Wisconsin-Madison. In 1928, Bardeen received his B.S. in electrical engineering from the University of Wisconsin-Madison, and in 1929 he received his M.S. in electrical engineering from the University of Wisconsin. In 1936 he received his Ph.D. in mathematical physics from Princeton University.

In 1938, Bardeen became an assistant professor at the University of Minnesota. In 1941, because of World War II, he took a leave of absence and worked for the Naval Ordnance Laboratory. He stayed there for four years. After the war, in 1945, he took a position at Bell Telephone Laboratories, which was just starting a solid-state division. There, he worked with William Shockley (1910-1989) and Walter Brattain (1902-1987). Their assignment was to seek a solid-state alternative to glass vacuum tube amplifiers. And on December 23, 1947, they succeeded in creating a point-contact transistor that achieved amplification.

In 1951, Bardeen left Bell Labs and took a position at the University of Illinois at Urbana-Champaign as Professor of Electrical Engineering and of Physics.

Bardeen shared the Nobel Prize in physics in 1956 with Shockley and Brattain for inventing the transistor. The transistor revolutionized the electronics industry, allowing the Information Age to occur, and made possible the development of almost every modern electronic device, from telephones to computers to missiles.

In 1972, Bardeen shared a second Nobel Prize in physics with Leon Neil Cooper (1930-) and John Robert Schrieffer (1931-) for a fundamental theory of conventional superconductivity known as the BCS theory. The BCS theory explains conventional superconductivity—the ability of certain metals at low temperatures to conduct electricity without electrical resistance. His developments in superconductivity are used in medical advances, such as computed axial tomography (CAT) scans and magnetic resonance imaging (MRI). Bardeen died of cardiac arrest at Brigham and Women's Hospital in Boston, Massachusetts on January 30, 1991.

1.   John Bardeen, Biography, NobelPrize.org, http://nobelprize.org/nobel_prizes/physics/laureates/ 1972/bardeen-bio.
2.   John Bardeen, Microsoft Encarta Encyclopedia Online, http://encarta.msn.com/ encyclopedia_761567167/bardeen_john 008.
3.   John Bardeen, Wikipedia, http://en.wikipedia.org/wiki/John_Bardeen.

Lev Landau (1908-1968)

Developed theories for condensed matter, especially liquid helium

Lev Davidovich Landau, the son of an engineer and physician, was born into a Jewish family in Baku, Azerbaijan on January 22, 1908. He began his college work at Baku State University in 1922.

In 1924, Landau moved to the Physics Department of Leningrad University, graduating in 1927. Landau subsequently enrolled for post-graduate study at the Leningrad Physico-Technical Institute, and at age 21 received a doctorate.

In 1929, Landau went to Copenhagen to work in the Institute for Theoretical Physics with Niels Bohr (1885-1962).

In 1932, he returned to the Soviet Union where he headed the department of theoretical physics at Kharkov Mechanics and Machine Building Institute.

Landau was arrested on April 27, 1938 by the Communist secret police and held in prison until his release on April 29, 1939 after his colleague Pyotr Kapitsa (1894-1984), an experimental low-temperature physicist, wrote a letter to Stalin, personally vouching for Landau's behavior.

Landau's accomplishments included the co-discovery of the density matrix method in quantum mechanics, the quantum mechanical theory of diamagnetism, the theory of superfluidity, the theory of second-order phase transitions, the **Ginzburg-Landau** theory of superconductivity, the explanation of **Landau damping** in plasma physics, the **Landau pole** in quantum electrodynamics, and the two-component theory of neutrinos.

Landau received the 1962 Nobel Prize in physics for his development of a mathematical theory of superfluidity that accounts for the properties of liquid helium at a temperature below 2.17 K. Superfluidity is an unusual state of matter noted only in liquid helium cooled to near absolute zero and characterized by apparently frictionless flow.

In 1943, Landau published a nine-volume set of books entitled *Course of Theoretical Physics*.

In 1962, Landau's car collided with an oncoming lorry, and he was severely injured, spending three months in a coma. He was several times considered near death, and suffered a severe impairment of memory. By the time of his death on April 1, 1968 he had made only a partial recovery.

1. Landau, Lev Davidovich, Microsoft Encarta Encyclopedia Online, http://au.encarta.msn.com/ encyclopedia_761558628/landau_lev_davidovich, 2007.
2. Lev Landau, Biography, NobelPrize.org, http://nobelprize.org/nobel_prizes/physics/ laureates/1962/landau-bio.
3. Lev Landau, Wikipedia, http://en.wikipedia.org/wiki/Lev_Landau.
4. Superfluidity, Merriam-Webster, www.merriam-webster.com/dictionary/superfluidity.

Victor Weisskopf (1908-2002)

Made theoretical contributions to quantum electrodynamics, nuclear structure, and elementary particle physics

Victor Frederick Weisskopf was born to a well-to-do Jewish family in Vienna, Austria on September 19, 1908. He attended the University of Vienna, and then received a doctorate in physics from the University of Gottingen in 1931. He then went to Leipzig to work under Werner Heisenberg (1901-1976), and then in the spring term of 1932 Berlin to work under Erwin Schrödinger (1887-1961).

For the academic year 1932-33, Weisskopf received a Rockefeller Fellowship to work in Copenhagen with Niels Bohr (1885-1962) and in Cambridge with Paul Dirac (1902-1984).

In 1937, Weisskopf accepted a lectureship at the University of Rochester, and immigrated to the United States, where he became a naturalized citizen in 1943.

During World War II Weisskoph worked as group leader in the Theoretical Division at Los Alamos on the Manhattan Project to develop the atomic bomb, and later campaigned against the proliferation of nuclear weapons. Weisskopf was a co-founder and board member of the Union of Concerned Scientists, which is an organization formed to promote the peaceful use of atomic energy. In the late 1940s, he was a vocal opponent of the plans of Edward Teller (1908-2003) and others to develop the hydrogen bomb.

In 1946, Weisskopf joined the staff of the Massachusetts Institute of Technology (MIT), remaining there until 1960 when he obtained a leave of absence after being appointed one of the Directorate of Five to assist the Director-general of the European Organization for Nuclear Research (CERN). The following year he became Director-general of CERN.

Weisskopf left CERN to return to MIT in 1965, where he became in 1967 head of the department of physics, a post he held until 1973. In parallel with his work at MIT, he was chairman of the high-energy physics advisory panel of the Atomic Energy Commission in the United States from 1967 to 1975.

Over the years, Weisskopf made valuable theoretical contributions to the quantum theory of radioactive transitions, the self-energy of the electron, the electrodynamic properties of the vacuum, and to the theory of nuclear reactions. He died in Cambridge, Massachusetts on April 22, 2002.

1. Frederick Victor Weisskoph, Biographical Memoirs, National Academy of Sciences, www.nap.edu/ readingroom/books/biomems/vweisskopf.
2. Victor Weisskopf (1908 - 2002), AtomicArchive.com, www.atomicarchive.com/Bios/Weisskopf.
3. Victor Weisskopf, Wikipedia, http://en.wikipedia.org/wiki/Victor_Weisskopf.

H. Thomas Milhorn, MD, PhD

Edwin Land (1909-1991)

Invented inexpensive filters for polarizing light and instant polaroid photography

Edwin Herbert Land was born on May 7, 1909 in Bridgeport, Connecticut. He studied chemistry at Harvard, but after one year he left Harvard for New York City.

In New York, Land invented the first inexpensive filters capable of polarizing light. Because he was not associated with an educational institution, he had to sneak into a laboratory at Columbia University late at night to use equipment. He also scoured the scientific literature at the New York City public library for prior work on polarizing substances. He came to realize that instead of attempting to grow a large single crystal of a polarizing substance, he could manufacture a film with millions of micrometer-sized polarizing crystals that were coaxed into perfect alignment with each other. He thus invented polaroid film.

In 1932, Land, with the financial backing of his instructor at Harvard, George Wheelwright III (1903-2001), established the Land-Wheelwright Laboratories to commercialize Land's polarizing technology. After early successes developing polarizing filters for sunglasses and photographic filters, Land obtained funding from Wall Street investors. The company was renamed the Polaroid Corporation in 1937, with Land as president and head of research. The company quickly found many additional applications, including color animation in the Wurlitzer jukebox, glasses in full-color stereoscopic (3-D) movies, controlling brightness of light through windows, and as a component of all liquid crystal displays.

During World War II, Land worked on military tasks, which included developing dark-adaptation goggles, target finders, the first passively guided smart bombs, and a special stereoscopic viewing system called the Vectrograph which revealed camoflauged enemy positions in aerial photography.

On February 21, 1947, Land demonstrated an instant camera and associated film. Called the Land Camera, it was in commercial sale less than two years later. Land also discovered a system for projecting the entire spectrum of hues with only two colors of light. He developed instant color photography with the SX-70 film and camera.

In 1957, Harvard University awarded Land an honorary doctorate. He died in Cambridge, Massachusetts on March 1, 1991.

1. Dr. Edwin H. Land (1909 - 1991) , The Rowland Institute at Harvard, www.rowland.harvard.edu/organization/land/index.php.
2. Edwin H. Land, Wikipedia, http://en.wikipedia.org/wiki/Edwin_Land.
3. Edwin Land, The Great Idea Finder, www.ideafinder.com/history/inventors/land.

Nikolay Bogoliubov (1909-1992)

Contributed to the microscopic theory of superfluidity, to theory of elementary particles, and to nonlinear mechanics

Nikolai Nikolaevich Bogoliubov was born in Nizhny Novgorod, Russia on August 21, 1909. His family moved to Kiev in 1921, where after graduation from high school he began independent study of mathematics and physics, participating in seminars at Kiev University.

In 1925, Bogoliubov entered the Ph.D. program at the Academy of Science of Ukrainian SSR, from which he graduated in 1929.

Bogoliubov was skilled in using sophisticated mathematics to attack concrete physical problems. His main interests were nonlinear mechanics, statistical physics, quantum field theory, and elementary particle theory.

Bogoliubov published papers on variational calculus, approximation methods of mathematical analysis, differential equations, equations of mathematical physics, asymptotic methods of nonlinear mechanics, theory of stability, theory of dynamical systems, and many other areas.

In 1947, Bogoliubov introduced kinetic equations in superfluidity theory. He built a new theory of scattering matrices, formulated concepts of microscopical causality, obtained important results in quantum electrodynamics, and investigated dispersion relations that have important meaning in elementary particle theory.

In 1958, Bogoliubov formulated a theory of superconductivity, using an analogy between superconductivity and superfluidity phenomena. He investigated a new synthesis of the Bohr theory of quasiperiodic functions, and developed methods of asymptotic integration of nonlinear differential equations which describe oscillating processes.

The **Bogoliubov transformation** is often used to diagonalize Hamiltonians, which yields the steady-state solutions of the corresponding Schrödinger equation. The solutions of BCS theory in a homogeneous system, for example, are found using a Bogoliubov transformation.

The **Krylov-Bogolyubov theorem** is a fundamental theorem in the study of dynamical systems. It is named after Bogolyubov and Nikolay Krylov (1879-1955).

The Joint Institute for Nuclear Research in Dubna, Russia set up the N. N. Bogoliubov prize for Young Scientists in memory of Bogoliubov.

Bogoliubov was awarded the Dirac Medal in 1992. He died on February 13 of that year in Moscow.

1. Bogoliubov prize for young scientists, CERN Courier, Faces and Places, March 29, 1999.
2. Nikolai Bogoliubov, Answers.com, ://www.answers.com/topic/nikolay-bogolyubov?cat=technology
3. Nikolai Bogoliubov, Wikipedia, http://en.wikipedia.org/wiki/Nikolai_Bogolubov.

William Shockley (1910-1989)
Did research on semiconductors and discovered the transistor effect

William Bradford Shockley was born in London, England to American parents on February 13, 1910, and was raised in California. His father was a mining engineer from Massechusettes and his mother had been a deputy mineral surveyer in Nevada.

Shockly received his B.S. from the California Institute of Technology in 1932 and his Ph.D. from the Massachusetts Institute of Technology in 1936. The title of his doctoral thesis was *Electronic Bands in Sodium Chloride.*

After receiving his doctorate, Shockley joined a research group headed by Clinton Davisson (1881-1958) at Bell Telephone Laboratories. When World War II broke out, he became involved in radar research at the labs in Whippany, New Jersey, and in 1942 he took leave from Bell Labs to become a research director at Columbia University's Anti-Submarine Warfare Operations Group. In 1944, he organized a training program for B-29 bomber pilots to use new radar bomb sights.

Shortly after the end of the war in 1945, Bell Labs formed a Solid State Physics Laboratory to seek a solid-state alternative to glass vacuum tube amplifiers. In December 1947, Shockley, John Bardeen (1907-1991), and Walter Brattain (1902-1987) succeeded in creating a point-contact transistor that achieved amplification.

On his own, Shockley worked out the critical ideas of drift and diffusion and the differential equations that govern the flow of electrons in solid state crystals.

In 1955, Shockley joined Beckman Instruments, where he was appointed the Director of Beckman's newly founded Shockley Semiconductor Laboratory division in Mountain View, California. His personality, described as domineering and increasingly paranoid, led to professional and personal problems.

Along with Bardeen and Brattain, Shockley was awarded the 1956 Nobel Prize in Physics for his work on transisters.

Shockley lectured at Stanford University beginning in 1958, and became professor of engineering science in 1963.

Late in his life, Shockley became intensely interested in questions of race, intelligence, and eugenics. He died on August 12, 1989.

1. William B. Shockley, Biography, NobelPrize.org, http://nobelprize.org/nobel_prizes/physics/laureates/1956/shockley-bio.
2. William Shockley, Microsoft Encarta Encyclopedia Online, http://encarta.msn.com/encyclopedia_761562270/william_shockley, 2007.
3. William Shockley, Wikipedia, http://en.wikipedia.org/wiki/William_Shockley.

Luis Alvarez (1911-1988)

Developed the liquid hydrogen bubble-chamber with which he found
atomic particles produced by high-energy nuclear events

Luis Walter Alvarez, the son of a physician, was born in San Francisco,
California on June 13, 1911. He attended the University of Chicago, where
he received his bachelor's degree in 1932, his master's in 1934, and his
Ph.D. in 1936.

During World War II, Alvarez was a key participant in the Manhattan
Project, including the Project Alberta on the dropping of the bomb. He
flew as a scientific observer of the atomic bombing of Hiroshima. He and
one of his students designed the exploding-bridgewire detonators for the
spherical implosives used on the Trinity and Nagasaki bombs.

Alvarez also did important work relating to radar and navigation
technologies. In particular, he developed the Ground Controlled Approach
system (GCA), which pilots use to land a plane in low visibility
conditions.

After the war, Alverez went on to invent the synchrotron, a cyclic
particle accelerator in which the magnetic field (to turn the particles so
they move in a circle) and the electric field (to accelerate the particles) are
carefully synchronized with the travelling particle beam. He also
developed the Berkeley 40-foot proton linear accelerator, which was
completed in 1947.

Alvarez was awarded the 1968 Nobel Prize in physics for the
discovery of a large number of resonance states, made possible through his
development of the hydrogen bubble chamber. This was significant
because it allowed scientists to record and study the short-lived particles
created in particle accelerators.

In 1980, Alvarez and his son, Walter Alvarez (1940-), presented the
asteroid-impact theory as an explanation for the presence of an unusual
abundance of iridium associated with the geological event referred to as
the K-T extinction boundary. Ten years after this initial proposal, evidence
of a huge impact crater called Chicxulub off the coast of Mexico strongly
supported their theory. Since that time, the concept of impact by a large
meteorite has become the most widely accepted explanation for the
extinction of the dinosaurs.

Alvarez died on September 1 1988.

1. Alvarez, Luis Walter, Microsoft Encarta Encyclopedia Online, http://au.encarta.msn.com/
    encyclopedia_761557105/Science, 2007.
2. Luis Alvarez, Biography, NobelPrize.org, http://nobelprize.org/nobel_prizes/physics/laureates/
    1968/alvarez-bio.
3. Walter Alvarez, Wikipedia, http://en.wikipedia.org/wiki/Luis_Walter_Alvarez.

Polykarp Kusch (1911-1993)

Precisely determined the magnetic moment of the electron

Polykarp Kusch, the son of a Lutheran clergyman, was born_in Blankenburg, Germany, on January 26, 1911. His family moved to the United States one year later. He became a naturalized citizen of the United States in 1923.

Originally interested in chemistry, Kusch switched his major to physics, and received his B.S. in 1931 from Case Institute of Technology and his Ph.D. from the University of Illinois in 1933. At Illinois, he worked on problems in the field of optical molecular spectroscopy.

Kusch worked at the University of Minnesota in the field of mass spectroscopy from 1936 to 1937. In 1937, he joined the physics department at Columbia University. There, he worked with Isidor Rabi (1898-1988), assisting him with research on atomic, molecular, and nuclear properties, using the method of molecular beams.

Kusch's own research dealt primarily with the small details of the interactions of the constituent particles of atoms and of molecules with each other and with externally applied fields. He established the existence of the magnetic moment of the electron, and precisely determinated its magnitude as part of an intensive program of postwar research with atomic and molecular beams. Later, he became interested in problems in chemical physics, and applied the molecular beam technique to their experimental study. Kusch spent most of his career at Columbia University, except for 1941 when he joined a team of researchers at the Westinghouse Laboratories in Bloomfield, Pennsylvania to work on the development of vacuum tubes, and for a later move to Bell Telephone Laboratories, where he worked on vacuum tubes and microwave generators.

In 1955, Kusch was jointly awarded the Nobel Prize for physics with Willis Lamb (1913-) for his accurate determination that the magnetic moment of the electron was greater than its theoretical value, thus leading to innovations in quantum electrodynamics. Lamb's share was for his discoveries concerning the fine structure of the hydrogen spectrum.

In his later years, Kusch became increasingly concerned with problems of education, especially that of educating the young to understand a civilization strongly affected by the knowledge of science and by the techniques that result from this knowledge. Kusch died on March 20, 1993.

1. Polycarp Kusch, BookRags, www.bookrags.com/biography/polycarp-kusch-wsd.
2. Polycarp Kusch, Wikipedia, http://en.wikipedia.org/wiki/Polykarp_Kusch.
3. Polycarp Kusch, Biography, NobelPrize.org, http://nobelprize.org/nobel_prizes/ physics/ laureates/1955/kusch-bio.

Chien-Shiung Wu (1912-1997)

Proved that parity is not conserved in nuclear beta decay

Chien-Shiung Wu was born on May 13, 1912 in Shanghai, China, but was raised in Liuhe, a city about 30 miles from Shanghai. Her father was a proponent of gender equality and founded Mingde Women's Vocational Continuing School. From 1930 to 1934 she studied in the Physics Department of National Central University. For two years after her graduation she did postgraduate study and worked as an assistant at Zhejiang University.

In 1936, Wu went to the United States, where she studied at the University of California, Berkeley, receiving her Ph.D. in 1940. Two years later she married Luke Chia-Liu Yuan, also a physicist and former classmate at Berkeley. The two then moved east. He worked on radar at Princeton and she taught at Smith College, Princeton University from 1942 to 1944 and Columbia University from 1944 to 1980.

At Columbia, Wu contributed to the Manhattan Project by developing a process to separate uranium isotopes by gaseous diffusion and by developing improved Geiger counters. After the war, she stayed at Columbia as a research scientist.

In 1956, Tsung-Dao Lee (1926-) and Chen Ning Yang (1922-pesent) proposed that parity was not conserved for weak interactions. Wu tested the proposal by observing the beta particles given off by cobalt-60. She observed that there was a preferred direction of emission, and therefore parity was indeed not conserved. However, she did not share Lee's and Yang's Nobel Prize—a fact widely blamed on sexism by the selection committee. She was given many other honors and awards.

Wu later conducted research into the molecular changes in the deformation of the hemoglobins that causes sickle-cell disease.

Wu was the first female instructor in the Physics Department of Princeton University, the first woman with a Princeton honorary doctorate, the first woman to receive the Research Corporation Award, the first woman to receive the Comstock Prize from the National Academy of Sciences, and the first female President of the American Physical Society.

Wu retired in 1981. She then lectured widely and encouraged the participation of young women in scientific careers. She died of a stroke on February 16, 1997.

1. Chien-Shiung Wu (1912 - 1997), Experimental Physcist, Featured Women in History,
2. Chien-Shiung Wu, Wikipedia, http://en.wikipedia.org/wiki/Chien-Shiung_Wu.
3. http://atdpweb.soe.berkeley.edu/quest/herstory/C-S_Wu.
4. National Women's Hall of Fame, Chien-Shiung Wu (1912-1997), www.greatwomen.org/women.php?action=viewone&id=174.

Edward Purcell (1912-1997)

Developed new methods for nuclear magnetic precision measurements

Edward Mills Purcell was born in Taylorville, Illinois on August 30, 1912. He received his B.E. in electrical engineering from Purdue University, followed by his M.A. and Ph.D. in physics from Harvard University.

After serving two years as instructor in physics at Harvard, he joined the Radiation Laboratory at Massachusetts Institute of Technology, which was organized in 1940 for military research and development of microwave radar. He became Head of the Fundamental Developments Group in the Radiation Laboratory, which was concerned with the exploration of new frequency bands and the development of new microwave techniques.

At the end of the war, Purcell returned to Harvard, and in 1945 he and his colleagues discovered nuclear magnetic resonance (NMR). NMR provides scientists with an elegant and precise way of determining chemical structure and properties of materials, and is widely used in physics and chemistry. It also is the basis of magnetic resonance imaging (MRI), one of the most important medical advances of the 20th century.

For their discovery of NMR, Purcell and Felix Bloch (1905-1983) of Stanford University shared the 1952 Nobel Prize in physics.

Purcell also made contributions to astronomy, being the first to detect radio emissions from neutral galactic hydrogen, the hyperfine transition in hydrogen. This afforded the first views of the spiral arms of the Milky Way. His observation helped launch the field of radio astronomy.

Purcell also made contributions to solid state physics, with studies of spin-echo relaxation, nuclear magnetic relaxation, and negative spin temperature, which was important in the development of the LASER.

As a consequence of a fall in 1996, Purcell died on March 7, 1997 of respiratory failure at his home in Cambridge, Massachusetts.

NMR revolutionized the field of chemistry and has become the most important spectroscopic technique in chemistry and biology. Chemists use NMR instruments to check the moisture content of food and check the quality of drugs and medicines. NMR also helps researchers probe the nature of ribonucleic acid (RNA) and deoxyribonucleic acid (DNA).

1.   E. M. Purcell, Biography, NobelPrize.org, http://nobelprize.org/nobel_prizes/physics/laureates/1952/purcell-bio.
2.   Edward Mills Purcell, Wikipedia, http://en.wikipedia.org/wiki/Edward_Mills_Purcell.
3.   Pound, Robert V., Edward Mills Purcell, Biographical Memoirs, National Academy of Sciences, http://books.nap.edu/html/biomems/epurcell.
4.   Edward M. Purcell, Microsoft Encarta Encyclopedia Online, http://encarta.msn.com/encyclopedia_761583316/purcell_edward_mills, 2007.

Willis Lamb (1913-)

Made discoveries concerning the fine structure of the hydrogen spectrum

Willis Eugene Lamb, Jr., the son of a telephone engineer, was born in Los Angeles, California on July 12, 1913. He received a B.S. in Chemistry from the University of California, Berkeley in 1934. Working under Robert Oppenheimer with theoretical work on scattering of neutrons by a crystal, he received his Ph.D. in physics in 1938.

Lamb joined the faculty of Columbia University in 1938. From 1943 to 1951 he was associated also with the Columbia Radiation Laboratory. In 1951 he went to Stanford University as Professor of Physics.

During 1953-1954 he was the Morris Loeb Lecturer at Harvard University. From 1956 to 1962 he was a Fellow of New College and Wykeham Professor of Physics at the University of Oxford, England.

In 1962 he became Henry Ford II Professor of Physics at Yale University, New Haven, Connecticut.

Over the years, Lamb's research involved the theory of the interactions of neutrons and matter, field theories of nuclear structure, theories of beta decay, range of fission fragments, fluctuations in cosmic ray showers, pair production, order-disorder problems, ejection of electrons by metastable atoms, quadrupole interactions in molecules, diamagnetic corrections for nuclear resonance experiments, theory and design of magneton oscillators, theory of a microwave spectroscope, study of the fine structure of hydrogen, deuterium and helium, and theory of electrodynamic energy level displacements.

In 1955, Lamb shared the Nobel Prize in Physics with Polykarp Kusch (1911-1993) for his discoveries concerning the fine structure of the hydrogen spectrum. Kusch's share was for his precision determination of the magnetic moment of the electron.

The phenomenon called the **Lamb shift** was named for Lamb. The Lamb shift is a small difference in energy between two energy levels of the hydrogen atom in quantum mechanics.

Lamb's work led physicists to rethink the basic concepts behind the application of quantum theory to electromagnetism. His work became one of the foundations of quantum electrodynamics, a key aspect of modern elementary particle physics.

1. Willis E. Lamb, Biography, NobelPrize.org, http://nobelprize.org/nobel_prizes/physics/laureates/1955/lamb-bio.
2. Willis E. Lamb, Jr., the Hydrogen Atom, and the Lamb Shift, OSTI, www.osti.gov/accomplishments/ lamb.
3. Willis Eugene Lamb, Jr., http://physics.rug.ac.be/Fysica/Geschiedenis/Mathematicians/Lamb.
4. Willis Lamb, Wikipedia, http://en.wikipedia.org/wiki/Willis_Lamb.

Wolfgang Paul (1913-1993)
Developed the ion trap technique

Wolfgang Paul was born in Lorenzkirch, Germany on August 10, 1913. He grew up in Munich, where his father was professor for pharmaceutic chemistry. In 1932, he enrolled in the Technical University of Munich, but changed to the Technical University of Berlin in 1934, obtaining his undergraduate degree in 1937. He then went to the University of Kiel, but his education was interrupted when he was drafted into the German airforce. He finished his Ph.D. in 1940 at the Technical University of Berlin. During World War II, Paul researched isotope separation, which is necessary to produce fissionable material for use in making nuclear weapons.

In 1952, Paul was appointed director of the physics institute at the University of Bonn, where he stayed for the remainder of his career. There, he experimented with electric and magnetic fields as a way to focus beams of atoms and to separate different particles. He found that a quadrupole electric field could be used to separate ions with different masses. This technique became the basis of an instrument eventually used in many laboratories to conduct analysis of elements by examining their individual spectrums.

Paul experimented further with both electric and magnetic fields, and developed a three-dimensional version of the mass filter, which became known as the **Paul trap**. The Paul trap suspends ions in a small area so that researchers can better study them.

Paul's other research included optical observation of the Lamb shift (a small shift in the energy levels of a hydrogen atom) and the creation of a superconducting storage ring for containing and studying various slow neutrons.

Paul is also credited with the development of high-energy physics in Germany through the construction of several particle accelerators in Bonn.

Paul and Hans Dehmelt (1913-1993) were awarded the 1989 Nobel Prize in physics for their development of a technique to isolate individual ions, making it possible to study them. They shared the Prize with Norman Ramsey (1915-), who invented a highly accurate technique for studying atomic-energy oscillations.

Paul died on December 7, 1993.

1. Wolfgang Paul, Autobiography, NobelPrize.org, http://nobelprize.org/nobel_prizes/physics/laureates/1989/paul-autobio.
2. Wolfgang Paul, Microsoft Encarta Encyclopedia Online, http://encarta.msn.com/encyclopedia_761583298/Wolfgang_Paul.
3. Wolfgang Paul, Wikipedia, http://en.wikipedia.org/wiki/Wolfgang_Paul.

Henry Primakoff (1914-1983)

Developed the theory of spin waves

Henry Primakoff was born in Odessa, Ukraine on February 12, 1914. His father was a physician and his mother was a pharmacist. The family immigrated to New York in 1922. Primakoff became a U.S. citizen in 1930. In 1931, Primakoff enrolled at Columbia University. After graduating, he went to graduate school at Princeton University. After a year he transferred to New York University (NYU), where he received his Ph.D. in 1938. He worked for a time at Brooklyn Polytechnic Institute and then at Queens College.

While still in graduate school, Primakoff, with T.D. Holstein, developed the theory of spin waves, which was based on a physical model and employed theoretical techniques. Spin waves are propagating disturbances in the ordering of magnetic materials.

After World War II, Primakoff accepted a joint physics and mathematics appointment at NYU. He subsequently joined the physics faculty of Washington University in St. Louis. Later, he published a paper on the photoproduction of neutral mesons in nuclear electric fields, which first described the process known as the **Primakoff Effect**, and ultimately led to a precise measurement of the very short mean life of the neutral pion (pi meson).

The Primakoff effect is the resonant production of neutral mesons by high-energy photons interacting with an atomic nucleus. It can be viewed as the reverse process of the decay of the meson into two photons. The Primakoff effect is thought to take place in stars, and could be a production mechanism of hypothetical particles, such as the axion. It was postulated in order to provide a natural solution to the strong CP problem. CP is the product of two symmetries: C for charge conjugation, which transforms a particle into its antiparticle, and P for parity, which creates the mirror image of a physical system.

In the 1950s, Primakoff turned to studying the nuclear and particle phenomena that manifest the weak interaction.

Primakoff moved to the University of Pennsylvania in 1960, where he became the leading world authority on muon capture, double beta decay, and the interaction of neutrinos and nuclei.

Primakoff died in Philadelphia, Pennsylvania on July 25, 1983 after a long battle with cancer.

1. APS Announces Primakoff Lectureship and Forum Award Endowment, American Physical Society, www.aps.org/publications/apsnews/199701/endowment.cfm.
2. Henry Primakoff, Wikipedia, http://en.wikipedia.org/wiki/Henry_Primakoff.
3. Primakoff effect, Wikipedia, http://en.wikipedia.org/wiki/Primakoff_effect.

Charles Townes (1915-)

Did fundamental work in the field of quantum electronics leading to the
development of the MASER and LASER

Charles Hard Townes was born in Greenville, South Carolina on July 28,
1915. He received his B.S. physics and a B.A. in modern languages from
Furman University in 1935, his M.A. in physics from Duke University in
1936, and his Ph.D from the California Institute of Technology in 1939.

After obtaining his Ph.D., Townes became a member of the technical
staff at Bell Telephone Laboratories, where he worked on the development
of a new radar system for aircraft in World War II.

After the war, Townes continued to work at Bell Labs, creating new
radar by experimenting with different radio wavelengths.

In 1948, Townes joined the physics faculty at Columbia University,
where he returned to experimental physics. The military, which was still
partially funding his research, wanted radar systems with wavelengths of
only a few millimeters. This led Townes and his colleagues to focus on
microwave research. In December 1953, Townes and his students
constructed a device that produced microwaves in a beam. They dubbed
the process Microwave Amplification by Stimulated Emission of
Radiation (MASER). The new device provided the basis for an atomic
clock. In the late 1950s, Townes and his associates improved upon the
MASER by creating solid-state models that could amplify ultra-weak
signals better than any other known means of amplification.

In 1958 Townes developed the concepts for the LASER (Light
Amplification by Stimulated Emission of Radiation), which delivers
infrared or visible light in a beam instead of microwaves. Two years later,
Theodore H. Maiman (1927-2007) built the first LASER.

Townes shared the 1964 Nobel Prize in physics with Nikolai Basov
(1922-2001) and Aleksandr Prokhorov (1916-2002) for their fundamental
work in the field of quantum electronics, which led to the construction of
oscillators and amplifiers based on the MASER-LASER principle.

Throughout the rest of his career, Townes's primary interest remained
quantum theory, but he also pursued research in radio and infrared
astronomy. In 1967, he went to the University of California, Berkeley,
where his program in radio and infrared astronomy led to the discovery of
ammonia and water molecules in the interstellar medium.

1. Charles H. Townes, Biography, NobelPrize.org, http://nobelprize.org/nobel_prizes/physics/
   laureates/ 1964/townes-bio.
2. Charles Hard Townes, Microsoft Encarta Encyclopedia Online, http://encarta.msn.com/
   encyclopedia_761583380/townes_charles_hard, 2007.
3. Charles Hard Townes, Wikipedia, http://en.wikipedia.org/wiki/Charles_Hard_Townes.

Clifford Shull (1915-2001)

Developed the neutron diffraction technique

Clifford Glenwood Shull was born in Pittsburgh, Pennsylvania on September 23, 1915. His father owned a hardware store and an associated home repair service. Shull graduated from Carnegie Institute of Technology in 1937. He then attended New York University, where he received his Ph.D. in 1941. He then worked in Beacon, New York with the research laboratory of The Texas Company for five years. His work there gave him the opportunity to learn about diffraction processes, crystallography, and the new field of solid state physics.

In 1946, Shull joined the staff at the Oak Ridge National Laboratory in Tennessee, and in 1995, he joined the faculty of the Massachusetts Institute of Technology.

Shull began studying neutron scattering in the 1940s using neutrons produced from nuclear reactors. At the time, scientists knew that beams of neutrons were scattered (diffracted) by simple crystals, such as sodium chloride. However, they did not yet realize the potential of these beams for providing information about the atomic structure of a substance. Because neutrons are uncharged, they pass readily into a substance, but when the neutrons strike atomic nuclei, they are scattered at various angles. By using this information, Shull studied the scattering characteristics of dozens of atomic nuclei. He found that different atoms produce different diffraction patterns. He also discovered that the results for an atom of a particular element are always the same, independent of the other elements to which the atom is bonded. For example, a sodium atom scatters neutrons in a certain way, whether it is found in sodium chloride or sodium bromide. In addition, the small magnetic properties of neutrons were helpful in studying the structure of magnetic materials.

Neutron scattering is now widely used in laboratories throughout the world to study and develop various materials, including polymers, semiconductors, and superconductors.

Shull shared the 1994 Nobel Prize in physics with Canadian physicist Bertram Brockhouse (1918-2003) for his development of the technique of neutron scattering, which is used to study the atomic structure and magnetic properties of various materials. Brockhouse's share was for his development of neutron spectroscopy. Shull died on March 31, 2001.

1. Clifford G. Shull, Autobiography, NobelPrize.org, http://nobelprize.org/nobel_prizes/ physics/ laureates/1994/shull-autobio.
2. Clifford G. Shull, Microsoft Encara Encyclopedia Online, http://encarta.msn.com/ encyclopedia_761583350/Shull_Clifford_G_, 2007.
3. Clifford Shull, Wikipedia, http://en.wikipedia.org/wiki/Bertram_Brockhouse.

H. Thomas Milhorn, MD, PhD

Norman Ramsey, Jr. (1915-)

Invented the separated oscillatory field method used to synchronize atomic clocks

Norman Foster Ramsey, Jr. was born in Washington, D.C. on August 27, 1915. His mother had been a mathematics instructor at the University of Kansas, and his father was an officer in the Army Ordnance Corps. Ramsey earned his B.A. in mathematics from Columbia University in 1935 and a second undergraduate degree in physics from Cambridge University. He received his Ph.D. in physics from Columbia University in 1940. His doctoral work was supervised by Isidor Rabi (1898-1988).

During the Second World War, Ramsey worked at the MIT Radiation Laboratory, where he headed a group developing radar at three centimeter wavelength. He then went to Washington, D.C. as a radar consultant to the Secretary of War. In 1943, Ramsey went to Los Alamos, New Mexico, to work on the Manhattan Project. After the war, Ramsey returned to Columbia University. In 1947, he joined the faculty of Harvard University, where he remained until his retirement in 1986.

Prior to Ramsey's work on atomic energy oscillations, these oscillations were measured by passing a beam of atoms through a constant magnetic field. Maintaining the magnetic field at a constant level over a large area was difficult. Ramsey used two separate magnetic fields, which vastly increased the accuracy. Called the separated oscillatory fields method, this technique used oscillations to synchronize the components of atomic clocks, providing them with remarkable accuracy.

The separated oscillatory fields method was used in the creation of the first atomic MASER, which is a focused beam of microwaves. One of the MASER's many applications was in space research, where it was used to track the space probe Voyager 2.

Ramsey was awarded the 1989 Nobel Prize in physics for the invention of the separated oscillatory field method, which had important applications in the construction of atomic clocks. He shared the prize with Hans Dehmelt (1922-) and Wolfgang Paul (1913-1993) for their development of the ion trap technique.

Ramsey's other research included an explanation of chemical shifts in nuclear-magnetic-resonance (NMR) experiments. This provided the basis for the magnetic-resonance-imaging (MRI) instruments used in medicine.

1. Norman F. Ramsey, Autobiography, NobelPrize.org, http://nobelprize.org/nobel_prizes /physics/laureates/1989/ramsey-autobio.
2. Norman Foster Ramsey, Jr., Wikipedia, ttp://en.wikipedia.org/wiki/Norman_Foster_Ramsey,_Jr.
3. Norman Foster Ramsey, Microsoft Encarta Encyclopedia Online, http://encarta.msn.com/ encyclopedia_761583319/Ramsey_Norman_Foster, 2007.

Robert Hofstadter (1915-1990)

Measured the size and shape of the proton and neutron

Robert Hofstadter was born in New York City on February 5, 1915. He graduated from the City College of New York in 1935, and then did his graduate work at Princeton University, where he earned his Ph.D. in 1938. His dissertation work was concerned with infrared spectra of simple organic molecules, and in particular with the partial elucidation of the structure of the hydrogen bond.

In 1938, Hofstadter was awarded a fellowship at Princeton University for postdoctoral work. There, he studied photoconductivity in willemite crystals. Willemite is a zinc silicate (Zn2SiO4) that occurs in different colors. In 1939, Hofstadter received a fellowship at the University of Pennsylvania, where he helped to construct a large Van de Graaff machine for nuclear research. During World War II, Hofstadter worked first at the National Bureau of Standards and later at the Norden Laboratory Corporation. At the end of the war, he became Assistant Professor of Physics at Princeton University, where he did research on crystal conduction counters, on the Compton effect, and on scintillation counters.

Hofstadter joined the faculty of Stanford University in 1950. After 1953, electron-scattering measurements became his principal interest. With students and colleagues he investigated the charge distribution in atomic nuclei and then the charge and magnetic moment distributions in the proton and neutron. He determined the size and surface thickness parameters of nuclei. He measured with great precision the size and shape of the proton and neutron. Both were found to consist of a positively charged, dense, point-like core surrounded by two intermingling layers of meson clouds.

Hofstadtler shared the 1961 Nobel Prize in physics with Rudolf Mössbauer (1929-) for his studies of electron scattering in atomic nuclei and for his discoveries concerning the structure of the nucleons. Mössbauer's share was for his researches concerning the resonance absorption of gamma radiation.

In later years, Hofstadter became interested in astrophysics and applied his knowledge of scintillators to the design of the Energetic Gamma Ray Experiment Telescope (EGRET). He retired from Stanford in 1985, and died on November 17, 1990.

1. Robert Hofstadter, Biography, NobelPrize.org, http://nobelprize.org/nobel_prizes/physics/ laureates/ 1961/hofstadter-bio.
2. Robert Hofstadter, Microsoft Encarta Encyclopedia Online, http://encarta.msn.com/ encyclopedia_761576518/robert_hofstadter, 2007.
3. Robert Hofstadter, Wikipedia, http://en.wikipedia.org/wiki/Robert_Hofstadter.

H. Thomas Milhorn, MD, PhD

Aleksandr Prokhorov (1916-2002)

Did fundamental work in the field of quantum electronics leading to the development of the MASER and LASER

Aleksandr Mikhailovich Prokhorov, the son of Russian immigrants, was born in Atherton, Queensland, Australia on July 11, 1916. His family had fled Russia in 1911 because of the Great October Revolution. They relocated back to the Soviet Union in 1923. Prokhorov graduated from Leningrad University with a B.S. in physics in 1939. Beginning in 1941, he served in the Russian army, was wounded twice, and discharged in 1944.

In 1948, Prokhorov obtained his Ph.D. in physics at the Soviet Academy of Sciences in Moscow. His dissertation was entitled *Coherent Radiation of Electrons in the Synchotron Accelerator.* Two years later, he joined the research staff at the Lebedev Institute of Physics at the Soviet Academy of Sciences, also in Moscow.

Prokhorov and Nikolai Basov (1922-2001), conducting research in quantum mechanics, deduced that manipulating quantum energies might permit them to amplify microwaves and light waves. They then developed the theoretical basis of Microwave Amplification by Stimulated Emission of Radiation (MASER). The MASER quickly found many applications for its ability to send strong microwaves in any direction, and resulted in improvements in radar. The MASER also provided the basis for an atomic clock.

Prokhorov later helped develop the Light Amplification by Stimulated Emission of Radiation (LASER), which delivers infrared or visible light instead of microwaves. Both the MASER and LASER can collect and amplify energy waves hundreds of times. They can also produce a beam with almost perfectly parallel light waves and little or no interference or static.

In 1964, Prokhorov, Basov, and Charles Townes (1915-) were awarded the Nobel Prize in physics for their pioneering work on MASERs and LASERs.

From 1973 to 2001, Prokhorov was a Chairman at the Moscow Institute of Physics and Technology. He died in Moscow on January 8, 2002.

1. Aleksandr Mikhailovich Prokhorov , Microsoft Encarta Encyclopedia Online, http://encarta.msn.com/ encyclopedia_761583315/prokhorov_aleksandr_mikhailovich, 2007.
2. Aleksandr Mikhailovich Prokhorov, Biography, NobelPrize.org, http://nobelprize.org /nobel_prizes/physics/laureates/1964/prokhorov-bio.
3. Aleksandr Mikhailovich Prokhorov, Wikipedia, http://en.wikipedia.org/wiki/ Aleksandr_Mikhailovich_Prokhorov.

Vitaly Ginzburg (1916-)

Contributed to the theory of superconductors

Vitaly Lazarevich Ginzburg, the son of Jewish parents, was born in Moscow, USSR on October 4, 1916. His father was an engineer engaged in purification of water and his mother was a physician. Ginzburg graduated from Moscow State University in 1938 with a degree in physics. He received his doctorate from the same institution in 1942.

Ginzburg began to work at the P. N. Lebedev Physical Institute in Moscow in 1940, and remained at the Institute for the duration of his career. In addition, in 1945 he also began teaching at Gorky State University.

Among Ginzburg's achievements are the **Ginzburg-Landau theory** (a mathematical theory used to model superconductivity), the theory of electromagnetic wave propagation in plasmas, and a theory of the origin of cosmic radiation. In the 1950s, he played a key role in the development of the Soviet hydrogen bomb.

By the early 1950s, scientists were aware of the phenomenon of superconductivity. At the time, however, theoretical explanations for superconductivity had not been developed. Ginzburg, with colleague Lev Landau (1908-1968), proposed key theories to explain the relation between electrons and the magnetic field inside superconductors.

Ginzburg and Landau noted that some superconductors repel a magnetic field, and they termed this class of superconductor Type I. Ultimately, Russian-born physicist Alexei Abrikosov (1928-), building on the Ginzburg-Landau theories, described more fully this second class of superconductor, now called Type II. Modern superconductors, including the ceramic varieties that function at higher temperatures, are Type II. Ongoing research in superconductivity is expected to find application in the design of generators and engines, in improved transmission of electrical power over long distances, and other uses.

Anthony Leggett (1938-), a British-born American physicist, made theoretical breakthroughs in describing a related phenomenon known as superfluidity

Ginsburg, Abrikosov, and Leggett shared the 2003 Nobel Prize in physics for their pioneering contributions to the theory of superconductors and superfluids.

1. Vitaly Ginzburg, Microsoft Encarta Encyclopedia Online, http://encarta.msn.com/ encyclopedia_701666357/Vitaly_Ginzburg, 2007.
2. Vitaly Ginzburg, Wikipedia, http://en.wikipedia.org/wiki/Vitaly_Ginzburg.
3. Vitaly L. Ginzburg, Autobiography, NobelPrize.org, http://nobelprize.org/nobel_prizes/ physics/ laureates/2003/ginzburg-autobio.

James Rainwater (1917-1986)

Showed that nuclear particles can vibrate and rotate so as to distort the shape of the nucleus

Leo James Rainwater was born in Council, Idaho on December 9, 1917. His parents operated a general store. His mother and he moved to Hanford, California after the death of his father in the great influenza epidemic of 1918. James received his bachelor's degree in physics from California Institute of Technology in 1939.

During World War II, Rainwater worked on the atomic bomb project, mainly doing pulsed neutron spectroscopy using the small Columbia University cyclotron. He received his Ph.D. at Columbia University in 1946. Afterward, he joined the physics faculty at Columbia.

In 1949, Rainwater began studying the nuclei of atoms. At that time, there were two basic ideas about how the nucleus behaved. Danish physicist Niels Bohr (1885-1962) had formulated the liquid-drop model for the nucleus in 1936. This model describes the nucleus as a drop of liquid, capable of changing shape. The shell model was developed by the German-American physicist Maria Goeppert-Mayer (1906-1972) and German physicist Hans Jensen (1907-1973). In this model, nucleons move in concentric orbits around the center of the nucleus. Both models had their shortcomings.

The liquid-drop model helped explain nuclear fission, but could not explain other important nuclear phenomena. The shell model also offered explanations for some things and failed to do so for others.

Rainwater solved the discrepancies produced by both models by thinking of the nucleus as oblong in some situations. His ideas were later tested and confirmed by the experiments of Aage Bohr (1922-) and Ben Mottelson (1926-).

Rainwater, Bohr, and Mottelson were awarded the Nobel Prize in physics in 1975 for their discovery of the connection between collective motion and particle motion in atomic nuclei and the development of the theory of the structure of the atomic nucleus based on this connection.

Rainwater also contributed to the scientific understanding of X-rays and participated in Atomic Energy Commission and naval research projects.

Rainwater died on May 31, 1986.

1. James Rainwater, Autobiography, NobelPrize.org, http://nobelprize.org/nobel_prizes/physics/ laureates/ 1975/rainwater-autobio.
2. James Rainwater, Microsoft Encarta Encyclopedia Online, http://encarta.msn.com/ encyclopedia_761583317/rainwater_leo_james, 2007.
3. James Rainwater, Wikipedia, http://en.wikipedia.org/wiki/James_Rainwater

Bertram Brockhouse (1918-2003)

Developed neutron spectroscopy

Bertram Neville Brockhouse was born in Lethbridge, Alberta, Canada on July 15, 1918. He grew up on a farm that had been homesteaded by his father in 1910. He served in the Royal Canadian Navy from 1939 to 1945. In 1944, he was enrolled in a six-month course in electrical engineering at Nova Scotia Technical College, and then assigned to the test facilities at the National Research Council in Ottawa as an electrical Sub-Lieutenant.

Brockhouse graduated with a B.A. in mathematics and physics from the University of British Columbia in 1947, an M.A. in physics from the University of Toronto in 1948, and a Ph.D. in physics from the same institution in 1950. His dissertation work involved solid state physics at both low and high temperatures.

From 1950 to 1962, Brockhouse carried out nuclear-reactor research at Canada's Chalk River Nuclear Laboratory. In 1962, he became professor at McMaster University in Canada, where he remained until his retirement in 1984.

Building on the work of American physicist Clifford Shull (1915-2001), Brockhouse designed a new instrument for neutron-scattering research. Neutron scattering is based on the fact that beams of neutrons can pass readily into a substance because they lack charge. The neutrons are then scattered (diffracted) by the atoms in the substance. Brockhouse's instrument, called a triple-axis spectrometer, could measure the energy and momentum of the neutrons as they entered and left the sample substance. This allowed him to gather data about vibrations and other movements of the atoms within the substance, ultimately providing information about the substance's physical properties. His instrument is still widely used for applications in biology, chemistry, materials science, and engineering. Brockhouse's work is also credited with helping to form the basis of modern solid-state physics (also known as condensed-matter physics).

Brockhouse shared the 1994 Nobel Prize in physics with Shull for his work in developing neutron scattering techniques for studying condensed matter. Shull's share was for the development of the neutron diffraction technique.

Brockhouse died on October 13, 2003.

1. Bertram Brockhouse, Wikipedia, http://en.wikipedia.org/wiki/Bertram_Brockhouse.
2. Bertram N. Brockhouse, Biography, NobelPrize.org, http://nobelprize.org/nobel_prizes/physics/laureates/1994/brockhouse-autobio.
3. Bertram N. Brockhouse, Curriculum Vitae, NobelPrize.org, http://nobelprize.org/nobel_prizes/physics/laureates/1994/brockhouse-cv.
4. Bertram N. Brockhouse, Microsoft Encarta Encyclopedia, http://encarta.msn.com/encyclopedia_761583128/Brockhouse_Bertram_N_, 2007.

Frederick Reines (1918-1998)

Detected the neutrino

Frederick Reines, the son of Jewish emigrants from Russia, was born in Paterson, New Jersey on March 16, 1918. His father owned a small country store. Reines was one of the two Nobel Laureates in physics to achieve the level of Eagle Scout. Robert Richardson (1937-) is the other one.

Reines attended Stevens Institute of Technology in Hoboken, New Jersey, where he earned his M.E. and M.S. degrees before receiving his Ph.D. from New York University in 1944. He then joined the Manhattan Project in Los Alamos, New Mexico, where he and a colleague, Clyde Cowan (1919-1974), developed a detection procedure for neutrinos.

In 1956, Reines and a team of researchers, colliding atomic particles in water, were the first to detect neutrinos. Neutrinos had been first proposed theoretically by Wolfgang Pauli (1900-1958) 20 years earlier to explain undetected energy that escaped when a neutron decayed into a proton and an electron.

Reines became the head of the physics department of Case Western Reserve University in 1959. There, he led a group that was the first to detect neutrinos created in the atmosphere by cosmic rays. Reines had a booming voice, and had been a singer since childhood. During this time, besides being chairman of the physics department, he sang in the Cleveland Orchestra Chorus. In 1966, Reines took most of his neutrino research team with him to California to become the founding dean of physical sciences at the University of California, Irvine.

In 1987, Supernova 1987A was observed exploding. When a supermassive star collapses and then explodes, it is thought that the resulting jets of neutrinos bombard the escaping masses to create the elements heavier than iron, up through uranium. Without these natural neutrino processes in exploding supermassive stars, these elements would not exist. Reines' work was instrumental in detecting these neutrinos.

Reines and American physicist Martin Perl (1927-) shared the 1995 Nobel Prize in physics for their experimental contributions to lepton physics. A lepton is a particle with spin 1/2 that does not experience the strong interaction. Reines' Prize was for detecting the neutrino. Reines died on August 26, 1998.

1.   Frederick Reines, Autobiography, NobelPrize.org, http://nobelprize.org/nobel_prizes/physics/ laureates/1995/reines-autobio.
2.   Frederick Reines, Microsoft Encarta Encyclopedia Onlilne, http://encarta.msn.com/ encyclopedia_761583320/Frederick_Reines, 2007.
3.   Frederick Reines, Wikipedia, http://en.wikipedia.org/wiki/Frederick_Reines.

Julian Schwinger (1918-1984)

Contributed to the development of quantum electrodynamics

Julian Seymour Schwinger was born in New York City on February 12, 1918. He entered City College of New York at the age of 14, but later transferred to Columbia University, where he received his B.S. in 1936. He stayed at Columbia and earned his Ph.D. in 1939 at the age of 21.

Upon graduation, Schwinger was a researcher at the University of California at Berkeley from 1939 to 1941. Then, from 1941 to 1943 he was a professor at Purdue University. During the war, from 1943 to 1945, he performed radar research at the Massachusetts Institute of Technology

At the end of the war, Schwinger joined the faculty of Harvard University, where he remained until 1972. During this time, he developed the concept of renormalization, which explained the Lamb shift in an electron's magnetic field. He also postulated in his study of particle physics that neutrinos should come in multiple varieties, associated with leptons like the electron and muon. This was later experimentally verified.

In 1975, Schwinger became a professor at the University of California, Los Angeles, where he remained until his retirement in 1988.

Earlier work by British physicist Paul Dirac (1902-1984) applied quantum mechanics to an analysis of the electromagnetic field. Dirac predicted that particles such as the electron would have an infinite quantity of energy, which led to other predictions that contradicted experimental observations. Schwinger reworked Dirac's mathematics so that those infinite quantities no longer appeared. This adjustment made Dirac's theory consistent with observation, and permitted physicists to predict the properties of particles and radiation. Sin-Itiro Tomonaga (1906-1979) and Richard Feynman (1918-1988) made similar adjustments in Dirac's theory, working independently of Schwinger and each other.

Schwinger, Tomonaga, and Feynman were awarded the 1965 Nobel Prize in physics for their work in quantum electrodynamics, with deep-ploughing consequences for the physics of elementary particles.

Schwinger also postulated the existence of two different neutrinos associated with the electron and the muon, and predicted the discovery that massive charged particles carry weak nuclear forces. After 1989, Schwinger took a keen interest in the research of low-energy nuclear fusion reactions. He died on July 16, 1994.

1. Julian Schwinge, Biography, NobelPrize.org, http://nobelprize.org/nobel_prizes/physics/ laureates/ 1965/schwinger-bio.
2. Julian Schwinger, Wikipedia, http://en.wikipedia.org/wiki/Julian_Schwinger.
3. Julian Seymour Schwinger, Microsoft Encarta Encyclopedia Online, http://encarta.msn.com/ encyclopedia_761583342/schwinger_julian_seymour, 2007.

H. Thomas Milhorn, MD, PhD

Kai Siegbahn (1918-2007)

Contributed to the development of high-resolution electron spectroscopy

Kai Manne Börje Siegbahn was born in Lund, Sweden on April 20, 1918. He was the son of Karl Siegbahn (1886-1978), who had received the 1924 Nobel Prize in physics. Kai attended the University of Stockholm, where he received his Ph.D. in 1944. He then taught at the Royal Institute of Technology and the University of Uppsala. Siegbahn was instrumental in developing high-resolution spectroscopy. Spectroscopy is the study of the electromagnetic radiation that a substance emits when exposed to a stimulus, such as radiation. The substance absorbs or emits some of the energy, thereby producing a spectrum that can be analyzed. Before the 1950s, electron spectroscopy had a number of limitations, so that when a substance was excited with radiation, the energy of the electrons it emitted could only be partially analyzed.

In 1954, Siegbahn developed a type of spectrometer that provided a more accurate analysis of the electron's energy, and therefore a more complete electromagnetic spectrum. In particular, he developed the method known as Electron Spectroscopy for Chemical Analysis (ESCA), now usually described as X-ray photoelectron spectroscopy. This tool enabled Siegbahn and his coworkers to study characteristics of electrons in almost all known chemical elements, and subsequently contributed to the development of new, high-tech materials, such as polymers and plastics. ESCA is now widely used for nondestructive assays of corrosion of metal surfaces, wear on prostheses, identification of contaminants on electrical circuits, and many other industrial applications.

Siegbahn shared the 1981 Nobel Prize in physics with Nicolaas Bloembergen (1920-) and Arthur Schawlow (1921-1999) for his work in spectroscopy. Bloembergen and Schawlow's share was for their contribution to the development of laser spectroscopy.

After brief interludes at the Nobel Institute of Physics and the Royal Institute of Technology in Stockholm, Siegbahn returned to Uppsala, where he spent the rest of his career. He officially retired in 1984, but continued to work in his laboratory until his death. He died of a heart attack on July 20, 2007 at his summer cabin in Angelholm in southern Sweden.

1. Kai Manne Borje Siegbahn Microsoft Encarta Encyclopedia Online, http://encarta.msn.com/ encyclopedia_761583351/Siegbahn_Kai_Manne_Borje, 2007.
2. Kai Siegbahn, Wikipedia, http://en.wikipedia.org/wiki/Kai_Siegbahn.
3. Kia M. Siegbahn, Curriculum Vitae, NobelPrize.org, http://nobelprize.org/nobel_prizes/ physics/laureates/1981/siegbahn-cv.
4. Maugh II, Thomas H, Kai Siegbahn, 89; Nobel-winning physicist invented electron spectroscopy for chemical analysism New York Times, August 8, 2007.

Richard Feynman (1918–1988)

Contributed to the development of quantum electrodynamics

Richard Phillips Feynman was born in New York City on May 11, 1918. He graduated with a bachelor's degree in physics from the Massachusetts Institute of Technology in 1939, and obtained a Ph.D. in physics from Princeton University in 1942.

During World War II, Feynman worked at what would become Los Alamos National Laboratory in central New Mexico. He was in charge of a group responsible for problems involving large-scale computations to predict the behavior of neutrons in atomic explosions.

After the war, Feynman moved to Cornell University, where he continued developing his own approach to quantum electrodynamics. In 1950, he moved to the California Institute of Technology—first as a visiting professor and then as professor of theoretical physics.

Feynman, Julian Schwinger (1918-1994), and Sin-Itiro Tomonaga (1906-1979) shared the 1965 Nobel Prize in physics for their work on quantum electrodynamics. Each of the three had independently developed methods for calculating the interaction between electrons, positrons, and photons.

In Feynman's space–time approach, he represented physical processes with collections of diagrams showing how particles moved from one point in space and time to another. Feynman had rules for calculating the probability associated with each diagram, and he added the probabilities of all the diagrams to give the probability of the physical process itself.

Feynman also made important contributions to the theory of quarks and superfluidity, a state of matter in which a substance flows with no resistance. He created a method of mapping out interactions between elementary particles that became a standard way of representing particle interactions, now known as **Feynman diagrams**. He also worked on the weak interaction, the strong force, and the composition of neutrons and protons.

In addition to his work in theoretical physics, Feynman is credited with pioneering the field of quantum computing and introducing the concept of nanotechnology. He was a member of the panel that investigated the Space Shuttle Challenger disaster. He died on February 15, 1988. He was said to have had a reputation as a notorious practical joker.

1. Richard Feynman, Microsoft Encarta Encyclopedia Online, http://encarta.msn.com/ encyclopedia_761570286/richard_feynman, 2007.
2. Richard Feynman, Wikipedia, http://en.wikipedia.org/wiki/Richard_Feynman.
3. Richard P. Feynman, Biography, NobelPrize.org, http://nobelprize.org/nobel_prizes/physics/ laureates/1965/feynman-bio.

Robert Pound (1919-)

Co-measured the gravitational red shift predicted by Einstein's theory of general relativity

Robert Vivian Pound was born in Ridgeway, Ontario, Canada on May 16, 1919. He later became a naturalized citizen of the United States. He is best known for measuring, with Glen Rebka, in 1960 the gravitational red shift predicted by Albert Einstein (1879-1955). The red shift is due to the fact that gravity affects a light wave's frequency and wavelength as it moves upward from the earth's surface. As light does so, it should shift to lower frequency due to gravitational pull, but the amount should be tiny in earth's modest gravity.

To measure this minuscule change, Pound and Rebka had to find a source of electromagnetic radiation whose frequency was known with enormous precision. That ingredient came in 1959 when Rudolf Mössbauer (1929-) at the Max Planck Institute in Heidelberg, Germany discovered that excited atomic nuclei can decay to the ground state by emitting a gamma ray. He found that if the nuclei are part of a high-quality crystal, then every gamma ray emerges with precisely the same energy. Another sample with nuclei in the ground state can then absorb the gamma rays, but only if there is no velocity difference between emitter and absorber. Any relative velocity will mean, because of the Doppler effect, that the absorber sees a gamma ray of the wrong frequency, and can't accept it.

Pound and Rebka placed an emitter at the top of a tower in the Jefferson Physical Laboratory at Harvard University and installed a detector 74 feet below. By measuring the detection rate as they moving the emitter up and down slightly, the researchers measured the velocity difference between source and detector that compensated for the gravitationally induced change of frequency.

They then reversed their experimental set-up to measure the frequency shift of gamma rays going up the tower. The difference between the up and down measurements was the gravitational effect. It matched Einstein's prediction to 10 percent accuracy. By 1964, Pound and Rebka had improved the agreement to within one percent.

The satellite-borne clocks of the GPS navigational system must be regularly corrected for changes induced by gravitational red shift. So relativity calculations keep every fighter jet on course.

1. Pound, Robert Vivian (1919-), Albert Weisstein's World of Biography, WolframResearch, http://scienceworld.wolfram.com/biography/Pound.
2. The Weight of Light, Physical Review Focus, http://focus.aps.org/story/v16/st1, July 12, 2005.

Wolfgang Panofsky (1919-2007)

Co-isolated the neutral pi meson

Wolfgang Kurt Hermann Panofsky was born on April 24, 1919. He was the son of renowned art historian, Erwin Panofsky. Wolfgang spent his early years in Hamburg, where his father taught.

After Hitler came to power and issued the first anti-Jewish decrees in 1934, the family left for Princeton, New Jersey, where Erwin Panofsky took a post at the Institute for Advanced Study.

Wolfgang enrolled in Princeton University at the age of 15, and receved his bachelor's degree in 1938. He obtained his Ph.D. from California Institute of Technology (Caltech) in 1942.

For a year Panofsky was director of Caltech's Office of Scientific Research and Development, but he was soon recruited to work at Los Alamos on the Manhattan Project, where he stayed until 1945.

From 1945 to 1951, Panofsky was on the faculty of the University of California, Berkeley

In 1951, Panofsky became Professor of Physics at Stanford University. He soon became director of the High Energy Physics Laboratory there. He managed to secure approval for the construction of a two-mile-long, linear accelerator, which was then the world's largest physics research project. It was named the Stanford Linear Accelerator Center (SLAC). Between 1961 and 1984, Panofsky was the director of the SLAC.

With Jack Steinberger (1925-), Panofsky was the first to isolate the neutral pi meson, one of the subatomic particles which had been predicted by theoretical scientists to account for the strong force which binds the nuclei of atoms.

Despite his own involvement with the Manhattan Project which developed the first atomic bomb, Panofsky, became in later life a fierce advocate of arms control. Panofsky was an adviser on arms control in the Kennedy and Johnson administrations who promoted stronger scientific ties to Russia and China as a deterrent to nuclear war.

Panofsky was 5ft-2in in height and enjoyed tinkering with old cars. At the SLAC he was called "Pief" as friendly nickname.

Panofsky died of a heart attack in Los Altos, California on September 24, 2007.

1. A Brief Biography of Wolfgang K. H. Panofsky, SLAC, www2.slac.stanford.edu/panofsky_fellow/ career.
2. Professor Wolfgang Panofsky, Telegraph, January 10, 2007.
3. Wolfgang K. H. Panofsky, Wikipedia, http://en.wikipedia.org/wiki/Wolfgang_K._H._Panofsky.

H. Thomas Milhorn, MD, PhD

Nicolaas Bloembergen (1920-)

Contributed to the development of LASER spectroscopy

Nicolaas Bloembergen, the son of a chemical engineer, was born in Dordrecht, Holland on March 11, 1920. He did his undergraduate work, and additionally earned the equivalent of a M.Sc degree at the University of Utrecht.

After the Nazi closed the University, Bloembergen spent the next two years hiding indoors.

In 1945, Bloembergen entered the Ph.D. physics program at Harvard University, where he was hired to work part-time as a graduate research assistant for Edward Purcell (1912-1997) at the MIT Radiation Laboratory. There, he worked on the first Nuclear Magnetic Resonance machine (NMR). His dissertation was entitled *Nuclear Magnetic Relaxation.*

Bloembergen did postoctoral work at the Kamerlingh Onnes Laboratorium at the University of Leiden (1947-1948). His work there involved the nuclear spin relaxation mechanism by conduction electrons in metals and by paramagnetic impurities in ionic crystals, the phenomenon of spin diffusion, and the large shifts induced by internal magnetic fields in paramagnetic crystals.

Bloembergen joined the Harvard faculty in 1949, and in 1958 he became a naturalized citizen of the United States.

Bloembergen is noted for his pioneering research in LASER spectroscopy, which is the study of the electromagnetic spectrum produced by a substance exposed to certain kinds of energy, such as radiation. The spectrum provides information about molecular energy levels, chemical bonds, and other features of the substance. He was especially interested in using LASERs to excite a substance, and then studying the relative amounts of energy the substance absorbs.

Although LASERs are intense beams of light waves, the traditional laws of optics do not apply. Bloembergen worked out new laws of optics for these situations, and used these laws to develop additional techniques for LASER spectroscopy. Applications for these techniques range from the analysis of biological substances to the study of combustion in jet engines.

Bloembergen shared the 1981 Nobel Prize in physics with Arthur Schawlow (1921-1999) and Kai Siegbahn (1918-2007) for their work in LASER spectroscopy.

1. Nicolaas Bloembergen, Wikipedia, http://en.wikipedia.org/wiki/Nicolaas_Bloembergen.
2. Nicolaas Bloembergen, Biography, NobelPrize.org,
    http://nobelprize.org/nobel_prizes/physics/laureates/1981/bloembergen-autobio.
3. Nicolaas Bloembergen, Microsoft Encarta Encyclopedia, Nicolaas Bloembergen, ,2007.

Owen Chamberlain (1920-2006)

Co-discovered the antiproton

Owen Chamberlain, the son of a prominent radiologist, was born in San Francisco, California on July 10, 1920. He studied physics at Dartmouth College, receiving his A.B. in 1941. He entered graduate school in physics at the University of California, Berkeley, but his studies were interrupted by the involvement of the United States in World War II.

During the War, from 1942 to 1946, Chamberlain served as a researcher on the Manhattan Project. Within the Manhattan Project, he worked under Emilio Segrè (1905-1989) in Berkeley, California and in Los Alamos, New Mexico, investigating nuclear cross sections for intermediate-energy neutrons and the spontaneous fission of heavy elements. His part in the Manhattan Project led him to study alpha particle decay, neutron diffraction, and high-energy nuclear reactions

In 1946, after the war, Chamberlain resumed his doctoral work at the University of Chicago under Enrico Fermi (1901-1954), receiving his Ph.D. in 1949. His dissertation research involved work on the diffraction of slow neutrons in liquids

In 1949, Chamberlain became a member of the physics department at the University of California, Berkeley, where he, Segrè, and other physicists investigated proton-proton scattering. In 1955, a series of proton scattering experiments led to the discovery of the antiproton.

For their collaborative discovery of the antiproton, Chamberlain and Segrè shared the 1959 Nobel Prize for physics.

For the next few years, Chamberlain and his colleagues studied the interactions of antiprotons with hydrogen, deuterium, and other elements, and used antiprotons to produce antineutrons.

In 1960, Chamberlain and others pioneered the development and use of polarized proton targets to study the spin dependence of a wide variety of high energy processes, including the scattering of pi mesons and protons on polarized protons, the determination of the parity of hyperons, and a test of time reversal symmetry in electron-proton scattering..

In 1985, Chamberlain was diagnosed with Parkinson's disease, and retired from teaching in 1989. He died of complications from the disease in Berkeley on February 28, 2006.,

1. Chamberlain, Owen, Microsoft Encarta Encyclopedia Online, http://au.encarta.msn.com/ encyclopedia_1481601713/chamberlain_owen2007.
2. Owen Chamberlain, Biography, NobelPrize.org, http://nobelprize.org/nobel_prizes/physics/ laureates/1959/chamberlain-bio.
3. Owen Chamberlain, Wikipedia, http://en.wikipedia.org/wiki/Owen_Chamberlain.

Yoichiro Nambu (1921-)

Proposed the color charge of quantum chromodynamics

Yoichiro Nambu was born on January 18, 1921. He received his B.S. in 1942 and his D.Sc. in 1952, both from Tokyo University. He then joined the physics faculty at Osaka City University.

After being a member at the Institute for Advanced Study in Princeton, he began a 50-year career at The University of Chicago, from which he retired in 1991. Nambu is best known for having proposed the color charge of quantum chromodynamics, which is a theory of the strong nuclear interactions among quarks. Quarks are regarded as fundamental constituents of matter. Quantum chromodynamics seeks to explain why quarks combine in certain configurations to form the observed patterns of subnuclear particles, such as the proton and pi meson.

Nambu also did early work on spontaneous symmetry breaking in particle physics. This is a situation in which the solution of a set of physical equations fails to exhibit a symmetry possessed by the equations themselves. The massless bosons arising in field theory with spontaneous symmetry breaking are sometimes referred to as **Nambu-Goldstone bosons**.

Nambu is considered one of the founders of string theory, having discovered that the dual resonance model could be explained as a quantum mechanical theory of strings.

String theory in particle physics treats elementary particles as infinitesimal, one-dimensional string-like objects, rather than dimensionless points in space-time. Different vibrations of the strings correspond to different particles.

Introduced in the early 1970s in attempts to describe the strong force, string theory became popular in the 1980s when it was thought that it might provide a fully self-consistent quantum field theory that could describe gravitation as well as the weak, strong, and electromagnetic forces.

Nambu's contributions to the quark model in the sixties and, later, his geometrical formulation of the dual resonance models as the dynamics of a relativistic string are said to be of fundamental importance.

The **Nambu-Goto action** in string theory is named after Nambu and Tetsuo Goto (1950-). It is the starting point of the analysis of string behavior, using the principles of Lagrangian mechanics.

1. Quantum Chromodynamics, Answers.com, www.answers.com/topic/quantum-chromodynamics?cat=technology.
2. String Theory, Wikipedia, http://en.wikipedia.org/wiki/String_theory.
3. Yoichiro Nambu, Answers.com, www.answers.com/topic/yoichiro-nambu.

Arthur Schawlow (1921-1999)
Contributed to the development of LASER spectroscopy

Arthur Leonard Schawlow was born in Mount Verner, New York on May 5, 1921. When he was three years old, the family moved to Toronto, Canada. He did his undergraduate work at the University of Toronto, and then enrolled in the graduate program in physics.

Schawlow's studies were interrupted by World War II, and he taught classes to armed service personnel at the University of Toronto until 1944. Then he worked on microwave antenna development at a radar factory. He completed his Ph.D. in 1949.

Afterward receiving his Ph.D, he did postdoctoral work with Charles Townes (1915-) in the physics department at Columbia University.

Townes and Schawlow demonstrated theoretically the feasibility of using the principle of Microwave Amplification by Stimulated Emission of Radiation (MASER) to develop a light-amplifying device—the Light Amplification by Stimulated Emission of Radiation (LASER).

In 1951, Scawlow accepted a position at Bell Telephone Laboratories, where he worked on superconductivity, with some studies of nuclear quadrupole resonance.

Schawlow left Bell Labs in 1961 to join the faculty at Stanford University, where he was chairman of the department of physics from 1966 to 1970. He remained at Stanford until he retired in 1996.

In the early 1970s, Schawlow began working on LASER spectroscopy for investigating the energy levels of electrons in atoms and molecules. He used the new generation of organic dye LASERs to make very accurate measurements of the energy levels of the electron in the hydrogen atom, establishing a very precise value of the Rydberg constant.

Schawlow and Nicolaas Bloembergen (1920-) shared the 1981 Nobel Prize in physics with and Kai Siegbahn (1918-2007) for their contribution to the development of laser spectroscopy. Siegbahn's part of the award was for his contribution to the development of high-resolution electron spectroscopy.

Schawlow also pursued investigations in the areas of superconductivity and nuclear resonance.

Schawlow died of leukemia in Palo Alto, California on April 28, 1999.

1. Arthur L. Schawlow, Biography, NobelPrize.org, http://nobelprize.org/nobel_prizes/ physics/laureates/1981/schawlow-autobio.
2. Robert Schawlow, Wikipedia, http://en.wikipedia.org/wiki/Arthur_Leonard_Schawlow.
3. Schawlow, Arthur Leonard, Microsoft Encarta Encyclopedia Online, http://uk.encarta.msn.com/ encyclopedia_781533114/Schawlow_Arthur_Leonard, 2007.

Jack Steinberger (1921-)

Demonstrated the doublet structure of leptons through the discovery of the muon neutrino

Jack Steinberger was born in the city of Bad Kissingen in Bavaria, Germany on May 25, 1921. His family left Germany when he was 13 years old due to the increasing anti-Semitism and the rise of the Nazi party. They moved to Chicago, Illinois, where his father acquired a small delicatessen store, which was the basis of a very marginal income.

Steinberger obtained his undergraduate degree in chemistry at the University of Chicago in 1942. Then, as a member of the United States Army he was assigned to the MIT Radiation Laboratory, which inspired his interest in physics. He returned to the University of Chicago for graduate work, and earned his Ph.D. in 1948.

Steinberger conducted research at the Institute for Advanced Study in Princeton, New Jersey, at the University of California at Berkeley, and at Columbia University. In 1968, he joined the staff of the European Organization for Nuclear Research (CERN) in Switzerland, where he spent the remainder of his career.

In the early 1960s, Steinberger, Melvin Schwartz (1932-2006), and Leon Lederman (1922-) devised a way to capture neutrinos. Using the powerful particle accelerator at the Brookhaven National Laboratory in New York, they created a beam of high-energy neutrinos. With a specialized detector, Steinberger and his colleagues were able to study the neutrinos, and in doing so discovered the muon neutrino, thus proving that neutrinos exist in more than one variety. Neutrinos are elementary particles that travel close to the speed of light, lack an electric charge, are able to pass through ordinary matter almost undisturbed, and are thus extremely difficult to detect. They are now believed to have a minuscule, but non-zero mass.

Steinberger, Schwartz, and Lederman were awarded the 1988 Nobel Prize in physics for the neutrino beam method and the demonstration of the doublet structure of the leptons through the discovery of the muon neutrino. A lepton is a particle with spin 1/2 that does not experience the strong interaction Steinberger gave his Nobel medal to New Trier High School of which he is an alumnus.

1. Jack Steinberger, Autobiography, NobelPrize.org, http://nobelprize.org/nobel_prizes/physics/laureates/1988/steinberger-autobio.
2. Jack Steinberger, Microsoft Encarta Encyclopedia Online, http://encarta.msn.com/encyclopedia_761583363/Jack_Steinberger, 2007.
3. Jack Steinberger, Wikipedia, http://en.wikipedia.org/wiki/Jack_Steinberger.

Aage Bohr (1922-)

Showed that nuclear particles can vibrate and rotate so as to distort the shape of the nucleus

Aage Niels Bohr, the son of Nobel Prize winning physicist Niels Bohr, was born in Copenhagen, Denmark on June 19, 1922. He began studying physics at the University of Copenhagen in 1940.

In 1943, his father had to flee Denmark to avoid arrest by the Nazis, and the whole family managed to escape to Sweden, and then to England. Bohr assisted his father on the atomic bomb project at Los Alamos, New Mexico during World War II. They were members of the British team. His official position was that of a junior scientific officer.

Returning to the University of Denmark, Bohr obtained his master's degree in 1946. He then became an associate at the Institute of Theoretical Physics at the University of Copenhagen, where he devoted his attention to the inner structure of the atom.

Bohr received his doctorate at the University of Copenhagen in 1954. His dissertation dealt with a collective motion theory of the atomic nucleus that he had developed with Ben Mottelson (1926-) at the suggestion of the James Rainwater (1917-1986). The theory helped to explain many nuclear properties by showing that nuclear particles can vibrate and rotate so as to distort the shape of the nucleus from the expected spherical symmetry into an ellipsoid.

Bohr worked with Mottelson and Rainwater to summarize the current knowledge of nuclear structure at the time in a monograph. The first volume *Single-Particle Motion* appeared in 1969, and the second volume, *Nuclear Deformations,* in 1975.

Bohr, along with Mottleson and Rainwater, was awarded the 1975 Nobel Prize in physics for discovering the connection between collective motion and particle motion in atomic nuclei and the development of the theory of the structure of the atomic nucleus based on this connection.

Bohr served as the director of the Institute for Theoretical Physics (now named the Niels Bohr Institute) from 1963 to 1970. He then resigned to devote more time to research.

In 1975, Bohr became director of the Nordic Institute of Theoretical Atomic Physics, which shares research and facilities with the Niels Bohr Institute.

1. Aage N. Bohr, Autobiography, NobelPrize.org, http://nobelprize.org/nobel_prizes/physics/ laureates/ 1975/bohr-autobio.
2. Aage N. Bohr, Wikipedia, http://en.wikipedia.org/wiki/Aage_Niels_Bohr.
3. Aage Niels Bohr, Microsoft Encarta Encyclopedia Onine, http://encarta.msn.com/ encyclopedia_761579172/bohr_aage_niels, 2007.

Chen-Ning Yang (1922-)

Investigated parity laws which led to important discoveries regarding the elementary particles

Chen-Ning Franklin Yang was born in Hefei, Anhui, China on October 1, 1922. His father was a Professor of Mathematics at Tsinghua University. Yang received his B.S. from National Southwestern Associated University in Kunming in 1942. Two years later, he received his M.S. at Tsinghua University, also in Kunming.

In 1945, Yang attended the University of Chicago, where he studied with Edward Teller (1908-2003), and received his Ph.D. in 1948. He taught physics at the University of Chicago from 1948 to 1949, and assisted Enrico Fermi (1901-1954) in his research.

In 1949, Yang moved to the Institute for Advanced Study at Princeton. He was made a permanent member of the institute in 1952 and full professor in 1955. In 1965, he became the Albert Einstein Professor of Physics of the State University of New York at Stony Brook.

Yang had an interest in two fields—statistical mechanics and symmetry principles. His B.S. thesis, *Group Theory and Molecular Spectra*; his M.S. thesis, *Contributions to the Statistical Theory of Order-Disorder Transformations*; and his Ph.D. thesis, *On the Angular Distribution in Nuclear Reactions and Coincidence Measurement*, were instrumental in introducing him to these fields.

With his associate, Tsung-Dao Lee (1926-), Yang proved experimentally that one of the basic quantum-mechanics laws, called the conservation of parity, is violated in the so-called weak nuclear reactions—those nuclear processes that result in the emission of beta or alpha particles.

In 1957, Yang and Tsung-Dao Lee received the Nobel Prize in physics for their theory that weak force interactions between elementary particles did not have parity. Lee and Yang were the first Chinese Nobel Prize winners.

Yang is also well known for his collaboration with Robert Mills (1927-1999) in developing a gauge theory of a new class widely known as the Yang-Mills theory. Such "Yang-Mills theories" are a fundamental part of the Standard Model of particle physics. Yang retired from Stony Brook in 1999, and returned to Tsinghua University.

1. Chen Ning Yang, Biography, NobelPrize.org, http://nobelprize.org/nobel_prizes/physics/laureates/1957/yang-bio.
2. Chen Ning Yang, Microsoft Encarta Encyclopedia Online, http://encarta.msn.com/encyclopedia_761557309/yang_chen_ning, 2007.
3. Chen Ning Yang, Wikipedia, http://en.wikipedia.org/wiki/Chen_Ning_Yang.

Hans Dehmelt (1922-)

Developed the ion trap technique

Hans Georg Dehmelt was born in Görlitz, Germany on September 9, 1922. After high school, in 1940, he volunteered for service in the German army, which in 1943 ordered him to attend the University of Breslau to study physics. After a year of study he returned to army service, and was captured during the Battle of the Bulge.

After release from an American prisoner of war camp in 1946, Dehmelt returned to his study of physics at the University of Göttingen. He completed his master's degree in 1948 and his Ph.D. in 1950. He then served as a research fellow at Hans Kopfermann's Institute in Göttingen.

Beginning in 1952, Dehmelt did postdoctoral work at Duke University. He then joined the faculty at the University of Washington in Seattle, Washington in 1955.

Although the electron's mass and charge could be measured precisely with existing methods, its spin and magnetism could not. Building on the work of Wolfgang Paul (1913-1993), Dehmelt experimented with a three-dimensional electric field, the Paul trap, to suspend ions in a small area. By adding a strong magnetic field to the device in 1971 he was able to drive electrons out of the trap until only a single one remained. He used this device to study the magnetic properties and spin states of electrons. He also developed a technique to cool the particles being studied to slow their movement, thus improving accuracy of measurement. His measurements were so precise they provided verification of fundamental theories in quantum theory.

In 1976, Dehmelt and his coworkers used their trap to observe the quantum jump of a single ion, again confirming a prediction of quantum mechanics.

Dehmelt continued to perfect his techniques for studying atomic particles. He improved the accuracy of his magnetic measurements in electrons, achieving an accuracy of just a few parts in a trillion. In 1980 he successfully isolated, cooled, and photographed a single ion.

For co-developing the ion trap technique, Dehmelt and Paul received the Nobel Prize in physics in 1989. They shared it with Norman Ramsy (1915-) for his invention of the separated oscillatory fields method and its use in the hydrogen MASER and other atomic clocks.

1. Hans Dehmelt Biography, BookRags, www.bookrags.com/biography/hans-dehmelt-woi.
2. Hans G. Dehmelt, Autobiography, NobelPrize.org, http://nobelprize.org/nobel_prizes/physics/laureates/1989/dehmelt-autobio.
3. Hans Georg Dehmelt, Microsoft Encarta Encyclopedia Online, http://encarta.msn.com/encyclopedia_761583157/Dehmelt_Hans_Georg, 2007.
4. Hans Georg Dehmelt, Wikipedia, http://en.wikipedia.org/wiki/Hans_Georg_Dehmelt.

Leon Lederman (1922-)

Demonstrated the doublet structure of leptons through the discovery of the muon neutrino

Leon Max Lederman was born in New York to a family of Jewish immigrants from Russia on July 15, 1922. His father operated a hand laundry. Leon received his bachelor's degree in chemistry from the City College of New York in 1943, and then spent three years in the U.S. Army, where he rose to the rank of Second Lieutenant in the Signal Corps. Lederman received his Ph.D. from Columbia University in 1951. He then joined the Columbia faculty and directed its Nevis Laboratory from 1961 to 1978. In 1956, he discovered the long-lived, neutral K-meson particle,

In the early 1960s Lederman, Melvin Schwartz (1932-2006), and Jack Steinberger (1921-) devised a way to capture neutrinos. Using the powerful particle accelerator at the Brookhaven National Laboratory in New York, they created a beam of high-energy neutrinos. With a specialized detector, Lederman and his colleagues were able to study the neutrinos, and in doing so discovered the muon neutrino, thus proving that neutrinos exist in more than one variety. Neutrinos are elementary particles that travel close to the speed of light, lack an electric charge, are able to pass through ordinary matter almost undisturbed and are thus extremely difficult to detect. They are now believed to have a minuscule, but non-zero mass.

In 1977, Lederman found evidence for yet another elementary particle, the bottom quark.

Lederman took an extended leave of absence from Columbia in 1979 to become the director of Fermi National Accelerator Laboratory (Fermilab) in Batavia, Illinois.

Lederman, Schwartz, and Steinberger, were awarded the 1988 Nobel Prize in physics for the neutrino beam method and the demonstration of the doublet structure of the leptons through the discovery of the muon neutrino. A **lepton** is a particle with spin 1/2 that does not experience the strong interaction.

Lederman resigned from Columbia and Fermilab in 1989 and taught briefly at the University of Chicago before moving to the Illinois Institute of Technology.

1. Leon M. Lederman, Autobiography, NobelPrize.org, http://nobelprize.org/ nobel_prizes/physics/ laureates/1988/lederman-autobio.
2. Leon M. Lederman, Wikipedia, http://en.wikipedia.org/wiki/Leon_M._Lederman.
3. Leon Max Lederman, Microsoft Encarta Encyclopedia Online, http://encarta.msn.com/ encyclopedia_761583237/Lederman_Leon_Max, 2007.

Nikolai Basov (1922-2001)

Did fundamental work in the field of quantum electronics leading to the
development of the MASER and LASER

Nikolai Gennadiyevich Basov was born in the small town of Usman, outside the city of Voronezh, Russia on December 14, 1922. His father was a professor of the Voronezh Forest Institute.

After finishing secondary school in 1941, Basov was called up for military service and directed to the Kuibyshev Military Medical Academy. In 1943 he left the Academy with the qualification of a military doctor's assistant. He served in the Soviet Army and took part in the Second World War in the area of the First Ukranian Front.

In 1945, on discharge from the army, Basov entered the Moscow Institute of Physical Engineers where he studied theoretical and experimental physics. Then he received his D.Sc. degree in 1956 from Lebedev Physical Institute. His dissertation research involved working on the creation of a molecular oscillator utilizing an ammonia beam.

Basov, together with his teacher Aleksandr Prokhorov (1916-2002), conducted research in quantum mechanics, deducing that quantum mechanics permits the amplification of microwaves and light waves by inducing atoms to release energy. This helped them construct the theoretical basis of Microwave Amplification by Stimulated Emission of Radiation (MASER). The MASER quickly found many applications for its ability to send strong microwaves in any direction and resulted in improvements in radar. It also provided the basis for an atomic clock.

In 1953, Basov became a researcher at the Lebedev Institute. He later helped develop the Light Amplification by Stimulated Emission of Radiation (LASER)), which delivers infrared or visible light instead of microwaves. Both the MASER and the LASER can collect and amplify energy waves hundreds of times. They can also produce a beam with almost perfectly parallel light waves with little or no interference or static.

Basov, Prokhorov, and Charles Townes (1915-) received the 1964 Nobel Prize in physics for their fundamental work in the field of quantum electronics that led to the development of LASER and MASER.

Basov became the Director of the Lebedev Physical Institute in 1973, and remained there until 1988. He died in Moscow, Russia on July 1, 2001.

1. Nicolay G. Basov, Biography, NobelPrize.org, http://nobelprize.org/nobel_prizes/physics/laureates/ 1964/basov-bio.
2. Nikolai Gennadiyevich Basov, Microsoft Encarta Encyclopedia Online, http://encarta.msn.com/encyclopedia_761583102/basov_nikolay_gennadiyevich, 2007.
3. Nikolay Basov, Wikipedia, http://en.wikipedia.org/wiki/Nikolay_Basov.

Calvin Quate (1923-)

Developed the scanning probe microscope

Calvin F. Quate was born in Baker, Nevada on December 7, 1923. He earned his bachelor's degree in electrical engineering from the University of Utah in 1944 and his Ph.D. from Stanford University in 1950.

In 1949, Quate joined the technical staff at Bell Telephone Laboratories at Murray Hill, New Jersey, and soon became Department Head, and then Associate Director of Electronics Research.

Quate transferred to Sandia Corporation in Albuquerque, New Mexico in 1959. In 1960, he was appointed Vice President and Director of Research. In 1961, he joined the electrical engineering faculty at Stanford.

Quate's early work led to an understanding of noise space charge waves on electron beams, practical systems of periodic focusing of electron beams, coupled helixes in traveling wave tubes, and the design of practical microwave amplifiers and oscillators.

Quate worked on acoustic amplifiers, interaction of acoustics with semiconductors, and acoustic correlators. He initiated activity in acoustics that led to the scanning acoustic microscope, which had a resolution greater than that of optical microscopes. His later work was on electron tunneling microscopy, which reveals the atomic scale structure of crystalline substances.

A Scanning Acoustic Microscope uses focused sound to investigate, measure, or image an object. It is commonly used in failure analysis and non-destructive evaluation. The semiconductor industry has found the Scanning Acoustic Microscope useful in detecting voids, cracks, and delaminations within microelectronic packages. It also has applications in biological and medical research.

Quate also worked on the atomic force microscope (scanning force microscope), which is a very high-resolution type of scanning probe microscope, with a resolution of fractions of a nanometer, more than 1000 times better than the optical diffraction limit.

In 1992, Quate was awarded The President's National Medal of Science for his contributions to microscopy, particularly the scanning acoustic microscope and the atomic force microscope. The award was presented by President Bush 41 at a White House ceremony.

1. Atomic Force Microscope, Wikipedia, http://en.wikipedia.org/wiki/Atomic_force_microscope.
2. Calvin F. Quate, IEEE, www.ieee.org/web/aboutus/history_center/biography/quate.
3. Scanning Acoustic Microscope, Wikipedia, http://en.wikipedia.org/wiki/Scanning_acoustic_microscope.

Jack Kilby (1923-2005)

Did basic work on information and communication technology

Jack St. Clair Kilby was born in Great Bend, Kansas. His father ran a small electric company. Kilby received his B.S. in electrical engineering from the University of Illinois at Urbana-Champaign in 1947, and in 1950, while simultaneously working at Centralab and taking evening classes, he earned an M.S. in electrical engineering from the University of Wisconsin-Milwaukee.

Kilby was hired by Texas Instruments, and in the summer of 1958 came to the conclusion that manufacturing circuit components in a single piece of semiconductor material was possible. He succeeded in etching transistors, resistors, and other electrical components on a single wafer of germanium less than half the size of a paper clip. He had painted the shapes of the desired components on the germanium with wax, and then used acid to eat away the extraneous, unprotected germanium.

That fall, Kilby demonstrated to the management of Texas Instruments the first integrated circuit—a piece of germanium connected to an oscilloscope. He threw a switch, and the oscilloscope screen showed a continuous sine wave, proving that his integrated circuit worked. A patent for a Solid Circuit made of Germanium, the first integrated circuit, was filed on February 6, 1959.

Leaving Texas Instruments in 1970, Kilby spent several years as an independent inventor.

Kilby later returned to Texas Instruments before officially retiring in 1983, although he continued working as a consultant.

From 1978 to 1984 Kilby taught electrical engineering at Texas A&M University.

Kilby shared the 2000 Nobel Prize in physics with Zhores Alferov (1930-) and Herbert Kroemer (1928-) for his part in the invention of the integrated circuit. Alferov and Kroemer's part was for developing semiconductor heterostructures used in high-speed- and opto-electronics.

In addition to the integrated circuit, Kilby patented the electronic portable calculator and the thermal printer used in data terminals. In total, he held about 60 patents.

Kilby died in Dallas, Texas on June 20, 2005 after a brief bout with cancer.

1. Jack Kilby, Wikipedia, http://en.wikipedia.org/wiki/Jack_Kilby.
2. Jack S. Kilby, Autobiography, NobelPrize.org, http://nobelprize.org/nobel_prizes/physics/laureates/2000/kilby-autobio.
3. Jack S. Kilby, Microsoft Encarta Encyclopedia Online, http://encarta.msn.com/encyclopedia_701502049/Kilby_Jack_S_, 2007.

James Zimmerman (1923-1999)

Co-invented the radio-frequency superconducting quantum interference
device (SQUID)

James Edward Zimmerman was born in Lantry, South Dakota on
February 19, 1923. He grew up on a ranch. In high school, he became
fascinated with radio and spent his spare time building receivers and
transmitters from scavenged parts.

After earning an electrical engineering degree from the South Dakota
School of Mines and Technology in 1943, Zimmerman joined the
Westinghouse Research Laboratory in Pittsburgh, Pennsylvania to work on
microwave radar as part of the war effort.

Zimmerman eventually took a job in Sydney, Australia, where he
worked on the Australian radar program.

After the war, Zimmerman returned to Pittsburgh, where in 1951 he
earned a Ph.D. in physics from the Carnegie Institute of Technology. His
dissertation concerned the thermal properties of materials at low
temperatures.

In 1953, Zimmerman joined the Smithsonian Institution and spent two
years measuring the solar constant at observatories in California and Chile.
Then, in 1955 he joined the low-temperature group at Ford Motor
Company's Scientific Laboratory in Dearborn, Michigan, where he
measured the thermal, electrical, and magnetic properties of metals and
alloys at low temperatures.

In the early 1960s, Zimmerman became involved in experiments on
quantum interference in superconductors. Before the end of 1964 he
demonstrated a two-junction interferometer consisting simply of two U-
shaped bits of niobium wire pressed together. Later, he refined this
approach by developing adjustable point-contact junctions that became the
staple of early interferometers.

In 1965, Zimmerman and Arnold Silver realized that quantum
interference could be observed in a superconducting loop with a single
junction if it was excited by a radio-frequency bias. The single-junction
circuit was an immediate success, yielding a fully practical
magnetometer/amplifier with exquisite sensitivity, limited only by the
uncertainty principle. Zimmerman suggested SQUID as an acronym for
superconducting quantum interference device, and the term stuck.

Zimmerman died on August 4, 1999 in Boulder, Colorado after
battling cancer for three years.

1. Kautz, Richard L. and Donald B. Sullivan, James Edward Zimmerman, Obituary, Physics Today
    Online, www.physicstoday.org/vol-53/iss-7/p70, 2000.
2. SQUID, Wikipedia, http://en.wikipedia.org/wiki/SQUID.

Philip Anderson (1923-)

Did theoretical investigations of the electronic structure of magnetic and disordered systems

Philip Warren Anderson was born in Indianapolis, Indiana on December 13, 1923. He grew up in Urbana, Illinois where his father was professor of plant pathology at the University of Illinois. Anderson earned his bachelor's, master's, and doctoral degrees from Harvard University over the years 1939 to 1949, except for the years 1943 to 1945, when he interrupted his education to work as a radio engineer in the United States Navy during World War II. In 1949, Anderson went to work at Bell Telephone Laboratories in New Jersey, where he investigated a wide variety of problems in condensed matter physics, which deals with the macroscopic physical properties of matter. During this period he discovered the concept of localization, the idea that extended states can be localized by the presence of disorder in a system; the **Anderson Hamiltonian**, which describes electrons in a transition metal; the Higgs mechanism for generating mass in elementary particles; and the pseudospin approach to the BCS theory of superconductivity.

Anderson, working with amorphous materials like amorphous silicon, demonstrated in 1958 that it is possible for an electron to get trapped in a small area. This phenomenon, known as **Anderson localization**, suggests that amorphous materials can be used in place of the crystalline semiconductors that are in use today. Anderson's discoveries led to the development of electronic switching and memory devices made from amorphous materials, such as glass.

From 1967 to 1975, Anderson was a visiting professor of theoretical physics at the University of Cambridge in England, where he worked with Nevill Mott (1905-1996) on electronic structure of magnetic and disordered systems.

In 1977, Anderson, Mott, and American physicist John van Vleck (1899-1980) were awarded the Nobel Prize in physics for their investigations into the electronic structure of magnetic and disordered systems, which allowed for the development of electronic switching and memory devices in computers.

Anderson retired from Bell Labs in 1984, and joined the faculty at Princeton University.

1. Philip W. Anderson, Autobiography, NobelPrize.org, http://nobelprize.org/nobel_prizes/physics/laureates/1977/anderson-autobio.
2. Philip W. Anderson, Microsoft Encarta Encyclopedia Online , http://encarta.msn.com/encyclopedia_761583097/anderson_philip_warren , 2007.
3. Philip Warren Anderson, Wikipedia, http://en.wikipedia.org/wiki/Philip_Warren_Anderson.

Val Fitch (1923-)

Discovered violations of fundamental symmetry principles in the decay of neutral K-mesons

Val Logsdon Fitch was born on a cattle ranch in Cherry County, Nebraska, not far from the South Dakota border, on March 10, 1923. The ranch was close to the Sioux reservation. His father spoke their language, and him an honorary chief.

Not long after Val's birth, his father was badly injured when a horse he was riding fell with him. His father subsequently had to give up running a ranch and moved the family to Gordon, Nebraska, a town about 25 miles away, where he entered the insurance business.

During World War II, Fitch was a United States soldier stationed as a technician on the Manhattan Project in Los Alamos, New Mexico where the atomic bomb was being designed and built. in approaching the measurement of new phenomena, he learned not just to consider using existing apparatus, but to invent new ways of doing the job

Stimulated by the research of the Manhattan Project, Fitch subsequently completed his undergraduate degree in electrical engineering at McGill University in Montréal, Québec, Canada in 1948. He earned his Ph.D. at Columbia University in New York City in 1954.

After receiving his Ph.D., Fitch joined the faculty at Princeton University, where he remained throughout his career. He eventually was appointed to the Cyrus Fogg Brackett Professorship of Physics, and in 1976 became chairman of the Physics Department.

At Princeton, Fitch teamed with James Cronin (1931-) to explore the characteristics and behavior of subatomic particles. Until the 1950s, physicists believed that there was perfect balance, or symmetry, between matter and antimatter. Fitch and Cronin found that this is not always true. In 1964, they observed that on rare occasions, the decay of neutral K-meson particles violate CP (charge, parity) symmetry. They modified the CP symmetry principle and confirmed CPT (charge, parity, time-reversal) symmetry, where there is a balance between matter and antimatter moving forward and backward in time, respectively.

For their discovery of the violations of fundamental symmetry principles in the decay of neutral K-meson particles, Fitch and Cronin shared the 1980 Nobel Prize in physics.

1. Val Fitch, Autobiography, NobelPrize.org, http://nobelprize.org/nobel_prizes/physics/laureates/1980/fitch-autobio.
2. Val Logsdon Fitch, Microsoft Encarta Encyclopedia, http://encarta.msn.com/encyclopedia_761583177/Fitch_Val_Logsdon, 2007.
3. Val Logsdon Fitch, Wikipedia, http://en.wikipedia.org/wiki/Val_Logsdon_Fitch.

Felix Boehm (1924-)

Contributed to the understanding of the weak interaction and fundamental symmetries in the nucleus

Felix Hans Boehm was born in Switzerland in 1924. He received his Ph.D. in physics from the Eidgenössische Technische Hochschule in Zurich in 1951. He then came to New York to work at Columbia University in 1952.

Boehm was a research fellow at the California Institute of Technology (Caltech) from 1953 to 1958, and Caltech Professor of Physics from 1958 to 1995.

Boehm's principal work was on nuclear structure and the nature and behavior of subatomic particles. He pioneered the use of nuclear physics techniques to explore fundamental questions concerning the weak interactions and the nature of neutrinos. He developed the beta-gamma curricular polarization correlation method as a means of establishing the spin and angular momentum relationships in beta decay. This experience allowed him to make the first observation of hadonic parity violation in a nuclear system.

In addition to searching for violations of time-reversal invariance in nuclei, Boehm also spearheaded experiments which used the electron antineutrinos from the Goesgen reactor to place limits on neutrino oscillations, providing some of the most stringent existing bounds on neutrino mass.

In 1992, Boehm and Petr Vogel published *The Physics of Massive Neutrinos* through Cambridge University Press.

Boehm received the 1995 Tom W. Bonner Prize in Nuclear Physics for his pivotal contributions to our understanding of the weak interaction and fundamental symmetries in the nucleus. In particular, the award was for his:

(1) measurements of positron polarization in beta decay and their impact on the development of the V-A theory of weak interactions,

(2) pioneering studies providing convincing evidence for parity violation in nuclear transitions, and

(3) frontier defining searches for violations of time-reversal invariance in nuclei and for neutrino oscillations

1. 1995 Tom W. Bonner Prize in Nuclear Physics Recipient, Felix Bohem, APS Physics, www.aps.org/programs/honors/prizes/prizerecipient.cfm?name=Felix%20Boehm&year=1995.
2. Felix Hans Boehm, The Caltech Institute Archives, http://archives.caltech.edu/search_catalog.cfm?results_file=Detail_View&recsPerPage=1&firstRecToShow=0&search_field=Hans%20Boehm&entry_type=&photo_id=&cat_series=

Georges Charpak (1924-)

Invented the multiwire proportional chamber particle detector

Georges Charpak was born to a Jewish family in Dabrowica, Poland on August 1, 1924. His family moved to Paris when he was five years old. During World War II, Charpak served in the resistance and was imprisoned in 1943. In 1944, he was deported to the Nazi concentration camp at Dachau, where he remained until the camp was liberated in 1945.

In 1945, Charpak enrolled in the Paris-based École des Mines. He graduated in 1948 with a degree in mining engineering. He became a French citizen during this time. After graduating, he took a job at the National Centre for Scientific Research.

Charpak received his doctorate in Nuclear Physics in 1954 from the Collège de France in Paris, where he worked in the laboratory of Frédéric Joliot-Curie (1900-1958) designing particle detection equipment.

In 1959, Charpak joined the staff of the European Organization for Nuclear Research (CERN) in Geneva, where he concentrated on the study of particle detection. In 1984, while remaining affiliated with CERN, Charpak was named the Joliot-Curie Professor at the École Supérieure de Physique et Chimie in Paris.

Charpak is best known for his invention of the multiwire proportional chamber particle detector, which consists of a number of wires in a chamber of ionized gas. The wires attract electrons, and a computer analyzes the current produced in the wires. His invention replaced the slower and less efficient technique of photographic analysis of nuclear particles, and greatly advanced the study of the nature of matter.

Charpak was awarded the 1992 Nobel Prize in physics for his invention of devices for detecting atomic particles, in particular the multiwire proportional chamber. Charpak's inventions freed scientists from dependence on film to detect and record subatomic matter, greatly advancing the study of nuclear processes.

Charpak in later years focused his attention on the field of medicine, analyzing the structure of proteins with superfast X-rays, and studying receptors in the brain.

Inspired by his own wrongful incarceration during World War II, Charpak founded CERN's SOS Committee, which represents scientists whose civil rights have been denied by repressive governments.

1. George Charpak, Microsoft Encarta Encyclopedia, http://encarta.msn.com/encyclopedia_761583137/George_Charpak.
2. Georges Charpak, Curriculum Vitae, NobelPrize.org, http://nobelprize.org/nobel_prizes/physics/laureates/1992/charpak-cv.
3. Georges Charpak, Wikipedia, http://en.wikipedia.org/wiki/Georges_Charpak.

Leo Esaki (1925-)

Made discoveries regarding tunneling phenomena in semiconductors

Leo Esaki was born in Osaka, Japan on March 12, 1925. He received his B.S. in physics in 1947, his M.S. in 1947, and his Ph.D. in 1959, all three from Tokyo University. Esaki worked at the Sony Corporation in Japan while a student.

Esaki is best is known for his invention of the **Esaki diode**, which made use of electron tunneling. A diode is an electronic device that allows the passage of current in only one direction. The first such devices were vacuum-tube diodes.

The Esaki diode is a type of semiconductor diode which is capable of very fast operation, well into the microwave region (GHz). It uses quantum mechanical effects as follows: In classical physics, an electric current cannot flow in a circuit interrupted by an insulating barrier. Since the 1930s, quantum mechanics had predicted that electrons might be able to tunnel through an insulating barrier if it were thin enough. Esaki developed a diode with electrical junctions only 10 billionths of a meter thick through which electrons could tunnel.

Esake moved to the United States in 1960 and joined IBM's T. J. Watson Research Center in Yorktown Heights, New York. He was made an IBM fellow, the company's highest research position, in 1965.

At IbM, Esake published a paper on semiconductor superlattice. Superlattices are synthetic crystals composed of extremely fine layers of different semiconductors. One of the potential uses for this material is in high-speed computers.

Esaki shared the Nobel Prize in physics in 1973 with Ivar Giaever (1929-) and Brian Josephson (1940-). Esaki and Giaver's part of the award was for their experimental discoveries regarding tunneling phenomena in semiconductors and superconductors, respectively. Esaki's research was done in 1957 at Sony while he was still a student. Josephson's part of the award was for his theoretical predictions of the properties of a supercurrent through a tunnel barrier, in particular those phenomena, which are generally known as the Josephson effects.

Esaki worked for IBM for 33 years, eventually becoming a director of the company. When he retired from IBM he returned to Japan, subsequently serving as President of various Japanese universities.

1. Leo Asaki, Biography, NobelPrize.org, http://nobelprize.org/nobel_prizes/physics/ laureates/ 1973/esaki-bio.
2. Leo Asaki, Microsoft Encarta Encyclopedia Online, http://encarta.msn.com/ encyclopedia_761583167/Leo_Esaki.html#461521555, 2007.
3. Leo Asaki, Wikipedia, http://en.wikipedia.org/wiki/Leo_Esaki.

H. Thomas Milhorn, MD, PhD

Roy Glauber (1925-)

Contributed to the quantum theory of optical coherence

Roy Jay Glauber was born in New York City on September 1, 1925. His father was a traveling salesman, and drove a company-owned car. The 1929 market crash had an immediate impact for Glauber. The company his father worked for failed, and the car the family had been using was repossessed.

As a child, Glauber was mesmerized by Jules Verne, Alexandre Dumas, and Sir Walter Scott. In high school he built his own telescope and used it to photograph a lunar eclipse. Later he built a spectroscope that won the city's science fair and was displayed at the 1939 World's Fair in New York.

After his sophomore year of undergraduate school at Harvard University, Glauber entered the military, and briefly taught physics for the military's specialized training program. He was then assigned at age 18 to the Manhattan Project, where his work involved calculating the critical mass for the atom bomb.

After two years at Los Alamos, Glauber returned to Harvard, receiving his B.S. in 1946 and his Ph.D. in 1949. He then joined the Harvard physics faculty.

In 1963, Glauber created a model for photodetection, and explained the fundamental characteristics of different types of light, such as LASER light and light from light bulbs. His theories are widely used in the field of quantum optics.

Glauber was awarded one-half of the Nobel Prize in Physics in 2005 for his contribution to the quantum theory of optical coherence. The other half was split between John L. Hall (1934-) and Theodor Hänsch (1941-) for their contributions to the development of laser-based precision spectroscopy, including the optical frequency comb technique.

Glauber's more recent research dealt with problems in a number of areas of quantum optics, a field which studies the quantum electrodynamical interactions of light and matter.

Glauber also worked on several topics in high-energy collision theory, including the analysis of hadron collisions, and the statistical correlation of particles produced in high-energy reactions. A hadron is any strongly interacting composite subatomic particle. All hadrons are composed of quarks.

1. Roy J. Glauber, Autobiography, NobelPrize.org, http://nobelprize.org/nobel_prizes/physics/laureates/2005/glauber-autobio.
2. Roy J. Glauber, NNDB, www.nndb.com/people/738/000138324.
3. Roy J. Glauber, Wikipedia, http://en.wikipedia.org/wiki/Roy_J._Glauber.

Simon van der Meer (1925-)

Contributed to the project which led to the discovery of the field particles W and Z

Simon van der Meer, the son of a schoolteacher, was born in The Hague, Netherlands on November 24, 1925. In 1952, he received a degree in physical engineering from the Higher Technical School in Delft, Netherlands. After graduating, he worked for the Philips Company Eindhoven, mainly on high-voltage equipment and electronics for electron microscopes. In 1956, van der Meer joined the European Organization for Nuclear Research (CERN) in Switzerland, where he spent most of his career.

In the 1950s, physicists began searching for evidence to support Einstein's unification theory, in which all of nature's forces— electromagnetic, gravity, strong nuclear, and weak nuclear—are related. One piece of necessary evidence was the existence of W and Z field particles, which would be found in the nucleus and expected to be 100 times heavier than the proton. These particles were thought to convey the weak force, a force that causes certain particles to decay or transform into other particles.

The existence of The W and Z particles had been predicted, but never confirmed. In an experiment designed by Carla Rubbia (1934-), researchers hoped to observe the particles by colliding a beam of protons with a beam of antiprotons. Van der Meer's most significant contribution to this effort was developing a way to create the concentrated beam of protons and antiprotons. He built a device that could generate and store antiprotons, an especially difficult task. In January of 1983 a team of more than 100 physicists confirmed for the first time the existence of W and Z particles.

From 1967 to 1976 van der Meer returned to more technical work when he was responsible for the magnet power supplies, first of the Intersecting Storage Rings and then of the 400 GeV synchrotron.

Van der Meer shared the 1984 Nobel Prize in physics with his colleague, Carlo Rubbia, for contributing to the project, which led to the discovery of the field particles W and Z, communicators of the weak interaction.

Van der Meer retired in 1990.

1. Simon van der Meer, Autobiography, NobelPrize.org, http://nobelprize.org/nobel_prizes/physics/laureates/1984/meer-autobio.
2. Simon van der Meer, Microsoft Encarta Encyclopedia Online, http://encarta.msn.com/ncyclopedia_761583384/Van_der_Meer_Simon, 2007.
3. Simon van der Meer, Wikipedia, http://en.wikipedia.org/wiki/Simon_van_der_Meer.

Abdus Salam (1926-1996)

Contributed to the theory of the unified weak and electromagnetic
interaction between elementary particles

Abdul Salam was born at Jhang Sadar, India (now in Pakistan) on January
29, 1926. His father was an official in the Department of Education in a
poor farming district. Salam received an M.A. from Government College,
Lahore in 1946, and then a B.A. in mathematic and physics at St. John's
College, Cambridge in 1949. He earned a Ph.D. in theoretical physics from
Cambridge in 1951. His research was on Quantum Electrodynamics.

Salam returned to the Government College, Lahore as a professor of
Mathematics in 1951, and then went back to Cambridge as a lecturer in
mathematics in 1954. In 1957, he became Professor of Theoretical Physics
at Imperial College, London.

During the early 1960s, Salam played a significant role in
establishing the Pakistan Atomic Energy Commission and the Space and
Upper Atmosphere Research Commission. He was also instrumental in
setting up five Superior Science colleges throughout Pakistan to further the
progress in science in the country.

In 1964, while still on the faculty of the Imperial College of London,
Salam became director of the International Center for Theoretical Physics
in Trieste, Italy. He founded the Third World Academy of Sciences, and
was instrumental in the creation of a number of international centers
dedicated to the advancement of science and technology.

In 1967, with American physicist Steven Weinberg (1933-), Salam
offered a unification hypothesis that incorporated the known facts about
the electromagnetic and weak interactions between subatomic particles.
Their work gave rise to the electro-weak theory, which is the mathematical
and conceptual synthesis of the electromagnetic and weak interactions.
The validity of the theory was later verified through experiments carried
out at the Super Proton Synchrotron facility at the European Organization
for Nuclear Research (CERN) in Geneva.

Salam and Weinberg, along with American physicist Sheldon
Glashow (1932-), were awarded the 1979 Nobel Prize in physics for their
work in electro-weak theory. Salam was the first Muslim Nobel Laureate.
He died after a long illness on November 21, 1996 in Oxford, England,
and was buried in Rabwah, Pakistan.

1. Abdus Salam, Biography, NobelPrize.org, http://nobelprize.org/nobel_prizes/physics/
    laureates/1979/salam-bio.
2. Abdus Salem, Microsoft Encarta Encyclopedia Online, http://encarta.msn.com/
    encyclopedia_761556972/Abdus_Salam, 2007.
3. Abdus Salem, Wikipedia, http://en.wikipedia.org/wiki/Abdus_Salam.

Alex Müller (1926-)

Discovered superconductivity in ceramic materials

Karl Alexander Müller was born in Basel, Switzerland on April 20, 1927. His family immediately moved to Salzburg, Austria for his father to study music. After several years there, he and his mother moved to Dornach, near Basel, to the home of his grandparents. Then they moved to Lugano, in the Italian-speaking part of Switzerland where Alex learned to speak Italian fluently. His mother died when he was 11. At the age of 19, he did basic military training in the Swiss army.

Müller enrolled in the Physics and Mathematics Department of the Swiss Federal Institute of Technology in Zürich, and obtained his doctorate in 1958. He then joined the Battelle Memorial Institute in Geneva, where he soon became the manager of a magnetic resonance group. During this time he became a Lecturer at the University of Zürich, which led to his accepting a position at the IBM Zürich Research Laboratory in 1963.

At IBM, Müller's research involved $SrTiO3$ and related perovskite compounds (mineral consisting of an oxide of calcium and titanium). He studied their photochromic properties when doped with various transition-metal ions; their chemical binding; ferroelectric and soft-mode properties; and the critical and multicritical phenomena of their structural phase transitions. In the early 1980s, Müller began searching for substances that would become superconductive (the ability of a material to carry an electrical current indefinitely without resistance) at higher temperatures than possible up to that time. In 1983, he recruited German physicist Johannes Bednorz (1950-) to IBM to help systematically test various oxides. A few recent studies had indicated those materials might superconduct. In 1986, the two succeeded in achieving superconductivity in a barium-lanthanum-copper oxide (BaLaCuO) at a temperature of 35 K.

Müller and Bednorz were awarded the 1987 Nobel Prize in physics for their discovery that copper oxide ceramic materials can achieve superconductivity at temperatures well above the extremely low temperatures once associated with this remarkable property.

Superconductors are now used in scientific and medical instruments, and may find applications in the electronics industry and in electric power transmission and storage.

1. K Alex Müller, NobelPrize.org, Autobiography, http://nobelprize.org/nobel_prizes/physics/ laureates/1987/muller-autobio.
2. Karl Alexander Müller, Mircrosoft Encarta Encyclopedia, http://encarta.msn.com/ encyclopedia_761583278/M%C3%BCller_Karl_Alex., 2007.
3. Karl Alexander Müller, Wikipedia, ttp//en.wikipedia.org/wiki/Karl_Alexander_M%C3%BCller.

Allan Bromley (1926-2005)

The father of heavy-ion physics

David Allan Bromley was born in Westmeath, Ontario, Canada on May 4, 1926. He received a B.S. in 1949 and an M.S. in 1950 from Queen's University. He earned a Ph.D. in nuclear physics in 1952 from the University of Rochester.

In 1952, Bromley joined the physics faculty at the University of Rochester. In 1955, he was hired by Atomic Energy of Canada.

In 1960, Bromley moved to the United States to accept a position at Yale University. He was Associate Director of the Heavy Ion Accelerator Lab from 1960 to 1963.

Bromley was the founder and, from 1963 to 1989, the Director of Yale's A. W. Wright Nuclear Structure Lab. He carried out pioneering studies on the structure and dynamics of atomic nuclei. He became a U.S. citizen in 1970.

In 1988, Bromley was awarded The National Medal of Science, which is an honor bestowed by the President of the United States to individuals in science and engineering who have made important contributions to the advancement of knowledge in the fields of behavioral and social sciences, biology, chemistry, engineering, mathematics, and physics. His award was for "seminal work on nuclear molecules, for development of tandem accelerators and semi-conductor detectors for charged particles, for his contributions to particle-gamma correlation studies, and for his role in founding the field of precision heavy-ion physics."

From 1989 to 1992, Bromeley served as George H. W. Bush's science advisor. He pushed for major increases in scientific research funding so that the United States could compete with Japan and Germany in manufacturing. He also supported the expansion of the high-speed network which eventually became the Internet.

In 1992, Bromley returned to Yale University to serve as Professor and Dean of the Faculty of Engineering.

Bromley died of a heart attack in New Haven, Connecticut on February 10, 2005. He is considered the father of modern heavy ion science.

1. D. Allan Bromley, 79, Physicist Who Devised National Science Policy for the First President Bush, Dies, New York Times, February 13, 2005.
2. D. Allan Bromley, Wikipedia, http://en.wikipedia.org/wiki/D._Allan_Bromley.
3. David Allan Bromley, The President's National Medal of Science: Recipient Details, www.nsf.gov/od/nms/recip_details.cfm?recip_id=56.

Ben Mottelson (1926-)

Showed that nuclear particles can vibrate and rotate so as to distort the shape of the nucleus

Ben Roy Mottelson was born in Chicago, Illinois on July 9, 1926. He spent time in the United States Navy during World War II. As part of his military service, he was sent to officer training school at Purdue University in West Lafayette, Indiana.

Mottelson re-enrolled at Purdue University at the end of the war, and received a Bachelor's degree from in 1947. He obtained a Ph.D. under Julian Schwinger (1918-1994) at Harvard University in 1950.

Receiving a Sheldon Traveling Fellowship from Harvard University, Mottelson spent the year 1950-1951 at the Institute for Theoretical Physics in Copenhagen, Denmark. A fellowship from the U.S. Atomic Energy Commission allowed him to continue working in Copenhagen for two more years.

Mottelson then accepted a research position in the European Organization for Nuclear Research. He then became a professor at the Nordic Institute for Theoretical Atomic Physics in Copenhagen.

In 1957, Mottelson began his association with Aage Bohr (1922-). Mottelson and Bohr helped prove the theories of James Rainwater (1917-1986) regarding the structure of atomic nuclei. They developed a collective model of the structure of the nucleus of an atom, combining concepts proposed in two earlier models—the liquid drop model (described nucleus behavior as liquid like) and the shell model (explained orbital behavior of nucleons within the nucleus).

In 1971, Mottelson became a naturalized Danish citizen. That same year, he and Bohr authored *Collective and Individual Particle Aspects of Nuclear Structure*. In 1969, they published the first volume of *Nuclear Structure*, a three-volume set.

Mottelson, Bohr, and Rainwater were awarded the 1975 Nobel Prize in physics for the discovery of the connection between collective motion and particle motion in atomic nuclei and the development of the theory of the structure of the atomic nucleus based on this connection.

In 1981, Mottelson replaced Bohr as director of the Nordic Institute for Theoretical Atomic Physics.

1. Ben R. Mottelson, Autobiography, NobelPrize.org, http://nobelprize.org/ nobel_prizes/ physics/laureates/1975/mottelson-autobio.
2. Ben R. Mottelson, Microsoft Encarta Encyclopedia Online, http://encarta.msn.com/ encyclopedia_761583276/mottelson_benjamin_roy, 2007.
3. Ben Roy Mottelson, Wikipedia, http://en.wikipedia.org/wiki/Ben_Roy_Mottelson.

Donald Glaser (1926-)

Invented the bubble chamber for detecting electrically charged particles

Donald Arthur Glaser, the son of a businessman, was born in Cleveland, Ohio on September 21, 1926. He was an accomplished violinist, becoming a member of a symphony orchestra at age 16. He studied composition at the Cleveland Institute of Music.

Glaser received his B.S. in physics and mathematics from Case Institute of Technology in 1946, and his Ph.D. in physics from the California Institute of Technology in 1949. His doctoral thesis research was an experimental study of the momentum spectrum of high energy cosmic rays and mesons at sea level.

In 1949, Glaser accepted a position as an instructor at the University of Michigan. His research involved short-lived elementary particles using a bubble chamber he had designed and built.

In creating the bubble chamber, Glaser built on the work of C. T. R. Wilson (1869-1959), the inventor of the cloud chamber, and C. F. Powell (1903-1969), who developed photographic techniques to capture images of charged nuclear particles on film. Both the cloud chamber and the emulsion technique were limited to the detection of low-energy nuclear particles. Glaser wanted to build a device that could reveal high-energy particles. He found that nuclear particles left a trail of bubbles as they moved through superheated ether.

Using the bubble chamber, scientists have since discovered new high-energy atomic particles. They have also used it to track neutral atomic particles, advancing their understanding of the mass, lifetime, and decay of these particles.

In 1959, Glazer joined the faculty of the University of California, Berkeley as a Professor of Physics.

Glazer was awarded the 1960 Nobel Prize in physics for his invention of the bubble chamber. With this device, he contributed greatly to the understanding of atomic function and provided the technology for the discovery of new atomic particles.

In 1962, Glazer changed his research to molecular biology, starting with a project on ultraviolet induced cancer. In 1964, he was given the additional title of Professor of Molecular Biology, and then in 1989 Professor of Physics and Neurobiology in the Graduate School.

1. Donald A. Glazier, Biography, NobelPrize.org, http://nobelprize.org/nobel_prizes/physics/laureates/1960/glaser-bio.
2. Donald A. Glazier, Wikipedia, http://en.wikipedia.org/wiki/Donald_A._Glaser.
3. Donald Arthur Glazier, Microsoft Encarta Encyclopedia Online, http://encarta.msn.com/encyclopedia_761583193/glaser_donald_arthur, 2007.

Henry Kendall (1926-1999)

Investigated deep inelastic scattering of electrons on protons and bound neutrons

Henry Way Kendall was born in Boston, Massechusetts on December 9, 1926. His parents had moved to the United States from Canada. In 1945, he entered the U.S. Merchant Marine Academy, and was in basic training when the first atom bombs were exploded over Japan. He spent the winter of 1945-46 on a troop transport on the North Atlantic, returning to the Academy for advanced training in the spring of 1946. He resigned in October of that year to start as a freshman at Amherst College, majoring in mathematics.

In 1950, Kendall entered graduate school at the Massachusetts Institute of Technology (MIT), where he received his Ph.D. in physics in 1954. After two years of postdoctoral research, he joined the faculty at Stanford University. He returned to MIT in 1961, remaining on the faculty there throughout the rest of his career. His colleague at Stanford, Jerome Friedman (1930-) had gone to MIT a year earlier.

Protons and neutrons were once thought to be the elementary particles of matter. In the 1950s, physicists began to search for more elementary particles, and in 1964 scientists predicted the existence of quarks.

In the late 1960s, using a new and powerful particle accelerator at the Stanford Linear Accelerator Center (SLAC), Kendall and Friedman, working with Richard Taylor (1929-), bombarded protons with electrons traveling at near the speed of light. The researchers expected the electrons to pass through the protons or to bounce off them. Many of the electrons rebounded at odd angles, which was an unexpected result. Particles within the protons apparently were causing this unusual electron behavior. Later experiments confirmed that those particles to be quarks. Six different types of quarks have since been found, and have been given the names up, down, charm, strange, top, and bottom.

Kendall, Friedman, and Taylor shared the 1990 Nobel Prize in physics for their investigations of deep inelastic scattering of electrons on protons and bound neutrons, which had importance for the development of the quark model in particle physics.

Kendall died in Wakulla Springs State Park, Florida on February 15, 1999 while photographing an underwater cave system.

1. Henry W. Kendall, Autobiography, NobelPrize.org, http://nobelprize.org/nobel_prizes/physics/laureates/1990/kendall-autobio.
2. Henry Way Kendall, Microsoft Encarta Encyclopedia, http://encarta.msn.com/encyclopedia_761583229/Kendall_Henry_Way, 2007.
3. Henry Way Kendall, Wikipedia, http://en.wikipedia.org/wiki/Henry_Way_Kendall.

Tsung-Dao Lee (1926-)

Proved experimentally that the law of conservation of parity does not hold true for weak nuclear reactions

Tsung-Dao Lee was born in Shanghai, China on November 24, 1926. He entered Zhejiang University in 1943, but his education was interrupted in 1944 y the Second World War. After a short delay, he continued his education at the National Southwestern Associated University Kunming in 1945.

After completing his sophomore year, he received a Chinese government fellowship for graduate study in the United States. Lee entered the University of Chicago in 1946, and received his Ph.D. under Enrico Fermi (1901-1954) in 1950. The title of his disertation was *Hydrogen Content of White Dwarf Stars.*

During the years 1950 to 1953, Lee worked as a research associate and lecturer at Yerkes Astronomical Observatory in Wisconsin; the University of California at Berkley; and the Institute for Advanced Study in Princeton, New Jersey. During this time, he worked on elementary particles, statistical mechanics, field theory, astrophysics, condensed matter physics, and turbulence.

In 1953, Lee joined the faculty of Columbia University as an assistant professor. There, he worked mainly in particle physics and field theory. His first work was on the renormalizable field theory model, better known as the **Lee Model**. He eventually became the Enrico Fermi Professor of Physics.

Over the years, Lee did research ranging from symmetry violations in weak interactions to fields of high energy neutrino physics and Relativistic Heavy Ion Collider (RHIC) physics. He then turned his attention to the bosonic nature of high Tc superconductivity, the neutrino mapping matrix, and new ways to solve the Schrödinger equation. Tc is the critical temperature at which superconductiviy occurs.

With his associate, Chen Ning Yang (1922-), Lee proved experimentally that the law of conservation of parity does not hold true for weak nuclear reactions. For this discovery, the two men were awarded the 1957 Nobel Prize in physics.

Beginning in 1981, Lee held professorships at a number of Chinese universities.

1. Tsung Dao Lee, Microsoft Encarta Encyclopedia Online, http://encarta.msn.com/ encyclopedia_761560348/lee_tsung-dao, 2007.
2. Tsung-Dao Lee, Biography, NobelPrize.org, http://nobelprize.org/nobel_prizes/physics/ aureates/1957/lee-autobio.
3. Tsung-Dao Lee, Wikipedia, http://en.wikipedia.org/wiki/Tsung-Dao_Lee.

Albert Crewe (1927-)

Developed the scanning transmission electron microscope

Albert Victor Crewe was born in Liverpool, England on February 18, 1927. He was the only child in a fairly-wealthy family. He graduated from the University of Liverpool with a degree in engineering, and then obtained his Ph.D. from the same institution in 1950.

After receiving his Ph.D., he moved to Chicago, Illinois to work on the Cyclotron particle accelerator at Argonne National Laboratories.

Crewe became the director of the Particle Accelerator Division at Argonne National Laboratories, and in 1961 he became Argonne Director. He became a naturalized United States citizen shortly afterward.

While at Argonne, Crewe directed the design and construction in 1963 of the Zero Gradient Synchrotron (ZGS). The ZGS generated a high-energy proton beam by accelerating protons in a giant underground ring to an energy of 12.5 billion electron-volts.

Crewe also developed high resolution electron microscopes, particularly scanning microscopes. Among them was the first modern scanning transmission electron microscope (STEM). The STEM is a type of transmission electron microscope in which the electrons pass through the specimen, but as in scanning electron microscopy the electron optics focus the beam into a narrow spot, which is scanned over the sample in a raster. The rastering of the beam across the sample makes the STEM suitable for analysis techniques, such as mapping by energy dispersive X-ray spectroscopy, electron energy loss spectroscopy, and annular dark-field imaging.

The first rudimentary STEM had been built in 1938 by Baron Manfred von Ardenne (1907-1997), working in Berlin for Siemens. However, the microscope was destroyed in an air raid in 1944.

Later, Crewe worked on the correction of spherical and chromatic aberrations, which are the limiting factors in all present high performance instruments.

Crewe left Argonne in 1967, becoming a professor at the University of Chicago, and later a dean of the physical sciences division.

1. Albert Crewe, Wikipedia, http://en.wikipedia.org/wiki/Albert_Crewe.
2. Albert V. Crewe, Experimental Applied Physics and Experimental Nuclear Physics, The University of Chicago Department of Physics, http://physics.uchicago.edu/ x_applied.html#Crewe.
3. Scanning Transmission Electron Microscopy, Wikipedia, http://en.wikipedia.org/ wiki/ Scanning_transmission_electron_microscopy.
4. Zero-Gradient Synchrotron begun 40 years ago; was the Advanced Photon Source of its time, Argonne News, www.anl.gov/Media_Center/Argonne_News/news99/an990628, June 29, 1999.

H. Thomas Milhorn, MD, PhD

Martin Perl (1927-)

Discovered the tau lepton

Martin Lewis Perl was born in New York City on June 24, 1927. His parents were Jewish immigrants from Russia. His father worked as a clerk and then a salesman in a company dealing in printing and stationary, and his mother worked as a secretary and then a bookkeeper in a firm of wool merchants.

Perl enrolled at Brooklyn Polytechnic Institute, but left to become an engineering cadet in the program at the Kings Point Merchant Marine Academy. Then he was drafted and spent a year at an army installation in Washington, D.C.

Perl returned to Brooklyn Polytechnic Institue and received a B.S. in chemical engineering in 1948. He went on to receive a Ph.D. from Columbia University in physics in 1950.

In 1950, Perl joined the faculty at the University of Michigan. And then in 1964 he moved to the Stanford Linear Accelerator Center (SLAC) in Palo Alto, California.

In 1973, Perl set out to study electrons and muons and to look for differences between the two particles. He did this by colliding subatomic particles. Occasionally a collision resulted in the production of electrons and muons. Since Perl was aware that muons were unstable particles that broke down into electrons and neutrinos, he theorized that only the decay of a more massive particle could result in the production of muons. In 1975, Perl announced his discovery of the tau lepton, a particle that is about 3500 times heavier than the electron.

The lepton is one of three groups of elementary particles, along with quarks and bosons. There are six known types of leptons, three that carry a negative electric charge and three with no charge. Of the negatively charged leptons, the electron was discovered in 1897 and the muon in the 1940s. Uncharged leptons are called neutrinos.

Before Perl's discovery, theoretical physicists had not predicted the existence of the tau, the third charged lepton.

Perl shared the 1995 Nobel Prize in physics with American physicist Frederick Reines (1918-1998) for experimental contributions to lepton physics. Perl's part of the award was for his discovery of the tau lepton, a type of fundamental particle that makes up part of the atom.

1. Martin L. Perl, Autobiography, NobelPrize.org, http://nobelprize.org/nobel_prizes/physics/ laureates/ 1995/perl-autobio.
2. Martin L. Perl, Microsoft Encarta Encyclopedia Online, http://encarta.msn.com/ encyclopedia_761583302/Perl_Martin_L_, 2007.
3. Martin Lewis Perl, Wikipedia, http://en.wikipedia.org/wiki/Martin_Lewis_Perl.

Alexei Abrikosov (1928-)

Contributed to the theory of superconductors

Alexei Alexeyevich Abrikosov was born in Moscow, USSR on June 25, 1928. Both his parents were physicians. He graduated from Moscow State University in 1948 with a M.Sc. degree.

Abrikosov received his Ph.D. in 1951 from the Institute for Physical Problems. His dissertaion work involved a theory of thermal diffusion in plasmas. He then obtained a Doctor of Physical and Mathematical Sciences degree in 1955. His research involved quantum electrodynamics at high energies.

From 1965 to 1988, Abrikosov worked at the Landau Institute for Theoretical Physics. He became a professor at Moscow State University in 1965.

In 1991, Abrikosov joined the staff at the Argonne National Laboratory outside Chicago, Illinois.

By the early 1950s, scientists were aware of the phenomenon of superconductivity. At the time, however, theoretical explanations for superconductivity had not been developed. Vitaly Ginzburg (1916-), with colleague Lev Landau (1908-1968), proposed key theories to explain the relation between electrons and the magnetic field inside superconductors.

Ginzburg and Landau noted that some superconductors repel a magnetic field, and they termed this class of superconductor Type I. Ultimately, Abrikosov, building on the Ginzburg-Landau theories, described more fully this second class of superconductor, now called Type II. Modern superconductors, including the ceramic varieties that function at higher temperatures, are Type II.

Ongoing research in superconductivity is expected to find application in the design of generators and engines, in improved transmission of electrical power over long distances, and other uses.

Anthony Leggett (1938-), a British-born American physicist, made theoretical breakthroughs in describing a related phenomenon known as superfluidity

Abrikosov, Ginzburg, and Leggett shared the 2003 Nobel Prize in physics for their pioneering contributions to the theory of superconductors and superfluids.

1.   Alexei Abrikosov, Microsoft Encarta Encyclopedia Online, http://encarta.msn.com/ encyclopedia_701666364/Alexei_Abrikosov, 2007.
2.   Alexei Alexeyevich Abrikosov, Autobiography, NobelPrize.org, http://nobelprize.org/ nobel_prizes/ physics/ laureates/2003/abrikosov-autobio.
3.   Alexei Alexeyevich Abrikosov, Wikipedia, http://en.wikipedia.org/wiki/ Alexei_Alexeyevich_Abrikosov.

Herbert Kroemer (1928-)

Did basic work on information and communication technology

Herbert Kroemer was born in Weimar, Germany on August 25, 1928. His father was a civil servant working for the city administration. In 1947, Kroemer entered the University of Jena, majoring in physics. He received a Ph.D. in theoretical physics in 1952 from the University of Gottingen, Germany. His dissertation was on hot-electron effects in the then-new transistor.

Kroemer worked in a number of research laboratories in Germany and the United States.

In the early 1960s, Kroemer and Russian physicist Zhores Alferov (1930-) working independently concluded that electrons and holes within certain heterostructures could be made to emit light very efficiently. By carefully controlling the composition and arrangement of the semiconducting layers, they were better able to control a key source of performance loss in a transistor—electrons flowing in one direction, leaving behind positively charged "holes" that constitute an opposing flow in the opposite direction. Heterostructures are less susceptible to this opposing flow. This improved efficiency made semiconductor LASERs that operate continuously at room temperature possible.

From 1968 to 1976 Kroemer taught electrical engineering at the University of Colorado. He joined the University of California, Santa Barbara faculty in 1976.

Kroemer and Alferov were awarded the 2000 Nobel Prize in physics for basic work on information and communication technology, specifically for developing semiconductor heterostructures used in high-speed- and opto-electronics. They shared the Prize with Jack Kilby (1923-2005) for his part in the invention of the integrated circuit.

The heterostructures pioneered by Kroemer and Alferov are used in satellite communications and in the base stations of mobile-phone networks.

LASER technology based on heterostructures can be found in DVD players and the bar-code readers used to check out books in libraries. A new generation of highly efficient light-emitting diodes, which are based on the same heterostructure principles as the LASER, are used in automobile brake lights and traffic lights.

1. Herbert Kroemer, Autobiography, NobelPrize.org, http://nobelprize.org/nobel_prizes/physics/laureates/2000/kroemer-autobio.
2. Herbert Kroemer, Wikipedia, http://en.wikipedia.org/wiki/Herbert_Kroemer.
3. Kroemer, Herbert, Microsoft Encarta Encyclopedia Online, http://encarta.msn.com/encyclopedia_701502047/Kroemer_Herbert, 2007.

Stanley Mandelstam (1928-)

Represented the analytic properties of scattering amplitudes in the form of double dispersion relations

Stanley Mandelstam was born in Johannesburg, South Africa in 1928. He received a B.Sc. from the University of Witwatersrand in 1952, a B.A. from Trinity College, Cambridge in 1954, and a Ph.D. from Birmingham University in 1956.

Mandelstam was professor of mathematical physics at the University of Birmingham from 1960 to 1963. He was on the faculty of the Université de Paris Sud in 1979-80 and 1984-85, and became professor of physics at the University of California, Berkeley in 1963.

Mandelstam introduced the relativistically invariant **Mandelstam variables** into particle physics in 1958 as a convenient coordinate system for formulating his double dispersion relations. The double dispersion relations were a central tool in the bootstrap program, which sought to formulate a consistent theory of infinitely many particle types of increasing spin.

After Gabriele Veneziano (1942-) constructed the first tree-level scattering amplitude to describe infinitely many particle types, Mandelstam made crucial contributions. He identified the Virasoro algebra as a geometrical symmetry of a world sheet conformal field theory, and showed that the conformal invariance allowed string amplitudes to be calculated on many worldsheet domains.

Mandelstam was among the first to apply path integral quantization methods to string theory. This work was generalized and extended by many others in the following years and now forms an integral part of the modern formulations.

Mandelstam explicitly constructed the fermion scattering amplitudes in the Ramond and Neveu-Schwarz sectors, and later gave arguments for the finiteness of string perturbation theory.

In field theory, Mandelstam showed that the two dimensional quantum Sine-Gordon model is equivalently described by a Thirring model whose fermions are the kinks. He also demonstrated that the 4d N=4 supersymmetric gauge theory is power counting finite, proving that this theory is scale invariant to all orders of perturbation theory. This was the first example of a field theory where all the infinities in Feynman diagrams cancel.

Mandelstam retired from the University of California, Berkeley in 1964, becoming Professor Emeritus.

1. Dirac Medalist, 1991, http://users.ictp.it/~sci_info/awards/Dirac/DiracMedallists/DiracMedal91.
2. Stanley Mandelstam, Wikipedia, http://en.wikipedia.org/wiki/Stanley_Mandelstam.

Ivar Giaever (1929-)

Made discoveries regarding tunneling phenomena in superconductors

Ivar Giaever, the son of a pharmacist, was born in Bergen, Norway on April 5, 1929. He earned a degree in mechanical engineering from the Norwegian Institute of Technology in 1952. He then served in the Norwegian Army from 1952 to 1953. In 1953, he accepted a position as a patent examiner for the Norwegian government.

Giaever immigrated to Canada in 1954, where he was employed as an electrical engineer by the Canadian division of General Electric. General Electric sent him to an advanced engineering program in its Schenectady, New York offices in 1956. He remained there, becoming an applied mathematician at the company's research and development center. In 1958, he became interested in physics, and took a position in a solid-state physics research group at General Electric. He then returned to school at night, and while working at General Electric earned a Ph.D. from Rensselaer Polytechnic Institute in 1964. He became a United States citizen the same year. Shortly afterward, he began his research on tunneling effects in semiconductors and superconductors

According to classical physics, an electric current cannot flow in a circuit interrupted by an insulating barrier. Since the 1930s, quantum theory had predicted that electrons might be able to tunnel through an insulating barrier if it were thin enough. Giaever discovered that cooling two semiconductors to near-absolute zero temperatures greatly enhanced superconductivity. As a result of these studies, he developed a technique that allowed physicists to easily observe and measure semiconductor properties. This technique allowed Giaever to confirm certain previously unproven theories about superconductivity.

Giaever and Leo Esaki (1925-) shared the Nobel Prize in physics in 1973 with and Brian Josephson (1940-) for for their experimental discoveries regarding tunneling phenomena in semiconductors and superconductors, respectively. Josephson's part of the award was for his theoretical predictions of the properties of a supercurrent through a tunnel barrier. During the 1970s, Giaever temporarily left General Electric to study biophysics at the University of Cambridge. When he returned to General Electric he began investigating the properties of cell membranes and how proteins interact with solid surfaces.

1. Ivar Giaever, Biography, NobelPrize.org, http://nobelprize.org/nobel_prizes/physics/laureates/1973/giaever-bio.
2. Ivar Giaever, Microsoft Encarta Encyclopedia Online, http://encarta.msn.com/encyclopedia_761583189/giaever_ivar, 2007.
3. Ivar Giaever, Wikipedia, http://en.wikipedia.org/wiki/Ivar_Giaever.

Murray Gell-Mann (1929-)

Developed a system for classifying subatomic particles

Murray Gell-Mann, the son of Jewish immigrants, was born in New York City on September 15, 1929. He earned a bachelor's degree in physics from Yale University in 1948 and a Ph.D. in physics from MIT in 1951.

Gell-Mann was a postdoctoral research associate in 1951 and a visiting research professor at University of Illinois at Urbana-Champaign, Illinois from 1952 to 1953.

After serving as Visiting Associate Professor at Columbia University from 1954 to 1955, he became a professor at the University of Chicago before moving to the California Institute of Technology, where he taught from 1955 until 1993.

Gell-Mann's work in the 1950s involved recently discovered cosmic ray particles that came to be called kaons and hyperons. Classifying these particles led him to propose a new quantum number called "strangeness."

Gell-Mann grouped related particles into an eightfold-way scheme that brought order out of the chaos created by the discovery of some 100 kinds of particles in collisions involving atomic nuclei. Gell-Mann subsequently found that all of those particles, including the neutron and proton, are composed of fundamental building blocks that he named quarks.

In 1963, Gell-Mann and George Zweig (1937-) independently advanced the quark theory. They hypothesized that quarks are the smallest particles of matter, and that they carry fractional electric charges. The quarks are permanently confined by forces coming from the exchange of gluons. Research in particle physics has thus far supported these theories.

Gell-Mann and others later constructed the quantum field theory of quarks and gluons, called quantum chromodynamics (QCD), which seems to account for all the nuclear particles and their strong interactions. QCD is a theory of the strong interaction called color force, which is a fundamental force describing the interactions of the quarks and gluons found in hadrons. A hadron is any strongly interacting composite subatomic particle. All hadrons are composed of quarks.

Gell-Mann's non-physics interests involved historical linguistics, archeology, natural history, the psychology of creative thinking, and other subjects connected with biological and cultural evolution and with learning.

1. Murray Gell-Mann, Biography, NobelPrize.org, http://nobelprize.org/nobel_prizes/physics/laureates/1969/gell-mann-bio.
2. Murray Gell-Mann, Wikipedia, http://en.wikipedia.org/wiki/Murray_Gell-Mann.
3. Murray Gill-Mann, Microsoft Encarta Encyclopedia Online, http://encarta.msn.com/encyclopedia_761569611/murray_gell-mann, 2007.

H. Thomas Milhorn, MD, PhD

Peter Higgs (1929-)

Proposed broken symmetry in electroweak theory, explaining the origin of
mass of elementary particles

Peter Ware Higgs was born on in Newcastle upon Tyne, England on May 29, 1929. Because of childhood asthma he missed some early schooling, and was taught at home. He attended King's College London, where he received a Ph.D. in physics. His dissertation topic was the vibration spectra of molecules.

Higgs then spent six years moving back and forth between Edinburgh, University College London and Imperial College before landing a permanent position at Edinburgh.

It was in Edinburgh in 1961 that Higgs first became interested in mass, developing the idea that particles were weightless when the universe began, acquiring mass a fraction of a second later as a result of interacting with a theoretical field now known as the **Higgs field**. Higgs postulated that this field permeates space, giving all elementary subatomic particles that interact with it their mass.

While the Higgs field is postulated to confer mass on quarks and leptons, it is thought to represent only a tiny portion of the masses of other subatomic particles, such as protons and neutrons. In these particles, gluons that bind quarks together are thought to confer most of the particle mass.

Higgs is best known for his proposal of broken symmetry in electroweak theory, explaining the origin of mass of elementary particles in general and of the W and Z bosons in particular. This **Higgs mechanism** predicts the existence of a new particle, the **Higgs boson**.

Although the Higgs boson has not yet been found in accelerator experiments, the Higgs mechanism is generally accepted as an important ingredient in the Standard Model of particle physics.

Stephen Hawkings (1942-), professor of mathematics at Cambridge University, raised some professional eyebrows when he put money on particle physicists failing to discover the Higgs boson with the Large Electron Positron (LEP) particle accelerator at the CERN nuclear laboratory in Geneva. Much to the disappointment of particle physicists, the LEP finally closed down without finding the elusive boson. This disagreement has been billed as the battle of the heavy weights.

1. Conner, Steve, Higgs v Hawking: a battle of the heavyweights that has shaken the world of theoretical physics, The Independent, http://web.archive.org/web/20031116195026/ http://millennium-debate.org/ind3sept023, September 3, 2002.
2. Peter Higgs: The Man Behind the Boson, PhysicsWorld.com, http://physicsworld.com/cws/ article/ print/19750, July 10, 2004.
3. Peter Higgs, Wikipedia, http://en.wikipedia.org/wiki/Peter_Higgs.

Richard Taylor (1929-)

Investigated deep inelastic scattering of electrons on protons and bound neutrons

Richard Edward Taylor was born in Medicine Hat, Alberta, Canada on November 2, 1929. In high school, during World War II, he developed an interest in explosives, and blew three fingers off his left hand. The atomic bomb that ended the war later that summer made him intensely aware of physics.

Taylor was educated at the University of Alberta and at Stanford University, where he received his Ph.D. His dissertation work involved the production of polarized gamma rays from the accelerator beam and the use of those gamma rays to study p-meson production.

Between sessions at the University, he spent two summers as a research assistant at the Defense Research Board installation near Medicine Hat.

After three years at École Normale Supérieure in Paris and a brief stay at the Lawrence Berkeley National Laboratory, Taylor joined the faculty of the Stanford Linear Accelerator Center (SLAC) in 1968 and remained there throughout his career.

Protons and neutrons were once thought to be the elementary particles of matter. In the 1950s, physicists began to search for more elementary particles, and in 1964 scientists predicted the existence of quarks.

In the late 1960s, using a new and powerful particle accelerator at SLAC, Taylor worked with Jerome Friedman (1930-) and Henry Kendall (1926-1999) to bombard protons with electrons traveling at near the speed of light. They expected the electrons to pass through the protons or to bounce off them. Many of the electrons rebounded at odd angles, which was an unexpected result. Particles within the protons apparently were causing this unusual electron behavior. Later experiments confirmed those particles to be quarks. Six different types of quarks have since been found, and have been given the names up, down, charm, strange, top, and bottom.

In 1990, Taylor shared the Nobel Prize for physics with collegues Friedman and Kendall for work with deep inelastic scattering of electrons on protons and bound neutrons, which were important for the development of the quark model in particle physics.

The Université de Paris-Sud conferred an honorary Doctorate on Taylor in 1980.

1. Richard E. Taylor, Autobiography, NobelPrize.org, http://nobelprize.org/nobel_prizes/physics/laureates/1990/taylor-autobio.
2. Richard Edward Taylor, Microsoft Encarta Encyclopedia, http://encarta.msn.com/encyclopedia_761583371/Taylor_Richard_Edward.

Rudolf Mössbauer (1929-)

Discovered the resonant and recoil-free emission and absorption of
gamma rays by atoms bound in a solid form

Rudolf Ludwig Mössbauer was born in Munich, Germany on January 31,
1929. He studied physics at the Technical University of Munich, where he
received his Ph.D. in 1958. He did his dissertation work with Heinz Maier-
Leibnitz (1911-2000.

As a graduate student, Mössbauer began to investigate the absorption
of gamma rays in matter. The emission and absorption of X-rays by gases
had been observed previously, and it was expected that a similar
phenomenon would be found for gamma rays. Gamma rays are created by
nuclear transitions, as opposed to X-rays, which are produced by electron
transitions. However, attempts to observe gamma-ray resonance in gases
failed because of energy being lost to recoil. Mössbauer was able to
observe resonance in solid iridium, which raised the question of why
gamma-ray resonance was possible in solids but not in gases. Mössbauer
proposed that for the case of atoms bound into a solid, under certain
circumstances, a fraction of the nuclear events could occur essentially
without recoil. He attributed the observed resonance to this recoil-free
fraction of nuclear events. This phenomenon has become known as the
Mössbauer effect.

Mössbauer spectroscopy is a spectroscopic technique based on the
Mössbauer effect. In its most common form, **Mössbauer Absorption
Spectroscopy**, a solid sample is exposed to a beam of gamma radiation
and a detector measures the intensity of the beam that is transmitted
through the sample, which changes depending on how many gamma rays
are absorbed by the sample.

Mössbauer joined the faculty at the California Institute of Technology
in 1960, where he continued his investigations of gamma absorption.
Three years later, he returned to the physics department of the Technical
University of Munich, this time as a full professor

Along with Robert Hofstadter (1915-1990), Mössbauer was awarded
the Nobel Prize in physics in 1961 for his 1957 discovery of the
Mossbauer effect—research he carried out as a Ph..D. student.
Hofstadter's part of the award was for his studies of electron scattering in
atomic nuclei and discoveries concerning the structure of the nucleons

1. Rudolf Mössbauer, Biography, NobelPrize.org, http://nobelprize.org/nobel_prizes/physics/
    laureates/1961/mossbauer-bio.
2. Rudolf Mössbauer, Wikipedia, http://en.wikipedia.org/wiki/Rudolf_M%C3%B6%C3%9Fbauer.
    Rudolf Ludwig Mössbauer, Microsoft Encarta Encyclopedia Online,
    http://encarta.msn.com/encyclopedia_761574647/m%C3%B6ssbauer_rudolf_ludwig, 2007.

Jerome Friedman (1930-)

Investigated deep inelastic scattering of electrons on protons and bound neutrons

Jerome Isaac Friedman was born in Chicago, Illinois on March 28, 1930. His parents were Russian immigrants. His parents were born in Carpatho-Ruthenia (currently in the Soviet Union). They immigrated to the United States in their teens, meeting in New York. His father worked for Singer Sewing Machine Co., and his mother worked in a garment factory until she married. Friedman developed an interest in physics in high school after reading Einstein's book *Relativity*.

Friedman studied physics at the University of Chicago, receiving his Ph.D. under Enrico Fermi (1901-1954) in 1956. After postdoctoral work at the University of Chicago and at Stanford University, he joined the physics faculty of the Massachusetts Institute of Technology (MIT) in 1960. His collegue at Stanford, Henry Kendall (1926-199), joined him at MIT year later.

Protons and neutrons were once thought to be the elementary particles of matter. In the 1950s, physicists began to search for more elementary particles, and in 1964 scientists predicted the existence of quarks.

In the late 1960s, using a new and powerful particle accelerator at the Stanford Linear Accelerator Center (SLAC), Friedman and Kendall and their colleague, Richard Taylor (1929-), bombarded protons with electrons traveling at near the speed of light. The researchers expected the electrons to pass through the protons or to bounce off them. Many of the electrons rebounded at odd angles, which was an unexpected result. Particles within the protons apparently were causing this unusual electron behavior. Later experiments confirmed that those particles to be quarks. Six different types of quarks have since been found, and have been given the names up, down, charm, strange, top, and bottom.

In 1980, Friedman became Director of the Laboratory for Nuclear Science at MIT and then served as Head of the Physics Department from 1983 to 1988.

Friedman, Taylor, and Kendall shared the 1990 Nobel Prize in physics for their work on scattering of electrons on protons and bound neutrons, which was important for the development of the quark model in particle physics.

1. Jerome I. Friedman, Autobiography, NobelPrize.org, http://nobelprize.org/nobel_prizes/physics/laureates/1990/friedman-autobio.
2. Jerome Isaac Friedman, Microsoft Encarta Encyclopedia Online, http://encarta.msn.com/encyclopedia_761583179/Friedman_Jerome_Isaac, 2007.
3. Jerome Isaac Friedman, Wikipedia, http://en.wikipedia.org/wiki/Jerome_Isaac_Friedman.

Joel Lebowitz (1930-)

Contributed to statistical mechanics and mathematical physics

Joel L. Lebowitz was born in Tacev, Czechoslovakia (now Ukraine) to Jewish parents on May 10, 1930. During World War II he and his family were deported to Auschwitz, where his father, mother, and younger sister were murdered.

After being liberated from the camp, he moved to United States where he enrolled at Brooklyn College. After receiving his undergraduate degree, he did graduate study at Syracuse University, where he obtained a Ph.D. in 1956. He then joined the faculty at Yale University.

In 1957, Lebowitz moved to the Stevens Institute of Technology in Hoboken, New Jersey and then to the Belfer Graduate School of Science of Yeshiva University in New York City in 1959.

In 1975, Lebowitz founded the Journal of Statistical Physics. He moved to Rutgers University in 1977.

Lebowitz made many important contributions to statistical mechanics and mathematical physics. He proved, along with Elliott Lieb (1932-), that the Coulomb interactions among charged particles obey the thermodynamic limit, which is the limit reached as the number of particles (atoms or molecules) in a system approaches infinity—or in practical terms, one mole (6.02×10^{23} particles).

Lebowitz also established the **Lebowitz inequalities** for the ferromagnetic Ising model. The Ising model, named after the physicist Ernst Ising (1900-1998), is a mathematical model in statistical mechanics which has been used to model diverse phenomena in which bits of information, interacting in pairs, produce collective effects.

The Boltzmann Medal for 1992 was awarded to Lebowitz for his many important contributions to equilibrium and non-equilibrium statistical mechanics and for his leadership role in the statistical physics community.

In later years, Lebowitz turned his attention to the arrow of time; that is, why physical phenomenon always move forward.

Lebowitz is the George William Hill Professor of Mathematics and Physics at Rutgers University. As someone who had known oppression during the Nazi period, he devoted his life to helping people throughout the world who had been imprisoned, persecuted, or otherwise deprived of their rights.

1. Ising Model, Wikipedia, http://en.wikipedia.org/wiki/Ising_model.
2. Joel Lebowitz, Wikipedia, http://en.wikipedia.org/wiki/Joel_Lebowitz.
3. Joel Lebowitz: Winner of the Boltzmann Medal, http://stp.clarku.edu/great_contributors/ Boltzmann_medal/lebowitz.

Leon Cooper (1930-)
Developed a theory of superconductivity

Leon Neil Cooper was born in New York City on February 28, 1930. He received a B.A. in 1951, an M.A. in 1953, and a Ph.D. physics in 1954 from Columbia University. His dissertation work was supervised by Robert Oppenheimer (1904-1967), who had directed the development of the atomic bomb.

In 1955, Cooper was working on quantum field theory at the Institute for Advanced Study in Princeton, New Jersey when John Bardeen (1908-1991) at the University of Illinois invited him to work on superconductivity with him. Superconductivity is a phenomenon in which certain metals, when cooled to a temperature that is near absolute zero, no longer resist the flow of electricity. Electrical devices can waste large amounts of energy simply in trying to overcome electrical resistance at normal temperatures, a problem that could be virtually eliminated if superconducting materials were used instead.

From 1955 to 1957 Cooper, Bardeen, and Robert Schrieffer (1931-) developed what would later become known as the BCS theory of superconductivity, which explains why certain materials can be superconductive. BCS is the combined initials of the men's last names.

Cooper contributed significantly to the development of a theory of superconductivity by discovering what are now called **Cooper pairs**—two electrons that, when situated a certain way among positive ions, no longer repel each other, but instead develop an attraction. These pairs then accumulate and move in the same direction, resulting in a superconducting metal because there is no resistance to the flow of electricity.

In 1957, Cooper left the University of Illinois to become an assistant professor of physics at Ohio State University. In 1958, he began teaching physics at Brown University in Providence, Rhode Island.

Cooper shared the 1972 Nobel Prize for physics with John Bardeen and Robert Schrieffer for his role in developing a theory of superconductivity.

In recent years, Cooper has devoted his efforts toward a better understanding of memory and other brain functions. He is Co-founder and Co-chairman of Nestor, Inc., an industry leader in applying neural-network systems to commercial and military applications.

1. Leon Cooper, Wikipedia, http://en.wikipedia.org/wiki/Leon_Cooper.
2. Leon N. Cooper, Biography, NobelPrize.org, http://nobelprize.org/nobel_prizes/physics/laureates/1972/cooper-bio.
3. Leon N. Cooper, Microsoft Encarta Encyclopedia Online, http://encarta.msn.com/encyclopedia_761583144/cooper_leon_n_, 2007.

Mildred Dresselhaus (1930-)

Contributed to the understanding of electronic properties of materials, especially novel forms of carbon

Mildred Dresselhaus (born Mildred Spiewak) was born in the Brooklyn, New York on November 11, 1930, but grew up in the poor section of the Bronx. She received her undergraduate degree at Hunter College in New York City. She was a Fulbright Fellow at the Cavendish Laboratory, Cambridge University in 1951-52, and obtained her master's degree at Radcliffe in 1953, and then received a Ph.D. at the University of Chicago in 1958. Her thesis was on *The Microwave Surface Impedance of a Superconductor in a Magnetic Field.*

Dresselhaus did a National Science Foundation postdoctoral fellowship at Cornell University in Ithaca, New York, where she continued her studies on superconductivity before she and her physicist husband, Gene Dresselhaus, moved to Lincoln Laboratory at the Massachusetts Institute of Technology in 1960.

At the Lincoln Laboratory, Dresselhaus switched from research on superconductivity to magneto-optics, and carried out a series of experiments that led to a fundamental understanding of the electronic structure of semimetals, especially graphite. She studied graphite intercalation compounds, where single graphene layers sandwiched between guest species could be explored within the context of low dimensional physics.

Further investigation of carbon fibers and liquid carbon led to the study of fullerenes and carbon nanotubes. Her work illuminated the unique electronic structure of carbon nanotubes and the use of spectroscopy to probe the geometric structure of individual nanotubes.

Carbon nanotubes are allotropes of carbon in which the length of the molecule is far greater than the diameter. An allotrope is one of many forms in which a chemical element occurs, each differing in physical properties.

Dresselhaus is particularly noted for her work on thermoelectrics, graphite, graphite intercalation compounds, and carbon nanotubes. Her group has made frequent use of band structure calculations, Raman scattering, and high-field magnetotransport methods.

In 2000-2001, Dresselhaus was director of the Office of Science at the U.S. Department of Energy.

1. Mildred Dresselhaus, 2008 Oliver E. Buckley Condensed Matter Prize Recipient, APS Physics, www.aps.org/programs/honors/prizes/prizerecipient.cfm?name=Mildred%.
2. Mildred Dresselhaus, Wikipedia, http://en.wikipedia.org/wiki/Mildred_Dresselhaus.
3. Mildred S. Dresselhaus, History that Matters, www.aip.org/history/historymatters/dresselhaus.

Zhores Alferov (1930-present)
Did basic work on information and communication technology

Zhores Ivanovich Alferov was born on March 15, 1930 in Vitebek, Belarus, in what was then part of the Union of Soviet Socialist Republic (USSR). At the age of eighteen his father immigrated to Saint Petersburg, in the year 1912. After graduating in 1952 from the V. I. Ulyanov Electrotechnical Institute in Saint Petersburg, he joined the staff of the A. F. Ioffe Physico-Technical Institute. His graduation thesis was devoted to the problem of obtaining the thin films and investigating the photoconductivity of bismuth telluride compounds.

Alferov earned his doctoral degree in physics and mathematics at the Institute in 1970. He became the Institute's director in 1987.

Alferov is also a Russian politician, and has been a member of the Russian State Parliament, the Duma, since 1995.

In the early 1960s, Alferov worked on creating faster transistors. At the time, conventional single-block semiconducting materials, such as silicon, faced a limitation—negatively charged electrons flow in one direction, leaving positively charged "holes" that flow in the opposite direction. These opposing flows reduce the transistor's efficiency.

Alferov tried a new method. Instead of working with a single block of semiconducting material, he experimented with structures made of layers of different semiconducting materials. By combining separate materials, such as gallium arsenide and aluminum gallium arsenide, in layers as thin as a few atoms, he vastly improved transistor performance. These layered semiconductors are called heterostructures. American physicist Herbert Kroemer (1929-present), working independently, arrived at the same conclusion.

In the winter of 1970-1971 and spring 1971, Alferov spent six months in the United States working in laboratory of semiconductor devices at the University of Illinois.

Alferov and Kroemer shared the 2000 Nobel Prize in physics with American engineer and inventor Jack Kilby (1923-2005) for their basic work on information and communication technology, specifically for developing semiconductor heterostructures used in high-speed- and opto-electronics. Kilby's part of the prize was for his part in the invention of the integrated circuit.

1. Zhores Alferov, Microsoft Encarta Encyclopedia, http://encarta.msn.com/encyclopedia_701502048/Zhores_Alferov, 2007.
2. Zhores Alferov, Wikipedia, http://en.wikipedia.org/wiki/Zhores_Ivanovich_Alferov.
3. Zhores I. Alferov, Autobiography, NobelPrize.org, http://nobelprize.org/nobel_prizes/physics/laureates/2000/alferov-autobio.

Burton Richter (1931-)

Discovered the heavy elementary particle J/ψ

Burton Richter was born in Brooklyn, New York on March 22, 1931. He received a bachelor's degree in physics from Massachusetts Institute of Technology (MIT) in 1952 and his Ph.D. in 1956. His dissertation involved making the relatively short-lived mercury-197 isotope by using the MIT cyclotron to bombard gold with a deuteron beam.

After receiving his Ph.D., he became a research associate at Stanford University, and was appointed to the faculty in 1960. He was director of the Stanford Linear Accelerator Center (SLAC) from 1984 to 1999.

At Stanford, Richter's research focused on the study of subatomic particles moving extremely fast. At MIT, the accelerators Richter had used were cyclotrons and synchrotrons. Both of these accelerators direct the high-energy particles at stationary atoms. However, the linear accelerator at SLAC works in a different manner. It generates both positively and negatively charged particles. Because the particles have opposite charges, the magnetic field used to contain and accelerate them affects them differently. The positively charged particles (positrons) move through the accelerator counterclockwise, while the negatively charged particles (electrons) move clockwise. Therefore, physicists create much higher-energy collisions between the two moving beams of particles than can with one moving beam directed at stationary atoms. When the positrons and electrons collided in Richter's experiments, they produced a burst of electromagnetic energy out of which other particles were produced.

In 1974, Richter's group was studying the rate at which such collisions produced hadrons, a class of particles related to the proton and neutron. They noticed that the production rate of hadrons peaked sharply at a particular energy, often a sign of a new particle. Richter gave the new particle the name ψ.

About the same time, Samuel Ting (1936-) had discovered the particle at MIT using a different technique. Ting had named the particle J, so the two names were joined into J/ψ. The discovery of the J/ψ particle gave experimental evidence of the theorized fourth elementary particle known as a quark. This fourth quark was called charm.

In 1976, Richter shared the Nobel Prize for physics with American physicist for their independent discoveries of the subatomic particle J/ψ.

1. Burton Richter, Autobiography, NobelPrize.org, http://nobelprize.org/nobel_prizes/physics/laureates/1976/richter-autobio.
2. Burton Richter, Microsoft Encarta Encyclopedia Online, http://encarta.msn.com/text_761583324___0/Burton_Richter, 2007.
3. Burton Richter, Wikipedia, http://en.wikipedia.org/wiki/Burton_Richter.

David Lee (1931-)

Discovered superfluidity in helium-3

David Morris Lee was born in Rye, New York on January 20, 1931. His father was trained as an electrical engineer and his mother was an elementary school teacher. They were the children of Jewish immigrants who had emigrated from England and Lithuania, respectively. As a young teenager Lee became very interested in meteorology. He kept his own weather records and subscribed to the daily weather map issued by the U.S. weather bureau.

Lee graduated from Harvard University with a physics major in 1952, and then joined the U.S. Army, serving for 22 months during the Korean War. He obtained a Masters degree from the University of Connecticut and a Ph.D. from Yale University in 1959.

Afterward, he joined the faculty at Cornell University. His responsibilities were to set up a research laboratory in low temperature physics and to teach courses in the physics department.

Superfluidity was discovered independently by two scientists in 1937 and in 1938 in a more abundant form of helium—helium-4. In a superfluid state atoms move together in such a way that fluid flows without resistance. For many years afterward, scientists believed that the isotopes, helium-4 and helium-3, were so dissimilar in their atomic structures that helium-3 would not exhibit superfluidity as had helium-4. In the late 1950s, however, physicists developed a new theory that suggested helium-3 should become a superfluid at very low temperatures. Many laboratories attempted to achieve this state, but none were successful.

In the early 1970s, Lee, Robert Richardson (1937-), and a graduate student, Doug Osheroff (1945-) discovered that helium-3 did indeed exhibit superfluidity at extremely low temperatures. They weren't looking for superfluidity, but cooled helium-3 to extremely low temperatures, within a few thousandths of a degree of absolute zero, to explore the magnetic properties of the isotope. The discovery launched vigorous investigations into helium-3 and its superfluid properties

Lee, Richardson, and Osheroff were awarded the 1996 Nobel Prize in physics for their discovery of superfluidity in helium-3. Lee also worked on the discovery of nuclear spin waves in spin polarized atomic hydrogen gas, and more recently studied impurity-helium solids.

1. David Lee (physicist), Wikipedia, http://en.wikipedia.org/wiki/David_Lee_%28physicist%29.
2. David M. Lee, Autobiography, NobelPrize.org, http://nobelprize.org/nobel_prizes/physics/laureates/1996/lee-autobio.
3. David M. Lee, Microsoft Encarta Encyclopedia Online, http://ca.encarta.msn.com/encyclopedia_761589272/Lee_David_M_, 2007.

James Cronin (1931-)

Discovered violations of fundamental symmetry principles in the decay of neutral K-mesons

James Watson Cronin was born in Chicago, Illinois on September 29, 1931. At the time, his father was a graduate student in classical languages at the University of Chicago. His mother had met his father in a Greek class at Northwestern University. His father later became Professor of Latin and Greek at Southern Methodist University in Dallas, Texas.

Cronin received his bachelor's degree in mathematics and physics at Southern Methodist University in 1951. He earned both his master's and doctoral degrees in physics in 1955 from the University of Chicago.

From 1955 to 1958 Cronin served as an assistant physicist at Brookhaven National Laboratory in Long Island, New York. When the violation of parity was discovered, Cronin began a series of electronic experiments to investigate parity violation in hyperon decays. In early 1958, the Brookhaven Cosmotron suffered a severe magnet failure. As a consequence, the experiments were moved to the Berkeley Bevatron.

Cronin taught at Princeton University from 1958 to 1971, and in 1971 he became professor of physics at the University of Chicago.

In 1964, Cronin and Val Fitch (1923-) were studying a subatomic particle called the K-meson (or kaon). At one time physicists thought that the universe followed three fundamental rules of symmetry: Charge conjugation symmetry (C), parity symmetry (P), and Time-reversal symmetry (T). Before Cronin and Fitch began their experiment, the rules for C and P had both been found to be untrue in some cases. However, it was believed that any reaction had to be symmetric over a combination of C and P. Cronin and Fitch disproved this rule by showing that K-mesons do not always preserve P, C, or CP symmetry. However, it was still believed that all reactions must be symmetric in the combination of charge conjugation, parity, and time (CPT). Cronin and Fitch's result showed that T symmetry has to fail when CP symmetry fails. That means reactions that do not preserve CP symmetry cannot run backward.

Cronin shared the 1980 Nobel Prize for physics with Val Fitch for demonstrating that, unlike as previously thought, symmetry is not always preserved when some elementary particles change in state from matter to antimatter.

1. James Cronin, Autobiography, NobelPrize.org, http://nobelprize.org/nobel_prizes/ physics/ laureates/1980/cronin-autobio.
2. James Cronin, Wikipedia, http://en.wikipedia.org/wiki/James_Cronin.
3. James W. Cronin, Microsoft Encarta Encyclopedia Online, http://encarta.msn.com/ encyclopedia_761583151/Cronin_James_Watson, 2007.

John Schrieffer (1931-)

Developed a theory of superconductivity

John Robert Schrieffer was born in Oak Park, Ilinois on May 31, 1931. His family moved in 1940 to Manhasset, New York, and then in 1947 to Eustis, Florida. In 1949, Schrieffer was admitted to the Massachusetts Institute of Technology, where for two years he majored in electrical engineering before switching to physics. He completed a bachelor's thesis on multiplets in heavy atoms in 1953. He attended graduate school at the University of Illinois at Urbana-Champaign, where he was hired as a research assistant to John Bardeen (1908-1991).

In his third year of graduate school, Schrieffer joined Bardeen and Leon Cooper (1930-) in developing the theory of superconductivity. Cooper had discovered that electrons in a superconductor are grouped in pairs, now called **Cooper pairs**, and that the motions of all Cooper pairs within a single superconductor function as a single entity. Schrieffer's mathematical contribution was to describe the behavior of all Cooper pairs at the same time, instead of each individual pair. Schrieffer showed his equations to Bardeen, who immediately realized they were the solution to the problem.

After completing his doctoral dissertation, Schrieffer spent the 1957-1958 academic year as a National Science Foundation fellow at the University of Birmingham in England and at the Niels Bohr Institute in Copenhagen, where he continued research into superconductivity.

Following a year as assistant professor at the University of Chicago, Schrieffer returned to the University of Illinois in 1959 as a faculty member. Two years later, he joined the faculty of the University of Pennsylvania in Philadelphia, and in 1964 he published a book on the BCS theory, *Theory of Superconductivity*.

In 1972, Schrieffer, Bardeen, Cooper were awarded the Nobel Prize in Physics for developing the BCS theory. BCS are first letter of the men's last names. During the remainder of his career, Schrieffer held positions at the University of California, Santa Barbara and Florida State University.

In 2005, Schrieffer was sentenced to two years in prison for causing a car crash that killed one person and injured seven others. At the time, his license was under suspension. He was incarcerated in R.J. Donovan Correctional Facility San Diego, California.

1. John Robert Schrieffer, Microsoft Encarta Encyclopedia Online, http://encarta.msn.com/encyclopedia_761583338/schrieffer_john_robert, 2007.
2. John Robert Schrieffer, Wikipedia, http://en.wikipedia.org/wiki/John_Robert_Schrieffer.
3. Robert Schieffer, Biography, NobelPrize.org, http://nobelprize.org/nobel_prizes/physics/laureates/1972/schrieffer-bio.

Kenneth Fowler (1931-)

Contributed to the theory of plasma physics and magnetic fusion

T. Kenneth Fowler was born in Tomaston, Georgia on March 27, 1931. He received a B.S. in engineering in 1953 and an M.S. in physics in 1955 from Vanderbilt University. He received a Ph.D. in Physics in 1957 from the University of Wisconsin.

Fowler joined the staff of Oak Ridge National Laboratory in 1957, and was appointed Group Leader of the Plasma Theory activity at the Laboratory in 1962. After serving with General Atomic in San Diego, California for two years, he joined the University of California Lawrence Livermore Laboratory as Group Leader of Plasma Theory in 1967. He was appointed Associate Director for Magnetic Fusion in 1970.

Fowler's research in plasma physics was aimed toward the development of controlled thermonuclear fusion. Plasma physics is the study of ultra-hot, electrically charged gases. Atomic fusion is a process whereby two atomic nuclei are fused, with the release of a large amount of energy. Through his theoretical investigations and analyses of experiments, he created new fusion concepts that have contributed to the search for solutions to controlled thermonuclear fusion power.

Fowler is the author of *Fusion Quest*, a book that argues that there has been important progress in recent years, and that the commercial exploitation of fusion power will be attainable, even if it takes 50 years or more.

So far, controlled hydrogen fusion—the continuous joining together of the nuclei of hydrogen atoms—has proved to be a far tougher to achieve than originally thought. In the Sun, enormous forces compress the entire star, which is the driving force of solar fusion of hydrogen, but an earthbound reactor is much smaller, so its fuel must be confined by something other than gravity, either a complex web of interlocking magnetic fields or the violent, machine-gun-like implosions of many pea-sized capsules of thermonuclear fuel.

Fowler's ongoing projects include the development of computer models of heat transport and gun injection in spheromaks. The spheromak is a very compact magnetic configuration to confine hot plasmas in a fusion reactor.

1. Browne, Malcome W., Fuel's Paradise, New York Times, September 21, 1997.
2. T. Kenneth Fowler, 1981 Award Recipients, Engineer's Day, College of Engineering, University of Wisconsin-Madison, www.engr.wisc.edu/eday/eday1981.html#Duffie.
3. T. Kenneth Fowler, Department of Nuclear Engineering, University of California, Berkeley, www.nuc.berkeley.edu/people/faculty/fowler.

Martinus Veltman (1931-)

Elucidated the quantum structure of electroweak interactions

Martinus Justinus Godefriedus Veltman was born in Waalwijk, Netherlands on June 27, 1931. His father was the head of the local primary school. In 1940, the Netherlands were overrun by the German army, and his father's school was requisitioned by the Germans for troop lodging. The young Veltman had electronics as a hobby. Unfortunately, there was practically no electronics material around. The Germans confiscated all radios, and there was a great scarcity of anything that could catch a radio signal. In the fall of 1944 they were liberated.

As an undergraduate, Veltman studied mathematics and physics at Utrecht University, and then obtained his Ph.D. in theoretical physics at the same institution in 1963.

Veltman was a fellow at the European Organization for Nuclear Research (CERN) in Geneva, Switzerland from 1963 to 1966. In 1966, he became professor of physics at the University of Utrecht.

In 1969, Veltman and his graduate student, Gerardus 't Hooft (1946-), were looking for a unifying theory that would link the four fundamental forces in nature—gravitation, electromagnetism, the strong nuclear force, and the weak nuclear force. Their results, called the non-abelian gauge theory of electroweak interaction, linked two of the fundamental forces, and gave the concepts of particle physics a firmer mathematical foundation.

The theory Veltman and 't Hooft developed explains how elementary particles interact with one another through the electromagnetic and weak forces. The electromagnetic force determines how particles with electric charge interact, while the weak force governs how particles radioactively decay or change into other particles. Together they constitute the electroweak force. Many of the theory's predictions have been confirmed experimentally, thus verifying its interpretation of electroweak interactions.

In 1981, Veltman left Utrecht University for the University of Michigan-Ann Arbor for what was said to be personal reasons.

Veltman and 't Hooft were awarded the 1999 Nobel Prize in physics for elucidating the quantum structure of electroweak interactions

In 2003, Veltman published a book about particle physics for a broad audience, entitled *Facts and Mysteries in Elementary Particle Physics*.

1. Martinus J. G. Veltman, Autobiography, NobelPrize.org, http://nobelprize.org/nobel_prizes/ physics/ laureates/1999/veltman-autobio.
2. Martinus Veltman, Microsoft Encarta Encyclopedia Online, http://ca.encarta.msn.com/ encyclopedia_461510950/Veltman_Martinus_J_G_, 2007.

Melvin Schwartz (1932-2006)

Demonstrated the doublet structure of the leptons through the discovery of the muon neutrino

Melvin Schwartz was born in New York City on November 2, 1932. He earned a B.A. in 1953 and Ph.D. in 1958 at Columbia University. His doctoral work was supervised by Jack Steinberger (1921-), who later became a research colleague. Afterward, Schwartz joined the faculty at Columbia.

In the early 1960s, Schwartz, Steinberger, and Leon Lederman (1922-) devised a way to capture neutrinos. Using the powerful particle accelerator at the Brookhaven National Laboratory in New York, they created a beam of high-energy neutrinos. With a specialized detector, Schwartz and his colleagues were able to study the neutrinos, and in doing so discovered the muon neutrino, thus proving that neutrinos exist in more than one variety. Neutrinos are elementary particles that travel close to the speed of light, lack an electric charge, are able to pass through ordinary matter almost undisturbed, and are thus extremely difficult to detect. They are now believed to have a minuscule, but non-zero mass.

In 1966, after 17 years at Columbia, Schwartz moved to Stanford University, where a new accelerator was just being completed. There, he was involved in investigating the charge asymmetry in the decay of long-lived neutral kaons and another project which produced and detected relativistic hydrogen-like atoms made up of a pion and a muon.

In 1979, Schwartz left teaching to found a business, Digital Pathways, which designed security systems for computer data networks.

Schwartz returned to Columbia University in 1991. He also became Associate Director of High Energy and Nuclear Physics at Brookhaven National Laboratory.

Schwartz, Lederman and Steinberger, were awarded the 1988 Nobel Prize in physics for the neutrino beam method and the demonstration of the doublet structure of the leptons through the discovery of the muon neutrino. A lepton is a particle with spin 1/2 that does not experience the strong interaction.

Schwartz retired in 2000 to Ketchum, Idaho. He died on August 28, 2006 in a nursing home at Twin Falls, Idaho after struggling with Parkinson's disease and hepatitis C.

1. Melvin Schwartz, Autobiography, NobelPrize.org, http://nobelprize.org/nobel_prizes/physics/ laureates/ 1988/schwartz-autobio.
2. Melvin Schwartz, Microsoft Encarta Encyclopedia Online, http://encarta.msn.com/ encyclopedia_761583339/Melvin_Schwartz, 2007.
3. Melvin Schwartz, Wikipedia, http://en.wikipedia.org/wiki/Melvin_Schwartz.

Pierre-Gilles de Gennes (1932-2007)

Discovered that methods developed for studying order phenomena in simple systems could be generalized to more complex forms of matter

Pierre-Gilles de Gennes was born in Paris, France on October 24, 1932. He graduated from École Normale Supérieure in 1955. Then he became a research engineer for Commissariat à l'Énergie Atomique, working mainly on neutron scattering and magnetism.

De Gennes received his Ph.D. in research science from the Centre d'Études Nucléaires de Saclay in 1959. In 1961, he became assistant professor in Orsay, and soon started the Orsay group on *supraconductors*. Later, in 1968 he switched to liquid crystals. He spent 27 months in the French Navy. Then, in 1961 he became an assistant professor at the University of Paris, Orsay.

In 1971, de Gennes became professor at the Collège de France in Paris, and was a participant of STRASACOL on *polymer physics*.

In 1976, de Gennes was appointed director of the École de Physique et Chimie in Paris, and in 1988 he became the science director for chemical physics at Rhone-Poulenc, one of the largest chemical companies in France.

De Gennes analyzed the characteristics of polymers by comparing them to simpler systems, like magnets and liquid crystals. He discovered that these simpler systems shared mathematical properties with the more complex polymers. Understanding the fundamental nature of these complex substances provided him insight into how molecules are arranged in other substances, ranging from super glue to liquid helium. His research also helped scientists manipulate important properties of these substances to get the best results when designing new molecules.

In 1980, de Gennes became interested in interfacial problems, in particular the *dynamics of wetting*.

De Gennes was awarded the 1991 Nobel Prize in physics for discovering that methods developed for studying order phenomena in simple systems could be generalized to more complex forms of matter, in particular to liquid crystals and polymers. After receiving the Nobel Prize, de Gennes decided to give talks on science, innovation, and common sense to high school students. He visited around 200 high schools during 1992-1994. De Gennes died in Orsay, France on May 18, 2007.

1. Pierre-Gilles de Gennes, Biography, NobelPrize.org, http://nobelprize.org/nobel_prizes/physics/laureates/1991/gennes-bio.
2. Pierre-Gilles de Gennes, Microsoft Encarta Encyclopedia Online, http://encarta.msn.com/encyclopedia_761583188/Gennes_Pierre-Gilles_de, 2007.
3. Pierre-Gilles de Gennes, Wikipedia, http://en.wikipedia.org/wiki/Pierre-Gilles_de_Gennes.

H. Thomas Milhorn, MD, PhD

Sheldon Glashow (1932-)

Contributed to the theory of the unified weak and electromagnetic interaction between elementary particles

Sheldon Lee Glashow was born in Manhatan, New York on December 5, 1932. His parents had immigrated to New York City from Bobruisk to find the freedom and opportunity denied to Jews in Czarist Russia. After years of struggle, his father became a successful plumber.

Glashow received a B.A. from Cornell University in 1954 and a Ph.D. in physics from Harvard University under Julian Schwinger (1918-1994) in 1958. His dissertation title was *The Vector Meson in Elementary Particle Decays.* He was then awarded a National Science Foundation postdoctoral fellowship to work at the Lebedev Institute in Moscow; however, his Russian visa was never approved. He spent the time in Copenhagen at the Niels Bohr Institute.

Glashow was a research fellow at the California Institute of Technology in 1960-1961, an assistant professor at Stanford University in 1961-1962, an associate professor at the University of California in Berkely from 1962 to 1966, and a professor at Harvard University from 1966 to 1982. About 1960, Glashow put forward a theory involving the unification of weak interactions and electromagnetism in the atomic nuclei, which Steven Weinberg (1933-) and Abdus Salam (1926-1996) also developed independently. The validity of the theory was later verified through experiments carried out at the Super Proton Synchrotron facility at the European Organization for Nuclear Research (CERN) in Geneva

Glashow, Weinberg, and Salam were awarded the 1979 Nobel Prize in physics for devising the electro-weak theory.

In collaboration with John Iliopoulos (1940-) and Luciano Maiani (1941-), Glashow also predicted the charm quark (J/ψ), which is a second-generation quark with a charge of +(2/3)e. It is the third most massive of the quarks. It was later experimentally verified in simultaneous discoveries by a group led by Burton Richter (1931-) and by a group led by Samuel Ting (1936-).

Glashow is said to be a skeptic of superstring theory due to its lack of experimentally testable predictions. The superstring theory is an attempt to explain all of the particles and fundamental forces of nature in one theory by modeling them as vibrations of tiny supersymmetric strings.

1. Glashow, Sheldon Lee, Microsoft Encarta Encyclopedia Online, http://uk.encarta.msn.com/encyclopedia_761574485/Glashow_Sheldon_Lee, 2007.
2. Sheldon Glashow, Autobiography, NobelPrize.org, http://nobelprize.org/nobel_prizes/physics/laureates/1979/glashow-autobio.
3. Sheldon Lee Glashow, Wikipedia, http://en.wikipedia.org/wiki/Sheldon_Lee_Glashow.

Claude Cohen-Tannoudji (1933-)
Developed methods to cool and trap atoms with LASER light

Claude Cohen-Tannoudji, the son of Jewish parents, was born on April 1, 1933 in Constantine, Algeria, when Algeria was part of France. His family, originally from Tangiers, had settled in Tunisia, and then in Algeria in the 16th century after having fled Spain during the Inquisition.

From 1953 to 1957 he attended the Ecole Normale Supérieure in Paris.

After graduating, because of the Algerian War, Cohen-Tannoudji did military service for 28 months. During this time, he was assigned to a group studying the upper atmosphere with rockets releasing sodium clouds at the sunset.

In 1960, Cohen-Tannoudji returned to École normale supérieure to work on a doctoral degree, which he obtained in 1962. He then accepted a teaching position at the University of Paris.

Cohen-Tannoudji joined the Centre National de la Recherche Scientifique (CNRS) in 1960. In 1973, while still associated with CNRS, he joined the Collége de France in Paris as professor of physics. He was instrumental in the creation of a research center devoted to atomic physics and optics.

At room temperature, atoms move at speeds of about 4000 km/h, too fast to be studied thoroughly. Slowing atoms down lowers their temperature. Cohen-Tannoudji was among the first to propose using LASERs to slow down atoms by bombarding them with LASER light. Packets of light wave energy strike the atoms in a way that is roughly the same as raindrops hitting a beach ball. Even though the photons have no mass, their extremely high speed produces enough momentum to slow the atoms due to their impacts.

The techniques Cohen-Tannoudji and others developed led to significant advancements in the study and manipulation of atoms, resulting in many applications, including more accurate atomic clocks and more precise devices for measuring gravity.

Cohen-Tannoudji shared the 1997 Nobel Prize for physics with two American scientists, Steven Chu (1948-) and William D. Philips (1948-), for development of methods to cool and trap atoms with laser light They had worked independently.

1. Claude Cohen-Tannoudji, Autobiography, NobelPrize.org, http://nobelprize.org/nobel_prizes/ physics/laureates/1997/cohen-tannoudji-autobio.
2. Claude Cohen-Tannoudji, Microsoft Encarta Encyclopedia Online, http://encarta.msn.com/ encyclopedia_761596407/Claude_Cohen-Tannoudji, 2007.
3. Claude Cohen-Tannoudji, Wikipedia, http://en.wikipedia.org/wiki/Claude_Cohen-Tannoudji.

H. Thomas Milhorn, MD, PhD

Heinrich Rorher (1933-)

Designed the scanning tunneling microscope

Heinrich Rohrer was born in St. Gallen, Switzerland on June 6, 1933. The family moved to Zurich in 1949. He enrolled in the Swiss Federal Institute of Technology in 1951, where he studied with Wolfgang Pauli (1900-1958). His doctoral dissertation dealt with measuring the length changes of superconductors at the magnetic-field-induced superconducting transition. He had to do most of his research at night after the city was asleep because his measurements were so sensitive to vibration.

After a stint in the Swiss mountain infantry, in 1961 Rorher did research on thermal conductivity of type-II superconductors and metals at Rutgers University in New Jersey. A type-II superconductor is characterised by its gradual transition from the superconducting to the normal state.

In 1963, Rorher joined the IBM Research Laboratory in Switzerland. In 1978, he began working with German physicist Gerd Binning (1947-) on a problem that required information on a microscopically small surface. Their resulting invention, the scanning tunneling microscope (STM), was based on electron tunneling, a quantum phenomenon characterized by the passage of electrons through barriers in a way that cannot be explained by classical physics.

The STM uses a probe with an extremely fine tip that slowly scans over a sample at a constant distance. Electrons tunnel between the probe tip and the sample's surface. The variations in distance are recorded, and a computer generates a three-dimensional map of the sample's surface. The measurements involved are on such a small scale that the map shows distinct atoms. The first atomic image that Binning and Rohrer produced was of the surface of gold.

The STM has been used to study biological samples, analyze industrial materials, and test miniaturized electronic circuits. It was a major advancement in the field of nanotechnology.

For inventing the STM, Rohrer and Binning shared the 1986 Nobel Prize in physics with German physicist Ernst Ruska, (1906-1988, who was honored for his invention of the electron microscope.

In 1974, Rohrer spent a sabbatical year at the University of California in Santa Barbara, California studying nuclear magnetic resonance.

1. Heinrich Rorher, Autobiography, NobelPrize.org, http://nobelprize.org/ nobel_prizes/physics/ laureates/ 1986/rohrer-autobio.
2. Heinrich Rorher, Microsoft Encarta Encyclopedia, http://encarta.msn.com/ encyclopedia_761583328/ Heinrich_Rohrer, 2007.
3. Heinrich Rorher, Wikipedia, http://en.wikipedia.org/wiki/Heinrich_Rohrer.

Jeffrey Goldstone (1933-)

Co-discovered bosons that appear in models with spontaneously broken symmetry

Jeffrey Goldstone was born on September 3, 1933. He received his B.A. from Cambridge University in 1954 and his Ph.D. from the same institution in 1958. At Cambridge, under the guidance of Hans Bethe (1906-2005), he worked on the theory of nuclear matter, and developed the use of Feynman diagrams for non-relativistic many-fermion systems.

Goldstone was a research fellow of Trinity College, Cambridge from 1956 to 1960, and held visiting research posts at Copenhagen, CERN, and Harvard. During this time, his research focus shifted to particle physics, and he investigated the nature of relativistic field theories with spontaneously broken symmetries.

From 1962 to 1976 Goldstone was a faculty member at Cambridge. In the early 1970s he, with others, worked out the light-cone quantization theory of relativistic strings. He moved to the United States in 1977 as Professor of Physics at the Massachusetts Institute of Technology (MIT), and was Director of the MIT Center for Theoretical Physics from 1983 to 1989.

Goldstone is best known for for the discovery of the **Nambu-Goldstone boson**. In particle and condensed matter physics, Nambu-Goldstone bosons are bosons that appear in models with spontaneously broken symmetry. They correspond to the broken symmetry generators, and are massless if the spontaneously broken symmetry is not also broken explicitly.

If the symmetry is not exact, that is, if it is explicitly broken as well as spontaneously broken, then the Goldstone bosons are not massless, though they typically remain light. These are called **pseudo-Nambu-Goldstone bosons**. **Goldstone's theorem** states that whenever a continuous symmetry is spontaneously broken, new massless (or light if the symmetry was not exact) scalar particles appear in the spectrum of possible excitation. There is one scalar particle, a Nambu-Goldstone boson, for each generator of the symmetry that is broken; that is, that does not preserve the ground state. Goldstone has published research in quantum field theory and on the quantum strong law of large numbers. In 1977, he began working on quantum computation algorithms.

1. Goldstone Boson/Goldstone's Theorem, Wikipedia, http://en.wikipedia.org/wiki/Goldstone%27s_theorem.
2. Jeffrey Goldstone, Faculty and Staff, Physics@MIT, http://web.mit.edu/physics/facultyandstaff/faculty/jeffrey_goldstone.
3. Jeffrey Goldstone, Wikipedia, http://en.wikipedia.org/wiki/Jeffrey_Goldstone.

Steven Weinberg (1933-)

Contributed to the theory of the unified weak and electromagnetic
interaction between elementary particles

Steven Weinberg, the son of Jewish parents, was born in New York City
on May 3, 1933. He graduated from Cornell University in 1954, attended
Copenhagen's Nordic Institute for Theoretical Atomic Physics for a year,
and obtained his doctorate from Princeton University in 1957. His
dissertation research was on the application of renormalization theory to
the effects of strong interactions in weak interaction processes.

Weinberg taught at Columbia University from 1957 to 1959 and the
University of California, Berkeley from 1959 to 1966. During this time, he
did research on a number of topics in particle physics, including the high
energy behavior of quantum field theory, symmetry breaking, pion
scattering, infrared photons, and quantum gravity.

In 1966, Weinberg accepted a lecturer position at Harvard, and in
1967 he was visiting professor at Massachusetts Institute of Technology
(MIT). It was at MIT that he and Abdus Salam (1926-1996) offered a
hypothesis that unified the known facts about the electromagnetic and the
weak interactions between atomic particles. The validity of the theory was
later verified through experiments carried out at the Super Proton
Synchrotron facility at the European Organization for Nuclear Research
(CERN) in Geneva

Weinberg continued his work in of particle physics, quantum field
theory, gravity, supersymmetry, superstrings, and cosmology. And in
1973, he proposed the confinement of quarks and gluons, and proposed the
theory of strong interations without the Higgs bosons. The Higgs boson is
a hypothetical massive elementary particle predicted to exist by the
Standard Model of particle physics. It is the only Standard Model particle
not yet observed.

Weinberg became professor of physics at Harvard University in 1973,
a position he held until 1982 when he moved to the University of Texas at
Austin.

In 1979, Weinberg and Salam shared the Nobel Prize in physics with
the American physicist Sheldon Glashow (1932-) for their contribution to
the understanding of the interactions of elementary particles. Glashow had
worked independently of the other two.

1. Steven Weinberg, Biography, NobelPrize.org, http://nobelprize.org/nobel_prizes/physics/
    laureates/1979/weinberg-autobio.
2. Steven Weinberg, Microsoft Encarta Encyclopedia Online, http://encarta.msn.com/
    encyclopedia_761577658/Steven_Weinberg, 2007.
3. Steven Weinberg, Wikipedia, http://en.wikipedia.org/wiki/Steven_Weinberg.

Carlo Rubbia (1934-)

Contributed to the discovery of the field particles W and Z

Carlo Rubbia was born in Gorizia, Italy on March 31, 1934. His father was an electrical engineer and his mother was an elementary school teacher. At the end of the World War II most of the province of Gorizia was overtaken by Yugoslavia, and his family fled to Venice first and then to Udine.

Rubbia received his Ph.D. from Scuola Normale in Pisa, Italy in 1958. He then went to the United States, where he spent 1-1/2 years at Columbia University performing experiments on the decay and the nuclear capture of mu mesons.

In 1961, Rubbia moved to Geneva, attracted by the newly founded European Organization for Nuclear Research (CERN), where he worked on the identification of W and Z bosons. These particles were believed to carry the weak force that causes radioactive decay in the atomic nucleus and controls the combustion of the Sun. It is also believed that the weak force has played a fundamental role in the nucleosynthesis of the elements.

In 1970, Rubbia was appointed Higgins Professor of Physics at Harvard University, where he spent one semester per year, while continuing his research activities at CERN.

An international team of more than 100 physicists, headed by Rubbia, was assembled to try to detect the W and Z bosons. To create energies high enough to create the particles, Rubbia felt they would need a beam of protons and a beam of antiprotons, counter rotating in the vacuum pipe of an accelerator and colliding head-on. As a result, they had to develop a number of techniques for creating antiprotons. These techniques were developed with Simon van der Meer (1925-).

In 1983, the team was successful in creating and identifying the W and Z bosons. The data on bosons confirmed the predictions included in the electro-weak theory of Sheldon Glashow (1932-), Steven Weinberg (1933-), and Abdus Salam(1926-1996) .

Rubbia and van der Meer were awarded the 1984 Nobel Prize in physics for their contributions which led to the discovery of the W and Z particles, communicators of weak interaction

In 1989, Rubbia was appointed Director-General of the CERN Laboratory.

1. Carlo Rubbia, Wikipedia, http://en.wikipedia.org/wiki/Carlo_Rubbia.
2. Carlo Rubbia, Autobiography, NobelPrize.org, http://nobelprize.org/nobel_prizes/physics/laureates/1984/rubbia-autobio.
3. Carlo Rubbia, Microsoft Encarta Encyclopedia, http://encarta.msn.com/encyclopedia_761560048/ Carlo_Rubbia, 2007.

David Thouless (1934-)
Co-developed the theory of topological phase transitions

David James Thouless was born in Bearsden, Scotland in 1934. He received a B.A. from Cambridge University in 1955. Working under Hans Bethe (1906-2005) on nuclear matter, he received a Ph.D. from Cornell University in 1958. He did postdoctoral work at the University of California, Berkeley and at the University of Birmingham.

After four years at Cambridge, in 1965 Thouless became Professor of Mathematical Physics at Birmingham University, where he collaborated with Michael Kosterlitz on the theory of topological phase transitions. He also worked on electron localization and the spin glass, which is a disordered material exhibiting high magnetic frustration.

In 1980, Thouless became Professor of Physics at the University of Washington in Seattle, where his main interests were the quantum Hall effect, vortices in superfluids, and other problems related to topological quantum numbers.

Thouless made many theoretical contributions to the understanding of extended systems of atoms, electrons, and nucleons. Areas that his work has impacted include superconductivity phenomena, properties of nuclear matter, and excited collective motions within nuclei.

The **Kosterlitz-Thouless transition** is the transition from the superfluid to the normal state in thin films of helium-3, which proceeds through the unbinding of vortices in the phase of the order parameter having opposite directions of rotation.

The **Thouless energy** is a characteristic energy scale of electrons in disordered conductors, derived by scaling Anderson localization. In stochastic processes, Anderson localization, also known as strong localization, is the absence of diffusion of waves in a random medium.

Thouless was the recipient of a number of prizes, including the 2000 Lars Onsanger Prize for the introduction, with Michael Kosterlitz, of the theory of topological phase transitions, as well as fundamental contributions to our understanding of electron localization and the behavior of spin glasses.

1. 2000 Lars Onsanger Prize Recipient, David James. David Thouless, www.aps.org/programs/honors/prizes/prizerecipient.cfm?name=David%20James% .
2. Anderson Localization, Answers.com, www.answers.com/topic/anderson-localization?cat=technology.
3. David J. Thouless, wikipedia, http://en.wikipedia.org/wiki/David_Thouless.
4. Kosterlitz-Thouless transition, Answers.com, www.answers.com/topic/kosterlitz-thouless-transition?cat=technology.
5. Kosterlitz–Thouless transition, Wikipedia, http://en.wikipedia.org/wiki/Kosterlitz-Thouless_transition.
6. Thousless Energy, Wikipedia, http://en.wikipedia.org/wiki/Thouless_energy.

James Bjorken (1934-)

Made theoretical contributions to elementary particle physics, including the concept of scaling and the introduction of charm

James Daniel Bjorken was born in 1934. He received a B.S. in 1956 from the Massachusetts Institute of Technology and a Ph.D. in 1959 from Stanford University. From 1959 to 1979, he was on the faculty of Stanford University

Bjorken discovered Light-Cone Scaling (**Bjorken Scaling**), a phenomenon in the deep inelastic scattering of light on strongly interacting particles (hadrons), like protons and neutrons. This discovery was critical to the recognition of quarks as fundamental particles, and led to the theory of strong interactions known as quantum chromodynamics (QCD).

QCD, a theory of the strong interaction, is known as color force, which is a fundamental force describing the interactions of the quarks and gluons found in hadrons such as protons, neutrons, and pions. QCD is a quantum field theory of a special kind called a non-abelian gauge theory. It is an important part of the Standard Model of particle physics.

Bjorken conceptualized that the quarks become point-like, observable objects at very short distances and high energies. Richard Feynman (1918-1988) subsequently reformulated this concept into the parton model, used by many physicists to understand the quark composition of hadrons when probed at high energies. It was later recognized that partons describe the same objects now more commonly referred to as quarks and gluons.

In the early 1970's, the predictions of Bjorken Scaling were confirmed in experiments at the Standard Linear Accelerator Center (SLAC) in which quarks were observed for the first time.

Bjorken and Curtis Callan of Princeton University were awarded the 2004 Dirac Medal for their work on the theory of the strong interaction.

From 1979 to 1989 Bjorken was at the Fermi Lab and from 1989 to 1998 he was at the SLAC. In 1998, he became Emeritus Professor with Stanford University.

Bjorken co-authored, with Sidney Drell, a classic companion volume textbook on Relativistic Quantum Mechanics and Quantum Fields that is still used by many practicing particle physicists.

1. James Bjorken, Wikipedia, http://en.wikipedia.org/wiki/James_Bjorken.
2. James Daniel Bjorken Professor (Emeritus), SLAC, www.slac.stanford.edu/slac/faculty/hepfaculty/bjorken.
3. Particle Theorists Win Dirac Medal, Science Forum Index, www.groupsrv.com/science/about51830, August 10, 2004.
4. Quantum Chromodynamics, Wikipedia, http://en.wikipedia.org/wiki/Quantum_chromodynamics.

John Hall (1934-)

Contributed to the quantum theory of optical coherence

John Lewis "Jan" Hall was born in Denver, Colorada on August 21, 1934. His father was trained as an electrical engineer and worked for the U.S. Bureau of Reclamation. His mother was an elementary school teacher and singer. Jan's had early interests in electricity and radio. His research on black-powder rockets was gradually replaced by social activities such as scouts and a church youth group

From Carnegie Institute of Technology in Pittsburg, Pennsylvanis, Hall obtained a B.S. in 1956, an M.S. in 1958, and a Ph.D. in physics in 1961. His dissertation research involved using a self-made electron microwave spin resonance spectrometer to study the hyperfine spectrum of interstitial hydrogen atoms in CaF_2 crystals.

Hall completed his postdoctoral studies at the Department of Commerce's National Bureau of Standards, and then worked there from 1962 until his retirement in 2004. He began lecturing at the University of Colorado at Boulder in 1967.

Hall developed a method to measure the frequency of light emitted from molecules and atoms to an accuracy of 15 digits. Lasers with extremely sharp colors can now be constructed, and with the frequency comb technique precise readings can be made of light of all colors. A frequency comb allows a direct link from radio frequency standards to optical frequencies. German physicist Theodor Hänsch (1941-) independently developed the same technique.

The work of Hall and Hänsch makes it possible to carry out studies of the stability of the constants of nature over time and to develop extremely accurate clocks and improved GPS technology.

Hall shared one-half of the 2005 Nobel Prize in phyics with Hänsch for contributing to the development of LASER-based precision spectroscopy, including the optical frequency comb technique. The other half of the Prize went to American physicist Roy Glauber (1925-) for contributing to the quantum theory of optical coherence.

A favorite gathering place for the Halls, their children, and their grandchildren is their mountain cabin at Marble, Colorado, in the mountains west of Aspen, which was built by the family during the 1980's.

1. John L. Hall, Biography, NobelPrize.org, http://nobelprize.org/nobel_prizes/physics/laureates/ 2005/hall-bio.
2. John L. Hall, Wikipedia, http://en.wikipedia.org/wiki/John_L._Hall.
3. Two Americans, a German Win 2005 Nobel Prize in Physics, www.physorg.com/news6948. October 4, 2005.

Ludvig Faddeev (1934-)

Made theoretical contributions in quantum field theory and mathematical physics

Ludvig Dmitrievich Faddeev was born in Leningrad, USSR on March 3, 1934. His father was professor of mathematics at Leningrad University. He attended undergraduate school at Leningrad University, and received a Ph.D. from St. Petersburg State University in 1959. His disertation title was *Properties of S-Matrix for the Scattering on a Local Potential.*

Faddeev is best known for the discovery of **Faddeev-Popov ghosts** and **Faddeev equations**. Faddeev-Popov ghosts are additional fields which need to be introduced in the realization of gauge theories as consistent quantum field theories. Gauge theories are a class of physical theories based on the idea that symmetry transformations can be performed locally as well as globally. Symmetry is a physical or mathematical feature of the system (observed or intrinsic) that is preserved under some change. Quantum field theory is a theoretical framework for constructing quantum mechanical models of field-like systems (many-body systems).

The Faddeev equations are equations that describe, at once, all the possible exchanges/interactions in a system of three particles in a fully quantum mechanical formulation.

Faddeev's work also led to the invention of quantum groups, which are certain noncommutative algebras that first appeared in the theory of quantum integrable systems.

Faddeev received a number of prizes for his work, including the Poincaré Prize for his "many deep and important contributions to the theory of quantum fields, quantization of non-commutative gauge theories, scattering in quantum mechanics and quantum field theory, and the theory of integrable systems." The Poincaré Prize was created to recognize outstanding contributions in mathematical physics. The prize is named after Henri Poincaré (1854-1912).

Faddeev is Editor-in-Chief of the Springer journal *Functional Analysis and its Applications* and a member of the Editorial Boards of *Letters in Mathematical Physics* and *Mathematical Physics, Analysis and Geometry.*

1. Faddeev Equations, Wikipedia, http://en.wikipedia.org/wiki/Faddeev_equations.
2. Faddeev-Popov Ghosts, Wikipedia, http://en.wikipedia.org/wiki/Faddeev-Popov_ghost.
3. Henri Poincar◆ Prize goes to Ludvig Faddeev, PsychCentral, http://psychcentral.com/news/archives/ 2006-09/s-hpp092506.
4. Ludvig Dmitrievich Faddeev, Mathematics Genealogy Program, www.genealogy.ams.org/id.php?id=46697.
5. Ludvig Faddeev, Wikipedia, http://en.wikipedia.org/wiki/Ludvig_Faddeev.
6. Quantum Group, Wikipedia, http://en.wikipedia.org/wiki/Quantum_group

Kenneth Wilson (1936-)
Developed a theory for critical phenomena in connection with phase transitions

Kenneth Geddes Wilson was born in Waltham, Massachusetts on June 8, 1936. His father was on the faculty in the Chemistry Department of Harvard University. Kenneth entered Harvard University at the age of 16, receiving his B.S. in mathematics and physics in 1956. He earned his Ph.D. in theoretical physics at the California Institute of Technology in 1961 under Murray Gell-Mann (1929-).

From 1959 to 1962, Wilson served as a Harvard junior fellow, followed by a year as a Ford Foundation fellow working in the field of elementary particles at the European Organization for Nuclear Research in Geneva, Switzerland. In 1963, he joined Cornell University's Department of Physics in Ithaca, New York, where he first became interested in the study of phase transitions.

Wilson is best known for his contribution to the understanding of how bulk matter undergoes phase transition; that is, a sudden and profound structural change resulting from variations in environmental conditions.

Applying a mathematical technique called renormalization, Wilson developed a description of substance behavior when it is close to the critical point. His theory offers an explanation for diverse phenomena, including ice melting and iron losing its magnetism at a certain critical temperature or pressure. It has application in fields ranging from chemistry to engineering.

Wilson has stated that one of the reasons he chose Cornell was because the school had a good folk-dancing group, a hobby he picked up in graduate school.

Wilson spent the 1969-1970 academic year at the Stanford Linear Accelerator Center, the spring of 1972 at the Institute for Advanced Study in Princeton, the fall of 1976 at Caltech as a Fairchild Scholar, and the academic year 1979-1980 at the IBM Zürich Laboratory.

In 1985, Wilson became the director of Cornell's Center for Theory and Simulation in Science and Engineering, and in 1988 he moved to Ohio State University.

Wilson was awarded the 1982 Nobel Prize in physics for developing his theory of second-order phase transitions in matter that accounts for the effects on neighboring molecules.

1. Kenneth G. Wilson, Autobiography, NobelPrize.org, http://nobelprize.org/nobel_prizes/physics/laureates/1982/wilson-autobio.
2. Kenneth G. Wilson, Wikipedia, http://en.wikipedia.org/wiki/Kenneth_G._Wilson.
3. Kenneth Geddes Wilson, Microsoft Encarta Encyclopedia, http://encarta.msn.com/encyclopedia_761583407/ Wilson_Kenneth_Geddes, 2007.

Samuel Ting (1936-)

Co-discovered the heavy elementary particle J/ψ

Samuel Chao Chung Ting was born in Ann Arbor, Michigan on January 27, 1936. His parents moved back to China when he was an infant. His father and mother later became professors of science and psychology, respectively, at the National Taiwan University in Taipei, Taiwan.

Ting studied one year at the National Cheng Kung University, Tainan City before returning to the United States in 1956 to study engineering, mathematics, and physics at the University of Michigan. In 1959, he was awarded B.S. degrees in both mathematics and physics, and in 1962 he earned a doctoral degree in physics.

In 1963, Ting worked in the European Organization for Nuclear Research in Switzerland. From 1965 to 1967 he taught at Columbia University and did research at Deutsches Elektronen-Synchrotron of Hamburg, Germany. In 1967 he joined the faculty of Massechusettes Institute of Technology.

Ting's research centered on high-energy particle physics. One way physicists study these particles is by smashing particles together and looking at what comes out of the collision. That is accomplished with particle accelerators designed to speed up charged particles to the desired velocity for the collision.

In 1971, Ting and a team of physicists at the Brookhaven National Laboratory in Upton, New York began searching for new particles, specifically relatively heavy particles with short lives. They used a proton accelerator that could hit a stationary target of beryllium.

In 1974, the group discovered that collisions of a given energy produced a large number of electrons and positrons. Ting concluded that these particles were the result of the decay of an unknown, short-lived particle, which they named particle J.

At about the same time, Burton Richter (1931-) at Stanford discovered the same particle, which he called ψ. The two physicists combined the names to call the particle J/ψ. Ting and Richter were awarded the Nobel Prize for physics in 1976 for their discovery of the subatomic particle J/ψ.

In 1995, Ting proposed the Alpha Magnetic Spectrometer, a space-borne cosmic ray detector, and subsequently began work on it. As of 2007, the cost was $1.5 billion.

1. Samuel C. C. Ting, Autobiography, NobelPrize.org, http://nobelprize.org/nobel_prizes/physics/laureates/1976/ting-autobio.
2. Samuel C. C. Ting, Wikipedia, http://en.wikipedia.org/wiki/Samuel_C._C._Ting.
3. Samuel Chao Chung Ting, Microsoft Encarta Encyclopedia Online, http://ca.encarta.msn.com/encyclopedia_761583376/Ting_Samuel_Chao_Chung, 2007.

Robert Richardson (1937-)

Discovered superfluidity in helium-3

Robert Coleman Richardson was born in Washington D.C. on June 26, 1937. He is one of the two Nobel Laureates in physic to achieve the level of Eagle Scout. Frederick Reines (1918-1998) is the other one.

Richardson attended Virginia Tech University, initially majoring in electrical engineering, and then switching to chemistry, and then to physics. He received a B.S. in physics in 1958. He also earned an M.S. in physics at the same institution in 1960. He received his Ph.D. from Duke University in 1966. He remained at Duke for another year as a research associate. Then he became a research assistant at Cornell University in Ithaca, New York. In 1975, he became a full professor, and in 1990 became director of the University's Laboratory of Atomic and Solid State Physics.

Superfluidity was discovered independently by two scientists in 1937 and in 1938 in a more abundant form of helium—helium-4. In a superfluid state atoms move together in such a way that fluid flows without resistance. For many years afterward, scientists believed that the isotopes, helium-4 and helium-3, were so dissimilar in their atomic structures that helium-3 would not exhibit superfluidity as had helium-4. In the late 1950s, however, physicists developed a new theory that suggested helium-3 should become a superfluid at very low temperatures. Many laboratories attempted to achieve this state, but none were successful.

A liquid in a superfluid state doesn't behave according to the same laws of physics that normal fluids obey. Instead, it is subject to the complex statistical rules of quantum mechanics. In the early 1970s, Richardson, David Lee (1931-), and a graduate student, Doug Osheroff (1945-), discovered that helium-3 did indeed exhibit superfluidity at extremely low temperatures. They weren't looking for superfluidity, but cooled helium-3 to extremely low temperatures, within a few thousandths of a degree of absolute zero, to explore the magnetic properties of the isotope. The discovery launched vigorous investigations into helium-3 and its superfluid properties

Richardson, Lee, and Osheroff were awarded the 1996 Nobel Prize in physics for their discovery of superfluidity in helium-3.

1. Robert C. Richardson, Autobiography, NobelPrize.org, http://nobelprize.org/nobel_prizes/ physics/ laureates/1996/richardson-autobio.
2. Robert C. Richardson, Microsoft Encarta Encyclopedia Online, http://encarta.msn.com/ encyclopedia_761589316/Richardson_Robert_C_, 2007.
3. Robert Coleman Richardson, Wikipedia, http://en.wikipedia.org/wiki/ Robert_Coleman_Richardson.

Albert Fert (1938-)
Discovered Giant Magnetoresistance

Albert Fert was born in Carcassonne, France on March 7, 1938. His father was a physicist and his mother was a teacher. Fert graduated in 1962 from the École Normale Supérieure in Paris with a degree in mathematics, received his master's degree in physics in 1963 at the University of Paris, and earned his Ph.D. in physics in 1970 at the Université Paris-Sud.

Fert joined the faculty of the University of Grenoble in 1962. In 1964, he moved to the Université Paris-Sud in Orsay, eventualy reaching the rank of professor and director of the Mixed Unit for Physics at CNRS/Thales in Orsay, France.

In 1988, Fert discovered an effect known as giant magnetoresistance (GMR). GMR is a quantum mechanical effect observed in thin film structures composed of alternating ferromagnetic and nonmagnetic metal layers. The effect manifests itself as a significant decrease in electrical resistance in the presence of a magnetic field. In the absence of an applied magnetic field the direction of magnetization of adjacent ferromagnetic layers is antiparallel due to a weak anti-ferromagnetic coupling between layers, and it decreases to a lower level of resistance when the magnetization of the adjacent layers align due to an applied external field.

The spins of the electrons of the nonmagnetic metal align parallel or antiparallel with an applied magnetic field in equal numbers, and therefore suffer less magnetic scattering when the magnetizations of the ferromagnetic layers are parallel. The effect is used commercially by manufacturers of hard disk drives and other memory devices.

GMR was simultaneously and independently discovered by German physicist Peter Grünberg (1939-) at the Université Paris-Sud in Orsay, France.

Fert and Grünberg shared the 2007 Nobel Prize in physics for their discovery of giant magnetoresistance.

The discovery of GMR is considered the birth of spintronics, which is an emerging technology that exploits the quantum spin states of electrons as well as making use of their charge state. The electron spin itself is manifested as a two state magnetic energy system. Spintronics is also known as magnetoelectronics.

1. Albert Fert, NNDB, www.nndb.com/people/967/000163478.
2. Albert Fert, Wikipedia, http://en.wikipedia.org/wiki/Albert_Fert.
3. Giantmagnetoresistance, wikipedia, http://en.wikipedia.org/wiki/Giant_magnetoresistive_effect.
4. Overbye, Dennis, Physics of Hard Drives Wins Nobel, New York Times, October 10, 2007.
5. Spintronics, Wikipedia, http://en.wikipedia.org/wiki/Spintronics.

Anthony Leggett (1938-)

Contributed to the theory of superfluids

Anthony James Leggett was born in Camberwell, South London, England on March 26, 1938. His father was as a secondary school teacher of physics, chemistry and mathematics. His mother also taught secondary school mathematics for a time, but gave it up when Anthony was born.

Leggett obtained his bachelor's degree in physics from the University of Oxford in 1961 and his doctoral degree in theoretical physics from the same institution in 1964.

After postdoctoral research at Oxford and at the University of Illinois in Urbana-Champaign, Leggett joined the physics faculty at the University of Sussex in England in 1967. He remained at Sussex until 1983, when he returned to the University of Illinois.

While at the University of Sussex in the 1970s Leggett became interested in the problem of superfluidity in liquid helium. Superfluids are liquids that exhibit strange properties, such as flowing without friction or internal resistance, when cooled to extremely low temperatures. A superfluid may climb the walls of a container and may even flow uphill against gravity.

Research in the 1930s had demonstrated that a certain isotope of helium, helium-4, becomes a superfluid when cooled to a temperature near absolute zero. Some 40 years later, three American physicists, David Lee (1931-), Douglas Osheroff (1945-), and Robert Richardson (1937-), succeeded in creating a superfluid from rarer helium isotope known as helium-3. Existing theories explained the superfluid state of helium-4, but the more complex atomic structure of helium-3 presented a challenge. Leggett succeeded in building a theory that explained how helium-3's atoms pair up and behave in the super cooled state.

Leggett's theories have since found broader application in the study of subatomic particles produced in high-energy accelerators. Investigators, studying the behavior of swirls, or vortices, in superfluids, such as helium-3, have learned much about the general concepts of turbulence and chaos.

Leggett shared the 2003 Nobel Prize in physics with Alexei Abrikosov (1928-) and Vitaly Ginzburg (1916-) for their pioneering contributions to the theory of superconductors and superfluids. Abrikosov and Ginzburg's part of the award was for their separate theoretical work on superconductors. Leggett was knighted by Queen Elizabeth II in 2004.

1. Anthony J. Leggett, Autobiography, NobelPrize.org, http://nobelprize.org/nobel_prizes/ physics/laureates/2003/leggett-autobio.
2. Anthony Leggett, Microsoft Encarta Encyclopedia Online, http://encarta.msn.com/ encyclopedia_701666361/Leggett_Anthony_J, 2007.

Daniel Tsui (1939-)

Discovered a new form of quantum fluid with fractionally charged
excitations

Daniel Chee Tsui, the son of illiterate parents, was born in a remote
village in Henan Province, China on February 28, 1939. His childhood was
filled with years of drought, flood, and war. His parents sent him to far
away to Hong Kong to begin his formal education.

In 1957, Tsui was admitted to the medical school of National Taiwan
University in Taiwan.

Tsui moved to the United States in 1958 to attend Augustana College
in Rock Island, Illinois on a full scholarship. He graduated from
Augustana in 1961 with a bachelor's degree in mathematics. He received
his doctorate in physics from the University of Chicago in 1966, and
remained there as a research associate until 1968.

In 1968, Tsui took a job at Bell Telephone Laboratories in New Jersey.
There, he and Horst Störmer (1939-) conducted experiments in 1982 using
very intenses magnetic fields and low temperatures. Electrons in these
conditions interact with each other, creating a new form of quantum fluid
in which electrons act together to form particle-like units called
quasiparticles. When electrons form these quasiparticles, they appear to
have only a fraction of their normal electric charge.

A quasiparticle can be thought of as a single particle moving through a
system, surrounded by a cloud of other particles that are being pushed out
of the way or dragged along by its motion.

Robert Laughlin (1950-) provided the theoretical analysis to explain
Stormer and Tsui's experimental discovery of this phenomenon, called the
fractional quantum Hall effect.

At particular magnetic fields, the electron gas condenses into a
remarkable state with liquid-like properties. A series of plateaus forms in
the Hall resistance.

In 1982, Tsui was appointed professor of Electrical Engineering at
Princeton University in New Jersey.

Tsui shared the 1998 Nobel Prize in physics with Störmer and
Laughlin for their discovery of a new form of quantum fluid with
fractionally charged excitations.

1. Daniel C. Tsui, Autobiography, NobelPrize.org, http://nobelprize.org/nobel_prizes/physics/
    laureates/1998/tsui-autobio.
2. Daniel C. Tsui, Microsoft Encarta Encyclopedia Online, http://encarta.msn.com/
    encyclopedia_1741500702/Tsui_Daniel_C_, 2007.
3. Daniel C. Tsui, Wikipedia, http://en.wikipedia.org/wiki/Daniel_C._Tsui.

Peter Grünberg (1939-)

Discovered Giant Magnetoresistance

Peter Andreas Grünberg was born on May 18, 1939 in Pilsen, Bohemia, which at the time was in a Nazi-occupied protectorate and is now in the Czeck Republic. Originally Grinberg, Peter's father changed his name to Grünberg in 1941. The Grünberg family, like the rest of the German-speaking majority of Pilsen, was expelled by the Czech government by means of the Beneš decrees after World War II. They came to Lauterbach, Hesse in central Germany.

Grünberg received his undergraduate degree in 1962 from the Johann Wolfgang Goethe University. He then attended the Darmstadt University of Technology in Germany, where he received his master's degree in 1966 and his Ph.D in phyics in 1969.

From 1969 to 1972, Grünberg did postdoctoral work at Carleton University in Ottawa, Canada. He later joined the Institute for Solid State Physics at the Jülich Research Center in Germany.

In 1988, Grünberg discovered the giant magnetoresistive effect (GMR). GMR is a quantum mechanical effect observed in thin film structures composed of alternating ferromagnetic and nonmagnetic metal layers. The effect manifests itself as a significant decrease in electrical resistance in the presence of a magnetic field.

In the absence of an applied magnetic field the direction of magnetization of adjacent ferromagnetic layers is antiparallel due to a weak anti-ferromagnetic coupling between layers, and it decreases to a lower level of resistance when the magnetization of the adjacent layers align due to an applied external field. The spins of the electrons of the nonmagnetic metal align parallel or antiparallel with an applied magnetic field in equal numbers, and therefore suffer less magnetic scattering when the magnetizations of the ferromagnetic layers are parallel.

The giant magnetoresistive effect is used commercially by manufacturers of hard disk drives and other memory devices. GMR was simultaneously and independently discovered by French physicist Albert Fert (1938-) at the Université de Paris Sud.

Grünberg and Fert were awarded the 2007 Nobel Prize in physics for their discovery of giant magnetoresistance.

The discovery of GMR is considered the birth of spintronics, which is an emerging technology that exploits the quantum spin states of electrons as well as making use of their charge state.

1. Giantmagnetoresistance, wikipedia, http://en.wikipedia.org/wiki/Giant_magnetoresistive_effect.
2. Overbye, Dennis, Physics of Hard Drives Wins Nobel, New York Times, October 10, 2007.
3. Peter Grünberg, Wikipedia, http://en.wikipedia.org/wiki/Peter_Gr%C3%BCnberg.

David Josephson (1940-)

Made theoretical predictions of the properties of a supercurrent through a tunnel barrier

Brian David Josephson was born in Cardiff, Wales, UK on January 4, 1940. He attended Trinity College at the University of Cambridge, where he earned a B.A. in 1960, an M.A. in 1964, and a Ph.D. in 1964, all three in physics. He published his first work while still an undergraduate. It dealt with certain aspects of the special theory of relativity and the Mössbauer effect.

In 1964, Josephson took a teaching position at the University of Cambridge. Then he spent a year at the University of Illinois as a visiting research professor before returning to the University of Cambridge in 1967.

Josephson was as a 22-year-old graduate student when he discovered the phenomenon of current flow across two weakly coupled superconductors separated by a very thin insulating barrier. Traditional quantum theory stated that only a small amount of current could tunnel through the nonconducting barrier. Josephson, extending the work of Leo Esaki (1925-) and Ivar Giaever (1929-), predicted that a much higher number of electrons would actually move across the insulator. He also noted that this current would be affected by an external magnetic field. The flow of electric current through nonconductive material became known as the **Josephson effect**.

This arrangement—two superconductors linked by a non-conducting barrier—is known as a **Josephson junction**; the current that crosses the barrier is the **Josephson current**. Josephson's discoveries have had practical applications in the development of miniature electronics.

Applying Josephson's discoveries, researchers at IBM Corporation had assembled by 1980 a computer switch structure which would permit switching speeds from 10 to 100 times faster than those possible with conventional silicon-based chips

For discovering the Josephson effect, Josephson was awarded the 1973 Nobel Prize for Physics, which he shared with Esaki and Giaever for their experimental discoveries regarding tunneling phenomena in semiconductors and superconductors.

1. Brian D. Josephson, Microsoft Encarta Encyclopedia Online, http://encarta.msn.com/
    encyclopedia_761583225/josephson_brian_david, 2007.
2. Brian David Josephson, Biography, NobelPrize.org, www.nobel-winners.com/Physics/
    brian_david_josephson.
3. Brian David Josephson, Wikipedia, http://en.wikipedia.org/wiki/Brian_David_Josephson.
4. Josephson Effect, Wikipedia, http://en.wikipedia.org/wiki/Josephson_effect.

Haim Harari (1940-)

Predicted the existence of the top and bottom quarks

Haim Harari was born in Jerusalem in 1940, a fifth-generation Israeli. His father was a member of the Knesset. Harari received his M.Sc. from Hebrew University of Jerusalem, and obtained his Ph.D. in physics from the same institution in 1965. He then became one of the first postdoctoral fellows in the newly established Stanford Linear Accelerator Center at Stanford University. Afterward, he joined the faculty of the Weizmann Institute of Science in Rehovot Israel .

Harari's main contributions to particle physics include the prediction for the top and bottom quarks in 1975 and the first complete statement of the standard six quark and six lepton model of particle physics. Quarks and leptons are currently believed to be the most fundamental particles in nature.

Harari also proposed the Rishon Model for a substructure of quarks and Leptons. The model has two kinds of fundamental particles, called rishons ("primary" in Hebrew). These are the T ("Third" for charge of 1/3 of the elementay charge "e") and V ("Vanishes" for charge zero). All leptons and all flavors of quarks are three-rishon combinations. These groups of three rishons have spin 1/2.

From 1979 to 1985, Harari served as Chairman of the Planning and Budgeting Committee of Israel's Council for Higher Education—the government body which distributes government funding for higher education and basic research, determines priorities in the higher education system, and approves the establishment of new schools, faculties, and programs.

Harari served as a Dean of the Graduate School of the Weizmann Institute from 1972 to 1978, and was a co-Founder of "Perach," a national tutoring program for underprivileged children in Israel. From 1988 to 2001 he was the President of the Institute.

In 1999, Harari established the Davidson Institute of Science Education at the Weizmann Institute. It is dedicated to numerous educational projects for teachers, students, and the general public in mathematics, science, and technology.

In 2004, Harari gave a speech entitled *A View from the Eye of the Storm* in which he presented his insight into the problems of the Middle East. He eventually turned it into a book by the same name.

1. Haim Harari, Summary of CV and Main Current Activities, Weizmann Institute of Science, www.weizmann.ac.il/home/harari/cv.
2. Ham Harari, Wikipedia, http://en.wikipedia.org/wiki/Haim_Harari.
3. Harari Rishon Model, Wikipedia, http://en.wikipedia.org/wiki/Harari_Rishon_Model.

Stanley Brodsky (1940-)

Contributed to theoretical understanding of high-energy physics

Stanley J. Brodsky was born in St. Paul, Minnesota on January 9, 1940. He obtained a B.S. in physics in 1961 and a Ph.D. in physics in 1964, both from the University of Minnesota. His dissertation topic was the higher order quantum electrodynamic radiative corrections to the hydrogen hyperfine splitting.

From 1964 to 1966 Brodsky was a research associate at Columbia University. From 1966 to1968, he was a research associate at the Stanford Linear Accelerator Center (SLAC). He joined the permanent staff at SLAC in 1968. From 1996 to 2002 he was the head of the Theoretical Physics Group at SLAC.

Brodsky's research interest have included high-energy theoretical physics, especially the quark-gluon structure of hadrons and novel effects in quantum chromodynamics (QCD); fundamental problems in atomic, nuclear, and high energy physics; precision tests of quantum electrodynamics; light-front quantization; nonperturbative and perturbative methods in quantum field theory; and applications of AdS/CFT (anti-de-Sitter space/conformal field theory) to quantum chromodynamics.

QCD is a theory of the strong interaction color force—a fundamental force describing the interactions of the quarks and gluons found in hadrons (such as the proton, neutron, and pion). It is a quantum field theory of a special kind called a non-abelian gauge theory. It is an important part of the Standard Model of particle physics.

Brodsky was a visiting professor at Princeton University in 1982 and Cornell University in 1985. In 1987, he was awarded the Senior United States Distinguished Scientist Award from the Alexander von Humboldt Foundation. In 1989, He became a Foreign Scientific Member and External Scientific Director of the Max Planck Institute for Nuclear Physics in Heidelberg, Germany.

Brodsky was a recipient of the 2007 J. J. Sakurai Prize in Theoretical Physics, awarded by the American Physical Society for "applications of perturbative quantum field theory to critical questions of elementary particle physics, in particular, to the analysis of hard exclusive strong interaction processes." In 2003, he was appointed the first Distinguished Fellow at the Thomas Jefferson Laboratory.

1. Plummer Brad, SLAC Theoretical Physicist Stanley J. Brodsky Awarded Sakurai Prize, SLAC Today, http://today.slac.stanford.edu/a/2006/10-09.
2. Stanley Brodsky, SLAC, 2007 J. J. Sakurai Prize for Theoretical Particle Physics Recipient, Theoretical Particle Physics, www.aps.org/programs/honors/prizes /prizerecipient.cfm?name=Stanley%20Brodsky&year=2007.
3. Stanley J. Brodsky, Professor, SLAC, www.slac.stanford.edu/slac/faculty/hepfaculty/brodsky.

David Gross (1941-)

Discovered asymptotic freedom in the theory of the strong interaction

David Jonathan Gross was born in Washington, D.C. on February 19, 1941. In his early adolescence, his family moved to Israel. He received his bachelor's and master's degrees from the Hebrew University of Jerusalem, Israel in 1962. He obtained his Ph.D. in physics from the University of California, Berkeley in 1966.

After being a Junior Fellow at Harvard University, Gross joined the faculty at Princeton University in 1969. In 1973, working with his first graduate student, Frank Wilczek (1951-), at Princeton, Gross discovered asymptotic freedom, which is an explanation for the behavior of the smallest building blocks of matter, the quarks, inside the atomic nucleus. Quarks never occur alone; they are always found in combination with other quarks in larger particles of matter. Six types of quarks exist—designated up, down, charm, strange, top, and bottom.

Scientists had isolated other subatomic particles in particle accelerator experiments, but it was not known why quarks had never been released and observed in isolation. How the strong force bound quarks in the atom's nucleus was unknown.

Gross and Wilczek determined that the force binding the pairs and trios of quarks inside the nucleus actually grows weaker when the quarks are close together. When the distance between the quarks grows, the attractive force grows stronger—much too strong for any particle accelerator to release quarks from within the nucleus. They concluded that, when close together, quarks act essentially like free particles, not subject to any force. They named this phenomenon asymptotic freedom.

In 1997, Gross joined the faculty of the University of California at Santa Barbara, where he became director of the Kavli Institute for Theoretical Physics.

For the discovery of asymptotic freedom in the theory of the strong interaction, Gross and Wilczek shared the 2004 Nobel Prize in physics with David Politzer (1949-), who independently formulated the same theory.

Gross, Wilczek, and Politzer significantly advanced the understanding of the strong nuclear force, one of four fundamental forces operating in the universe.

1. David Gross, Microsoft Encarta Encyclopedia Online, http://encarta.msn.com/encyclopedia_701704001/Gross_David_J_, 2007.
2. David Gross, Wikipedia, http://en.wikipedia.org/wiki/David_Gross.
3. David J. Gross, Autobiography, NobelPrize.org, http://nobelprize.org/nobel_prizes/physics/laureates/2004/gross-autobio.

Theodor Hänsch (1941-)

Contributed to the development of LASER-based precision spectroscopy

Theodor Wolfgang Hänsch was born in Heidelberg, Germany on October 30, 1941. His father was a businessman engaged in the export of farming machinery. Hänsch grew up during and after the Second World War. At times, he and his family had to huddle together in the basement bomb shelter of their home, listening to the sound of air raid sirens. After the war, his family, having lost its estate, had to share a small apartment with some war refugees as subtenants.

Hänsch started spending his weekly allowance in pharmacies willing to sell substances, like fuming nitric acid or white phosphorous, to a young boy. After an accident with bomb-making materials, his interests moved from chemistry to physics and electronics.

Hänsch received his undergraduate degree in physics from the University of Heidelberg in 1966, and his doctorate in physics from the same institution in 1969, Hänsch was a professor at Stanford University, California from 1975 to 1986. In 1986, he returned to Germany to head the Max-Planck-Institut für Quantenoptik.

In 1970, Hänsch invented a new type of LASER which generated light pulses with an extremely high spectral resolution; that is all the photons emitted from the laser had nearly the same energy. Using this device, he succeeded in measuring the transition frequency of the Balmer line of atomic hydrogen with a much higher precision than before.

During the late 1990's, Hänsch and his coworkers developed a new method to measure the frequency of LASER light to an even higher precision using a device called the optical frequency comb generator. A frequency comb allows a direct link from radio frequency standards to optical frequencies.

The invention was then used to measure the Lyman line of atomic hydrogen to an extraordinary precision of one part in a hundred trillion. At such a high precision, it became possible to search for possible changes in the fundamental physical constants of the universe over time. American physicist John Hall (1934-) independently developed the same technique.

Hänsch shared one-half of the 2005 Nobel Prize in phyics with Hall for contributing to the development of LASER-based precision spectroscopy, including the optical frequency comb technique. The other half of the Prize went to American physicist Roy Glauber (1925-) for contributing to the quantum theory of optical coherence.

1. Theodor W. Hänsch, Autobiography, NobelPrize.org, http://nobelprize.org/nobel_prizes/physics/
    laureates/2005/hansch-autobio.
2. Theodor W. Hänsch, Wikipedia, http://en.wikipedia.org/wiki/Theodor_W._H%C3%A4nsch.

Gabriele Veneziano (1942-)

One of the fathers of string theory

Gabriele Veneziano was born in Florence, Italy in 1942. His undergraduate education took place at the University of Florence, where he received an undergraduate degree in physics in 1965. He received his Ph.D. in physics from the Weizmann Institute of Science in Rehovot, Israel in 1967. Veneziano was a research associate and then a visiting professor at the Massachusetts Institute of Technology from 1968 to 1972. He then returned to the Weizmann Institute as a professor of physics.

In 1978, Veneziano became a permanent member of the Theoretical Physics Division at the European Organization for Nuclear Research (CERN) in Geneva, Switzerland. His pioneering discoveries in dual resonance models developed into string theory and a basis for the quantum theory of gravity.

In 1968, while working a CERN, Veneziano discovered that a 200-year-old formula, the Euler Beta function, used as a scattering amplitude (**Veneziano amplitude**) has many features that are useful for explaining physical properties of strongly interacting particles. This amplitude is now interpreted as the scattering amplitude for four open string tachyons. A tachyon is any hypothetical particle that travels at superluminal speed (faster than the speed of light).

Veneziano's work led to intense activity aimed at explaining strong nuclear interactions based on a field theory of strings with a length scale of fermi. A fermi is one million billionth of a meter.

String theory is a mathematical approach to theoretical physics whose building blocks are one-dimensional extended objects called strings, rather than the zero-dimensional point particles that form the basis for the Standard Model of particle physics. By replacing the point-like particles with strings, an apparently consistent quantum theory of gravity emerges, which has not been achievable under the Standard Model. String theory may be a way to unify the known natural forces (gravitational, electromagnetic, weak nuclear, and strong nuclear) by describing them with the same set of equations.

Veneziano was awarded the 2004 Dannie Heineman Prize for Mathematical Physics for his pioneering discoveries in dual resonance models which, partly through his own efforts, have developed into string theory and a basis for the quantum theory of gravity.

1. Gabriele Veneziano, 2004 Dannie Heineman Prize for Mathematical Physics Recipient, APS Physics, www.aps.org/programs/honors/prizes/prizerecipient.cfm?name=Gabriele%.
2. Gabriele Veneziano, Wikipedia, http://en.wikipedia.org/wiki/Gabriele_Veneziano.
3. String Theory, Wikipedia, http://en.wikipedia.org/wiki/String_theory.

Klaus von Klitzing (1943-)

Discovered the quantized Hall Effect

Klaus von Klitzing was born in Schroda, Germany (now part of Poland) on June 28, 1943. He fled westward to Lutten, Germany with his family to escape the invasion of the Soviet Army into Germany at the end of World War II. He earned his B.S. in physics at the Technical University in Brunswick in 1969 and his Ph.D. at the University of Wurzbürg in 1972.

From 1975 to 1976, von Klitzing was affiliated with the University of Oxford in England, where he conducted research using very powerful superconducting magnets. In 1978, he returned to the University of Würzburg before moving to the High-Field Magnet Laboratory of the Intstitut Max von Laue-Paul Langevin in Grenoble, France He became professor at the Technical University of Munich in 1980.

The Hall effect, discovered by American physicist Edwin Hall (1855-1938) is a phenomenon which occurs when an electrical current is passed through a conducting material while a magnetic field is applied at right angles to the current. This results in an accumulation of electrons, called the Hall voltage, along one edge of the conductor.

Von Klitzing discovered that by using extremely powerful magnetic fields and temperatures close to absolute zero, the Hall voltage varies in a step pattern as the magnetic field or the electric current is varied. The Hall conductivity takes on the quantized values $\sigma = ve^2/h$, where "e" is the elementary charge, "h" is Planck's constant, and "v" is an integer. In the quantum Hall effect, also known as the integer quantum Hall effect, v takes on the integer values ($v = 1, 2, 3$, etc.).

The quantum Hall Effect has allowed the definition of a new practical standard for an electrical resistance unit, h/e^2, which is roughly equal to 25812.8 ohms. It is referred to as the von Klitzing constant R_K. Von Klitzing's finding allows more accurate testing of theories about electronic movements within atoms.

For his discovery of the Integer Quantum Hall Effect, von Klitzing was awarded the 1985 Nobel Prize in physics. In that same year, he became a director of the Max Planck Institute for Solid State Research in Stuttgart, Germany.

More recently, von Klitzing's research has focused on the properties of low dimensional electronic systems, typically in low temperatures and in high magnetic fields.

1. Klaus von Klipzing, Microsoft Encarta Encyclopedia Online, http://encarta.msn.com/ encyclopedia_761583391/von_Klitzing_Klaus-Olaf.
2. Klaus von Klipzing, Wikipedia, http://en.wikipedia.org/wiki/Klaus_von_Klitzing.
3. Quantum Hall Effect, Wikipedia, http://en.wikipedia.org/wiki/Quantum_Hall_Effect.

H. Thomas Milhorn, MD, PhD

Chris Quigg (1944-)

Contributed to theoretical understanding of high-energy collisions and the fundamental interactions of elementary particles

Chris Quigg was born on December 15, 1944. He grew up in Bethlehem, a steel town in eastern Pennsylvania. He graduated from Yale University with a B.S. in physics in 1966 and received his Ph.D. in physics in 1970 from the University of California, Berkeley.

From 1971 to 1974 Quigg was at the Institute for Theoretical Physics, State University of New York in Stony Brook. From 1974 to 1987 he was in the Department of Theoretical Physics at Fermi National Accelerator Laboratory in Batavia, Illinois. From 1977 to 1987 he was the head of the Theoretical Physics Department at Fermilab.

From 1987 to 1989 Quigg was Deputy Director for Operations with the Superconducting Super Collider Central Design Group. He helped lead the team that designed the Superconducting Super Collider, an enormous accelerator scientists planned to build in the area around Waxahachie, Texas until Congress cancelled the funding. It was planned to have a ring circumference of 87 kilometers and an energy of 20eV per beam, potentially enough energy to create a Higgs boson, a particle predicted by the Standard Model, but not yet detected.

Quigg and colleagues derived a formula for a bound on the mass of the Higgs boson, the particle that physicists believe gives elementary particles their mass.

In 1984, Quigg coauthored *Supercollider Physics*, which strongly influenced the quest for future discoveries at hadron colliders. A hadron is any strongly interacting composite subatomic particle. All hadrons are composed of quarks.

Quigg has made many other contributions, including the study of the spectroscopy of heavy-light mesons, signatures for the production of heavy quarks and quarkonium, and the study of ultrahigh-energy neutrino interactions. Quarkonium designates a flavorless meson whose constituents are a quark and its own antiquark.

More recently, Quigg has been studying the newly-discovered particle known as X(3872). Though scientists observed the particle at the KEK accelerator in Japan and confirmed it at Fermilab, they haven't reached a consensus on what this particle actually is.

1. Chris Quigg, Curriculum Vitae, http://lutece.fnal.gov/CV/CV.
2. Chris Quigg, Physics Central, www.physicscentral.com/people/2004/quigg.
3. Chris Quigg, Wikipedia, http://en.wikipedia.org/wiki/Chris_Quigg.

Douglas Osheroff (1945-)

Discovered superfluidity in helium-3

Douglas Dean Osheroff was born in Aberdeen, Washington on August 1, 1945. His father was a physician and his mother was a nurse. Osheroff earned his bachelor's degree in 1967 from the California Institute of Technology, where he was a student of Richard Feynman (1918-1988). He received a Ph.D. from Cornell University in 1973.

From 1972 to 1987, Osheroff did research at the American Telephone and Telegraph Bell Laboratories at Murray Hill, New Jersey. Then he joined the departments of physics and applied physics at Stanford University. His research focused on phenomena that occur at extremely low temperatures. From 1993 to 1996, he served as Physics Department chair.

Superfluidity was discovered independently by two scientists in 1937 and in 1938 in a more abundant form of helium—helium-4. In a superfluid state, atoms move together in such a way that the fluid flows without resistance. For many years afterward, scientists believed that the isotopes, helium-4 and helium-3, were so dissimilar in their atomic structures that helium-3 would not exhibit superfluidity as had helium-4. In the late 1950s, however, physicists developed a new theory that suggested helium-3 should become a superfluid at very low temperatures. Many laboratories attempted to achieve this state, but none were successful.

In the early 1970s, Osheroff (a graduate student at the time), David Lee (1931-), and Robert Richardson (1937-) discovered that helium-3 did indeed exhibited superfluidity at extremely low temperatures. They weren't looking for superfluidity, but cooled helium-3 to extremely low temperatures, within a few thousandths of a degree of absolute zero, to explore the magnetic properties of the isotope. The discovery launched vigorous investigations into helium-3 and its superfluid properties

Osheroff, Lee, and Richardson were awarded the 1996 Nobel Prize in physics for their discovery of superfluidity in helium-3.

Osheroff served on the Space Shuttle Columbia investigation panel. He is also an avid photographer and introduces students at Stanford to medium-format film photography in a freshman seminar titled *The Technical Aspects of Photography*.

1. Douglas D. Osheroff, Autobiography, NobelPrize.org, http://nobelprize.org/nobel_prizes/physics/laureates/1996/osheroff-autobio.
2. Douglas D. Osheroff, Microsoft Encarta Encyclopedia Online, http://encarta.msn.com/encyclopedia_761589300/ Osheroff_Douglas_D_, 2007.
3. Douglas D. Osheroff, Wikipedia, http://en.wikipedia.org/wiki/Douglas_D._Osheroff.

H. Thomas Milhorn, MD, PhD

Gerardus 't Hooft (1946-)

Elucidated the quantum structure of electroweak interactions

Gerardus 't Hooft was born in Den Helder, Netherlands on July 5, 1946. When he was eight, his family moved for a 10 month period to London, England, where for the first time Gerardus was forced to master a foreign language—English. His grand-uncle, Frits Zernike (1888-1966), had earned the 1953 Nobel Prize for work that had led him to the invention of the phase contrast microscope.

'T Hooft was introduced to the piano at age 10, and subsequently took lessions—, Chopin, Debussy, Mendelssohn and many others. He still plays.

'T Hooft studied physics and mathematics at the University of Utrecht, Netherlands, completing his undergraduate work in 1966 and his Ph.D. in theoretical physics in 1972.

As a 22-year-old graduate student, 't Hooft studied high-energy physics with Martinus Veltman (1931-), who was then a professor of physics at the University of Utrecht. Veltman and 't Hooft were looking for a theory that would link the four fundamental forces in nature—gravitation force, electromagnetic force, the strong nuclear force, and the weak nuclear force. They were able to link the electromagnetic force and the weak nuclear force, which they called the non-abelian gauge theory of electroweak interaction. The electromagnetic force governs how particles with electric charge interact, and the weak force governs how particles radioactively decay, or transform into different particles. Together they constitute the electroweak force.

Veltman and 't Hooft's theory explains how elementary particles interact with one another, primarily through the electromagnetic and weak nuclear force. Many of the theory's predictions have been experimentally verified. Their work gave the concepts of particle physics a firmer mathematical foundation.

From 1972 until 1974, 't Hooft was a fellow at the European Organization for Nuclear Research (CERN) in Geneva, Switzerland. He returned to the University of Utrecht in 1976 as an assistant professor and became full professor of physics in 1977.

'T Hooft and Veltman were awarded the 1999 Nobel Prize in physics for elucidating the quantum structure of electroweak interactions in physics.

1. Gerardus 't Hooft, Autobiography, NobelPrize.org, http://nobelprize.org/nobel_prizes/physics/laureates/1999/thooft-autobio.
2. Gerardus 't Hooft, Microsoft Encarta Encyclopedia Online, http://encarta.msn.com/ncyclopedia_461510949/%E2%80%99t_Hooft_Gerardus, 2007.

Gerd Binning (1947-)

Designed the scanning tunneling microscope

Gerd Karl Binning was born in in Frankfurt, West Germany on July 20, 1947. His childhood was influenced by the Second World War, which had only just ended. He and the other children had fun playing among the ruins of the demolished buildings

Binning earned his undergraduate degree at J. W. Goethe University in Frankfort. In 1978, he obtained his Ph.D. from the same institution. His doctoral dissertation work was a study of superconductivity.

Binning joined the IBM Research Laboratory near Zürich, Switzerland in 1978, and began working with Swiss physicist Heinrich Rohrer (1933-) on a problem that required information on a microscopically small surface. Their resulting invention, the scanning tunneling microscope (STM), was based on electron tunneling, a quantum phenomenon characterized by the passage of electrons through barriers in a way that cannot be explained by classical physics.

The microscope uses a probe with an extremely fine tip that slowly scans over a sample at a constant distance. Electrons tunnel between the probe tip and the sample's surface. The variations in distance are recorded and a computer generates a three-dimensional map of the sample's surface. The measurements involved are on such a small scale that the map shows distinct atoms. The first atomic image that Binning and Rohrer produced was of the surface of gold.

The STM has been used to study biological samples, analyze industrial materials, and test miniaturized electronic circuits. It was a major advancement in the field of nanotechnology.

For inventing the STM, Binning and Rohrer shared the 1986 Nobel Prize in physics with German physicist Ernst Ruska, (1906-1988), who was honored for his invention of the electron microscope.

That same year, Binning developed the first atomic force microscope (AFM), further expanding the array of tools available to researchers seeking a better understanding of materials on an atomic scale.

In 1994, Binning founded Definiens, an image intelligence company that analyzed and interprets images on every scale, from microscopic cell structures to satellite images.

1. Gerd Binning, Autobiography, NobelPrize.org, http://nobelprize.org/nobel_prizes/physics/laureates/1986/binnig-autobio.
2. Gerd Binning, Wikipedia, http://en.wikipedia.org/wiki/Gerd_Binnig.
3. Gerd Karl Binning, Magnet Lab, www.magnet.fsu.edu/education/tutorials/pioneers/binnig.
4. Gerd Karl Binning, Microsoft Encarta Encyclopedia Online, http://ca.encarta.msn.com/encyclopedia_761583113/Binnig_Gerd_Karl, 2007.

Nathan Isgur (1947-2001)

Contributed to understanding the quark structure of baryon resonances and discovered a new symmetry of nature that describes the behavior of heavy quarks

Nathan Isgur was born in Sam Houston, Texas on May 25, 1947. Initially interested in biology, he switched to physics and graduated with a B.Sc. from the California Institute of Technologhy in 1968. He began work on his doctorate at the University of California, Berkeley, but received a draft notice during his first year. Denied a draft deferment, he went to Toronto to avoid serving in the Vietnam War, a war he disagreed with on moral and political grounds.

Isgur enrolled in the graduate program at the University of Toronto and received a Ph.D. in particle theory in 1974. He stayed at Toronto as a post-doctoral candidate, and in 1976 he was hired as an assistant professor in the Department of Physics.

His first important publication was in 1975. It was on the mixing angle of pseudo-scalar mesons due to annihilation into gluons, an explanation that is now standard.

Isgur eventually took out Canadian citizenship. He was unable to travel to the United States because of his position as a war resister until President Jimmy Carter issued a blanket amnesty for all draft resisters.

Isgur served on the faculty at Toronto until 1990 when he joined the faculty of the College of William and Mary in Williamsburg, Virginia and became the Theory Group Leader at Jefferson Lab.

He is best known for his work on the excited states of the proton and for his role in the discovery of a new symmetry of nature which describes the behavior of heavy quarks.

Isgur received a number of awards, including the 2001 J. J. Sakurai Prize for Theoretical Particle Physics for "the construction of the heavy quark mass expansion and the discovery of the heavy quark symmetry in quantum chromodynamics, which led to a quantitative theory of the decays of c and b flavored hadrons."

Isgur was diagnosed with multiple myeloma in 1996, and died in Williamsburg, Virginia on July 24, 2001.

1. Dr. Nathan Isgur, Jefferson Lab, www.jlab.org/div_dept/dir_off/directors/NIbio.
2. Gabriel Karl and Pekka Sinervo, Nathan Isgur, 1947-2001, In Memoriam, www.cap.ca/pic/archives/57.5(2001)/memoriam.
3. Nathan Isgur, 2001 J. J. Sakurai Prize for Theoretical Particle Physics Recipient, APS Physics, www.aps.org/programs/honors/prizes/prizerecipient.cfm?name=Nathan%20Isgur&year=2001.
4. Nathan Isgur, Wikipedia, http://en.wikipedia.org/wiki/Nathan_Isgur.

William Phillips (1948-)

Developed methods to cool and trap atoms with LASER light

William D. Phillips was born in Wilkes-Barre, Pennsylvania on November 5, 1948. His father was a carpenter and his mother operated a boarding house. Phillips earned a B.S. in physics at Juniata College in Huntingdon, Pennsylvania in 1970. In 1976, he earned his Ph.D. in physics from Massachusetts Institute of Technology (MIT). His dissertation research involved the measurement of the magnetic moment of the proton in H_2O. In 1978, after a post-doctoral fellowship at MIT, he joined the National Bureau of Standards.

Phillips began conducting experiments in trapping atoms using a magnetic device to slow the atoms. Meanwhile, Steven Chu (1948-) and a team at Bell Telephone Laboratories in Holmdel, New Jersey began using LASERs to capture atoms. Chu successfully used LASERs to cool atoms to 240 millionths of a Celsius degree above absolute zero.

At room temperature, atoms move at speeds of about 4000 km/h, much too fast for scientists to study them. Lowering the temperature of the sample of atoms slows the atoms' motion, making them easier to study. Chu and Phillips both bombarded atoms with finely tuned LASER beams. The photons of the beam strike the atoms in a way that is roughly like raindrops hitting a beach ball. Although photons have no mass, they travel at the speed of light, and therefore have enough momentum when they hit the atoms to slow them down.

By 1998, Phillips and his research team had cooled atoms to 40 millionths of a Celsius degree above absolute zero, lower than scientists thought was theoretically possible at the time. Philips method for capturing atoms at regular intervals was termed an optical lattice.

By 1995, in France, Claude Cohen-Tannoudji (1933-) and his team, working independently, used similar techniques to lower the temperature of a sample of atoms to 0.2 millionths of a Celsius degree.

For developing methods to cool and trap atoms with LASER light, Phillips shared the 1997 Nobel Prize for physics with Chu and Cohen-Tannoudji.

Trapped atoms have increased the accuracy of atomic clocks, which increases the accuracy of other instruments that use atomic clocks, such as navigation systems. It also helps calibrate instruments used to measure the force of gravity at spots on the Earth.

1. William D. Phillips, Autobiography, NobelPrize.org, http://nobelprize.org/nobel_prizes/physics/laureates/1997/phillips-autobio.
2. William D. Phillips, Microsoft Encarta Encyclopedia Online, http://encarta.msn.com/encyclopedia_761596429/Philips_William_D_,2007.

David Politzer (1949-)

Discovered asymptotic freedom in the theory of the strong interaction

Hugh David Politzerwas born in New York City on August 31, 1949. He received his B.S. from the University of Michigan in 1969, and his Ph.D. in 1974 from Harvard University. He then joined the faculty of the California Institute of Technology.

In his first published article, which appeared in 1973, Politzer described the phenomenon of asymptotic freedom, which is an explanation for the behavior of the smallest building blocks of matter, the quarks, inside the atomic nucleus. Quarks never occur alone; they always are found in combination with other quarks in larger particles of matter. Six types of quarks exist—designated up, down, charm, strange, top, and bottom.

Scientists had isolated other subatomic particles in particle accelerator experiments, but it was not known why quarks had never been released and observed in isolation. How the strong force bound quarks in the atom's nucleus was unknown.

Politzer determined that the force binding the pairs and trios of quarks inside the nucleus actually grows weaker when the quarks are close together. When the distance between the quarks grows, the attractive force grows stronger—much too strong for any particle accelerator to release quarks from within the nucleus. The principle can be compared to a rubber band—as the band is stretched more and more, an ever-increasing amount of energy is required to continue the stretching.

Politzer concluded that, when close together, quarks act essentially like free particles, not subject to any force. They named this phenomenon asymptotic freedom.

In 1989, Politzer appeared in a minor role as Manhattan Project physicist Robert Serber in the movie *Fat Man and Little Boy*, which starred Paul Newman as General Leslie Groves. Politzer, who did not even own a television at the time, was very reluctant to appear in the movie. The director spoke with him for many hours before Politzer relented to taking the part.

For the discovery of asymptotic freedom in the theory of the strong interaction, Politzer shared the 2004 Nobel Prize in physics with David Gross (1941-) and Frank Wilczek (1951-), who independently formulated the same theory.

1. H. David Politzer, Microsoft Encarta Encyclopedia Online, http://encarta.msn.com/ encyclopedia_701711812/politzer_h_david, 2007.
2. H. David Politzer, NNDB, http://www.nndb.com/people/716/000140296.
3. H. David Politzer, Wikipedia, http://en.wikipedia.org/wiki/H._David_Politzer.

Horst Störmer (1949-)

Discovered a new form of quantum fluid with fractionally charged excitations

Horst Ludwig Störmer was born in Frankfurt, Germany on April 6, 1949. His father owned an interior decoration shop and his mother was a school teacher. In 1968, he enrolled in physics at the Goethe University in Frankfurt, and transferred to physics the following year.

In 1977, Störmer received a doctorate in physics from the University of Stuttgart. He then took a job at Bell Telephone Laboratories. In 1983, he was promoted to the head of the department for Electronic and Optical Properties of Solids. In 1991, he was promoted to director of the Physical Research Laboratory, heading some 100 researchers in eight departments.

Störmer and Daniel Tsui (1939-) conducted experiments in 1982 at Bell Laboratories using very intense magnetic fields and low temperatures. Electrons in these conditions interact with each other, creating a new form of quantum fluid, in which electrons act together to form particle-like units called quasiparticles. When electrons form these quasiparticles, they appear to have only a fraction of their normal electric charge. A quasiparticle can be thought of as a single particle moving through a system, surrounded by a cloud of other particles that are being pushed out of the way or dragged along by its motion.

Robert Laughlin (1950-) provided the theoretical analysis to explain Störmer and Tsui's experimental discovery of this phenomenon, called the fractional quantum Hall effect. At particular magnetic fields, the electron gas condenses into a remarkable state with liquid-like properties. A series of plateaus forms in the Hall resistance.

Störmer shared the 1998 Nobel Prize in physics with Tsui and Laughlin. The three shared the prize for their discovery of a new form of quantum fluid with fractionally charged excitations.

Störmer was appointed professor of applied physics at Columbia University in 1998 after having worked for 20 years in the Research Area of Bell Laboratories. He remained Adjunct Physics Director at Bell Labs, part-time. His area of expertise continued to be in condensed matter physics with an emphasis on semiconductors.

Currently, Störmer studies electronic transport, emphasizing nanosized structures, such as graphene, electron sheets in semiconductors and molecules

1. Horst Stoermer, Microsoft Encarta Encyclopedia Online, http://es.encarta.msn.com/ encyclopedia_961545705/Horst_Stoermer, 2007.
2. Horst Störmer, Autobiograpy, NobelPrize.org, http://nobelprize.org/nobel_prizes/ physics/laureates/1998/stormer-autobio.

Steven Chu (1949-)

Developed methods to cool and trap atoms with LASER light

Steven Chu, the son of Chinese immigrants, was born in St. Louis, Missouri in 1948. He received his bachelor's degree from the University of Rochester in 1970 and his doctoral degree from University of California, Berkeley in 1976.

After receiving his doctoral degree, Chu remained at Berkeley for two years doing postdoctoral work before joining Bell Telephone Laboratories in 1978. In 1983, he became head of the Quantum Electronics Research Department and moved to another branch of Bell Labs at Holmdel, New Jersey.

At room temperature, atoms move at speeds of about 4000 km/h, much too fast for scientists to easily study them. Since the speed of atoms is related to their temperature, slowing a sample of atoms makes the atoms cooler and easier to study.

In the mid-1980s, Chu pioneered a technique for slowing and cooling atoms that uses LASERs to immerse the atoms in photons. The photons strike the atoms in a way that is roughly analogous to raindrops hitting a beach ball. Although photons have no mass, because they move at the speed of light they have momentum and can therefore affect a small mass, such as an atom. In Chu's trap, the impact of photons hitting atoms slowed the atoms so much that they seemed to be stuck, so Chu's team named the process "optical molasses."

Chu's research helped pave the way for important discoveries in atomic physics. It also led to the development of many new practical applications, including more accurate atomic clocks and more precise devices for measuring the pull of gravity.

Chu became professor of physics at Stanford University in 1987. He served as the chair of the Physics Department from 1990 to1993 and from 1999 to 2001. He shared the 1997 Nobel Prize in physics with Claude Cohen-Tannoudji (1933-) and American physicist William Philips (1948-), who made separate and complementary advancements in the field.

In 2004, Chu was appointed director of the Lawrence Berkeley National Laboratory.

Besides his scientific career, he also developed serious interest in various sports, including baseball, swimming, and cycling.

1. Claude Cohen-Tannoudji, Microsoft Encarta Encyclopedia Online, http://encarta.msn.com/ encyclopedia_761596407/Claude_Cohen-Tannoudji, 2007.
2. Steven Chu, Autobiography, NobelPrize, http://nobelprize.org/nobel_prizes/physics/ laureates/1997/ chu-autobio.
3. Steven Chu, Wikipedia, http://en.wikipedia.org/wiki/Steven_Chu.

Georg Bednorz (1950-)

Discovered superconductivity in ceramic materials

Johannes Georg Bednorz was born in Neuenkirchen, North Rhine-Westphalia, Germany on May 16, 1950. His father was a primary school teacher and his mother was a piano teacher. In 1968, Bednorz entered the University of Münster to study chemistry, but soon changed his major to cristallography, a field of mineralogy which is located between chemistry and physics.

In 1972, Bednorz went to the IBM Zürich Research Laboratory for three months as a summer student. The physics department to which he was assigned was headed by Alex Müller (1926-).

Bednorz went on obtain his Ph.D. at the Federal Institute of Technology in 1982. His dissertation work was crystal growth and characterization of Strontium titanate (SrTiO3).

In 1982, Bednorz was hired by IBM to work in its Zurich laboratories. There, he rejoined Müller, who was investigating superconducting materials that had been discovered in 1911 by Dutch physicist Kamerlingh Onnes (1853-1926). Superconductivity is a phenomenon occurring in certain materials at extremely low temperatures, characterized by zero electrical resistance.

In 1983, Bednorz and Müller began a systematic study of the electrical properties of ceramics formed from transition metal oxides. In 1986, by systematically testing nickel-containing and copper-containing oxides, they found a material, barium lanthanum copper oxide (BaLaCuO), that starts to become superconducting at 35K, a temperature higher than previously thought possible. All superconductors known up until that time lost their electrical resistance and entered the superconducting state at temperatures barely above absolute zero.

In 1987, Bednorz and Müller were jointly awarded the Nobel Prize in physics for discovering that copper oxide ceramic materials can achieve superconductivity at temperatures well above the extremely low temperatures once associated with this property.

Superconductors are now used in scientific and medical instruments, and may find applications in the electronics industry and in electric-power transmission and storage.

1. Georg Bednorz (1950-Present), Magnet Lab, www.magnet.fsu.edu/education/tutorials/pioneers/bednorz.
2. J. Georg Bednorz, Autobiography, NobelPrize.org, http://nobelprize.org/nobel_prizes/ physics/laureates/1987/bednorz-autobio.
3. Johannes Georg Bednorz, Microsoft Encarta Encyclopedia Online, http://encarta.msn.com/encyclopedia_761583105/Bednorz_Johannes_Georg, 2007.
4. Johannes Georg Bednorz, Wikipedia, http://en.wikipedia.org/wiki/Johannes_Georg_Bednorz.

Robert Laughlin (1950-)

Discovered a new form of quantum fluid with fractionally charged
excitations

Robert Betts Laughlin was born in Visalia, California on November 1,
1950. His father was a lawyer in the Tulare County District Attorney's
office. His mother, a school teacher, was the daughter of a local physician.
Laughlin entered the University of California, Berkeley in 1968 as an
electrical engineering student. He graduated with a degree in mathematics
in 1972. He then entered the military as an enlisted man. At the end of
basic training, he was ordered to Fort Sill, Oklahoma to learn how to fire
Pershing missiles. After missile school he spent the rest of his tour in
southern Germany.

Laughlin entered the Massachussetts Institute of Technology graduate
program in physics in 1974, and directed his attention toward solid state
physics. He completed his Ph.D. work in 1979, and then went to work at
Bell Telephone Laboratories in New Jersey. There, he learned about
semiconductors. In 1982, he became a research physicist at Lawrence
Livermore National Laboratory in California.

Laughlin, Daniel Tsui (1939-), and Horst Stormer (1949-) collaborated
on the discovery that electrons can act together to form particle-like units
called quasiparticles. When electrons form these quasiparticles, they
appear to have only a fraction of their normal electric charge. A
quasiparticle can be thought of as a single particle moving through a
system, surrounded by a cloud of other particles that are being pushed out
of the way or dragged along by its motion.

Laughlin provided the theoretical analysis to explain Stormer and
Tsui's experimental discovery of this phenomenon, called the fractional
quantum Hall effect. At particular magnetic fields, the electron gas
condenses into a remarkable state with liquid-like properties. A series of
plateaus forms in the Hall resistance.

Laughlin became an associate professor of physics at Stanford
University in California in 1985 and a professor of physics at Stanford in
1989. He shared the 1998 Noble Prize in physics with Tsui and Stormer
for discovering a new form of quantum fluid with fractionally charged
excitations.

From 2004 to 2006, Laughlin served as the president of KAIST, a
research university in Daejeon, South Korea.

1. Robert B. Laughlin, Autobiography, NobelPrize.org, http://nobelprize.org/nobel_prizes/physics/
    laureates/1998/laughlin-autobio.
2. Robert B. Laughlin, Microsoft Encarta Encyclopedia Online, http://encarta.msn.com/
    encyclopedia_1741500700/Laughlin_Robert_B_, 2007.

Carl Wieman (1951-)

Achieved Bose-Einstein condensation in dilute gases of alkali atoms

Carl Edwin Wieman was born in Corvallis, Oregon, home of Oregon State University, on March 26, 1951. His father worked as a sawyer in a sawmill. Much of Carl's youth was spent wandering around in the forests of towering Douglas fir trees.

Wieman earned a B.S. in 1973 from Massachusetts Institute of Technology and a Ph.D. from Stanford University in 1977.

After receiving his Ph.D. Wieman took a position as an assistant research scientist at the University of Michigan. In 1984, he accepted a faculty position at the University of Colorado in Boulder. Eric Cornell (1961-) joined him in 1990. Together, they worked on Bose-Enstein Condensation.

Bose-Einstein condensation is a state of matter of bosons confined in an external potential and cooled to temperatures very near to absolute zero. Under such super cooled conditions, a large fraction of the atoms collapse into the lowest quantum state of the external potential, at which point quantum effects become apparent on a macroscopic scale. This state of matter was first predicted by Satyendra Nath Bose (1984-1974) in 1925. Bose submitted a paper to the Zeitschrift für Physik, but it was turned down by the peer review. He then took his work to Einstein, who recognized its merit and had it published under the names Bose and Einstein, hence the acronym.

In 1995, Wieman and Cornell produced the first Bose-Einstein condensate using a gas of rubidium atoms cooled to 0.000000170 K. Working independently, German physicist Wolfgang Ketterle (1957-) achieved similar results.

Wieman was awarded an honorary Doctor of Science from the University of Chicago in 1997.

Wieman and Cornell shared the 2001 Nobel Prize in physics with Ketterle for achieving Bose-Einstein condensation in dilute gases of alkali atoms, and for their early fundamental studies of the properties of the condensates.

Wieman joined the University of British Columbia in 2007, but retained a 20 percent appointment at University of Colorado, Boulder to head the science education project he founded in Colorado.

1. Bose-Einstein Condensate, Wikipedia, http://en.wikipedia.org/wiki/ Bose%E2%80%93Einstein_condensate.
2. Carl E. Wieman, Autobiography, NobelPrize.org, http://nobelprize.org/nobel_prizes/physics/ laureates/ 2001/wieman-autobio.
3. Carl Wieman, Wikipedia, http://en.wikipedia.org/wiki/Carl_Wieman.

Edward Witten (1951-)

Contributed to manifold theory, string theory, and the theory of
supersymmetric quantum mechanics

Edward Witten was born into a Jewish family in Baltimore, Maryland on August 26, 1951. His father, Louis Witten, was a physicist specializing in gravitation and general relativity. Edward received his bachelor's degree in history, with a minor in linguistics, from Brandeis University in Waltham, Massachusetts, and planned to become a journalist. He attended the University of Wisconsin-Madison for one semester as an economics graduate student before dropping out. He then enrolled in applied mathematics at Princeton University, where he received an M.A. in 1971 and his Ph.D in physics in 1976. Witten then worked at Harvard University as a Junior Fellow before joining the physics faculty at Princeton University in 1980. He was at the California Institute of Technology from 1999 to 2001, before becoming the Charles Simonyi Professor of Mathematical Physics at the Institute for Advanced Study in Princeton, New Jersey.

Witten's wife, Chiara Nappi, is a physics professor at Princeton. His brother is Matt Witten, who is a producer of television shows, including L.A. Law and House.

Witten is regarded as one of the world's leading researchers in superstring theory. He made extensive contributions to theoretical physics, and in 1990 he was awarded the Fields Medal for his influence on the development of mathematics.

In 1995, Witten suggested the existence of M-theory, and used it to explain a number of previously observed dualities, sparking a flurry of new research in string theory, called the second superstring revolution. M-theory is a proposed master theory that unifies the five superstring theories.

Superstring theory is an attempt to explain all the particles and the four fundamental forces of nature in one theory by modeling them as vibrations of tiny supersymmetric strings. It is considered one of the most promising theories of quantum gravity. Superstring theory is shorthand for supersymmetric string theory because, unlike bosonic string theory, it is the version of string theory that incorporates fermions and supersymmetry.

Witten made Time Magazine's list of 100 most influential people of 2004.

1. Edward Witten, Big Ideas, www.thirteen.org/bigideas/witten.
2. Edward Witten, Biography Center, www.biography-center.com/biographies/3741-Witten_Edward.
3. Edward Witten, Wikipedia, http://en.wikipedia.org/wiki/Edward_Witten.

Frank Wilczek (1951-)

Discovered asymptotic freedom in the theory of the strong interaction

Of Polish and Italian origin, Frank Anthony Wilczek was born in Mineola, New York on May 15, 1951. His grandparents had emigrated from Europe in the aftermath of World War I. His father's side came from Poland and his mother's side from Italy.

Frank received his B.S. in mathematics at the University of Chicago in 1970, an M.A. in mathematics at Princeton University in 1972, and a Ph.D. in physics at Princeton University in 1974. He then stayed on, first as an instructor in physics, ultimately becoming a full professor in 1980.

In 1973, Wilczek, a graduate student working with David Gross (1941-) at Princeton University, discovered asymptotic freedom, which is an explanation for the behavior of the smallest building blocks of matter, the quarks, inside the atomic nucleus. Quarks never occur alone; they always are found in combination with other quarks in larger particles of matter. Six types of quarks exist—designated up, down, charm, strange, top, and bottom.

Scientists had isolated other subatomic particles in particle accelerator experiments, but it was not known why quarks had never been released and observed in isolation. How the strong force bound quarks in the atom's nucleus was unknown.

Wilczek and Gross determined that the force binding the pairs and trios of quarks inside the nucleus actually grows weaker when the quarks are close together. When the distance between the quarks grows, the attractive force grows stronger—much too strong for any particle accelerator to release quarks from within the nucleus. They concluded that, when close together, quarks act essentially like free particles, not subject to any force. They named this phenomenon asymptotic freedom.

In 1981, Wilczek moved to the University of California at Santa Barbara, and then in 1989 he became a professor at the Institute for Advanced Study in Princeton. In 2000, he moved to the Center for Theoretical Physics at the Massachusetts Institute for Technology in Cambridge. For the discovery of asymptotic freedom in the theory of the strong interaction, Wilczek shared the 2004 Nobel Prize in physics with Gross and David Politzer (1949-), who independently formulated the same theory.

1. Frank Wilczek, Autobiography, NobelPrize.org, http://nobelprize.org/nobel_prizes/physics/laureates/2004/wilczek-autobio.
2. Frank Wilczek, Microsoft Encarta Encyclopedia Online, http://encarta.msn.com/encyclopedia_701711813/Frank_Wilczek, 2007.
3. Frank Wilczek, Wikipedia, http://en.wikipedia.org/wiki/Frank_Wilczek.

Ralph Merkle (1952-)

Invented the encryption technology that allows secure translations over the internet

Ralph Charles Merkle was born on February 2, 1952, and grew up in Livermore, California. He obtained a B.A. from the University of California, Berkeley in 1974 and his M.S. in 1977. In 1979, he was awarded a Ph.D. in electrical engineering from Stanford University. His dissertation was entitled *Secrecy, authentication and public key systems*.

In 1980, Merkle became the manager of compiler development at Elxsi, a minicomputer manufacturing company established in the late 1970s. In 1988, he became a research scientist at Xerox PARC, a research and development company in Palo Alto, California that began as a division of Xerox Corporation. There, he pursued research in computational nanotechnology. In 1999, he began working as a nanotechnology theorist for Zyvex in Richardson, Texas. Zyvex was the first nanotechnology company.

Merkle's interest in Nanotechnology involves the design, modeling, and manufacture of systems that can inexpensively fabricate most products that can be specified in molecular detail. This includes molecular logic elements connected in complex patterns to form molecular computers, positional devices (molecular robotic arms or Stewart platforms) able to position individual atoms or clusters of atoms under programmatic control, and a wide range of other molecular devices.

In 2003, Merkle became a Distinguished Professor at Georgia Tech. In addition to his work at Georgia Tech, Merkle is a director of the Alcor Life Extension Foundation of Scottscale, Arizona. The company researches, advocates for, and performs cryonics, the preservation in liquid nitrogen of humans after legal death, with hopes of restoring them to full health when new technology is developed in the future.

Merkle was co-recipient of the 1998 Feynman Prize for Nanotechnology Theory.

Merkle co-invented the **Merkle-Hellman public key cryptosystem**, which was one of the earliset public key cryptosystems, and the **Merkle-Damgård construction**, a method to build cryptographic hash functions. A cryptographic hash function is a transformation that takes an input and returns a fixed-size string, which is called the hash value.

Merkle also invented the **Merkle trees**, a type of data structure which contains a tree of summary information about a larger piece of data.

1. Ralph C. Merkle, KurzweilAI.net, www.kurzweilai.net/bios/bio0043.
2. Ralph Merkle, Wikipedia, http://en.wikipedia.org/wiki/Ralph_Merkle.
3. Ralph Merkle's Home Page, www.merkle.com.

Eric Drexler (1955-)

Popularized the potential of molecular nanotechnology

Kim Eric Drexler was born in Oakland, California on April 25, 1955 and raised in Oregon. His father was a management consultant and his mother was an audiologist and speech pathologist. He received a B.S. from the Massachusetts Institute of Technology (MIT) in 1977. During his freshman year at MIT, Drexler first started thinking seriously about molecular machines.

In 1986, Drexler published *Engines of Creation*, an examination of the possibilities of nanotechnology. Molecular nanotechnology is the concept of engineering functional mechanical systems at the molecular scale.

Drexler received his M.S. from MIT in 1979 and his Ph.D. in Molecular Nanotechnology (the first of its kind) in 1991. His doctoral dissertation was revised and published by Doubleday as a book *Nanosystems Molecular Machinery Manufacturing and Computation,* which received the Association of American Publishers award for Best Computer Science Book of 1992.

Drexler is best known for popularizing the potential of molecular nanotechnology in the 1970s and 1980s. His 1981 paper in the *Proceedings of the National Academy of Sciences* established fundamental principles of molecular design, protein engineering, and productive nanosystems. One area in which he sees these super-smart mechanical computers playing a vital role is medicine. Nanomachines with artificial intelligence would bring undreamed of health benefits. Surgical precision, for instance, would be brought down to the molecular level.

Drexler was one of the founders of the Foresight Institute, a nonprofit educational organization established to help prepare for advanced technologies.

Drexler is mentioned in the science fiction book *The Diamond Age* by Neal Stephenson as one of the heroes of a future world, where nanotechnology is ubiquitous. In the science fiction novel *Newton's Wake* by Ken Macleod a "drexler" is a nanotech assembler of pretty much anything that can fit in the volume of the particular machine—socks to starships. Drexler is also mentioned in the science fiction book *Decipher* by Stel Pavlou.

Drexler serves as Chief Technical Advisor to Nanorex, a company developing design software for structural DNA nanotechnologies.

1. Berry, Michael, The Creater, An Interview with nanotechnologist Eric Drexler, www.sff.net/people/mberry/nano, 1991.
2. K. Eric Drexler, Ph.D., e-drexler.com, http://e-drexler.com/p/idx04/00/0404drexlerBioCV.
3. K. Eric Drexler, Wikipedia, http://en.wikipedia.org/wiki/K._Eric_Drexler.

Nathan Seiberg (1956-)

Contributed to the development of string theory

Nathan Seiberg was born in Israel on September 22, 1956. He received a B.S. from Tel Aviv University in 1977, and served in the Israeli Army from 1977 to 1982. He received his Ph.D. in physics from the Weizmann Institute of Science in 1982.

Seiberg was a scholar in physics at the Weizmann Institute from 1982 to 1986, and a teacher in physics at the Institute from 1986 to 1989. He became a professor of physics at Rutgers University in 1989 and a professor of mathematical physics at Princeton University's Institute for Advanced Study in 1997. He holds dual Israeli-American citizenship.

Seiberg has worked in field theory, gauge theory, Matrix theory, particle physics phenomenology, string theory, and supersymmetry. Supersymmetry is a symmetry that relates elementary particles of one spin to another particle that differs by half a unit of spin and are known as superpartners.

Seiberg helped develop string theory, which argues that the universe is comprised not of individual particles but of elongated energy in tiny loops that resemble strings. Working with Edward Witten (1951-), he devised a series of partial differential equations that simplified the classification of four-dimensional manifolds.

Seiberg has won a number of awards, including the Mifal Hapais Prize in 1979, Michael Landau Prize in 1981, J.F. Kennedy Prize in 1982, Oskar Klein Medal in 1995, New Jersey Pride Award in 1996, and Dannie Heineman Prize in 1998.

The Dannie Heineman Prize was for "decisive advances in elucidating the dynamics of strongly coupled supersymmetric field and string theories. The deep physical and mathematical consequences of the electric-magnetic duality they exploited have broadened the scope of Mathematical Physics."

During recent years, Seiberg has found, with various collaborators, exact solutions of supersymmetric quantum field theories and string theories. These solutions have applications to mathematics and to the dynamics of quantum field theories and string theory, leading to many new and unexpected insights. One of the insights is the fundamental role played by the duality between electricity and magnetism in these theories.

1. Faculy and Emeriti, Institute for Advanced Study, ://www.ias.edu/about/faculty-and-Emeriti/ seiberg.
2. Nathan Seiberg, 1998 Dannie Heineman Prize for Mathematical Physics Recipient, www.aps.org/programs/honors/ prizes/prizerecipient.cfm?name=Nathan%20Seiberg&year=1998.
3. Nathan Seiberg, Big Ideas, www.thirteen.org/bigideas/seiberg.
4. Nathan Seiberg, Curriculum Vitae, http://www.sns.ias.edu/~seiberg/vitae.pdf.
5. Nathan Seiberg, NNDB, www.nndb.com/people/539/000163050.

Wolfgang Ketterle (1957-)

Achieved Bose-Einstein condensation in dilute gases of alkali atoms

Wolfgang Ketterle was born in Heidelberg, Germany on October 21, 1957. When he was three years old, the family moved from Heidelberg to the village of Eppelheim, three miles away. His father first joined an oil and coal distribution company as an apprentice and retired as a director. As a student, Wolfgang liked to play soccer and basketball and enjoyed trying out the various disciplines within track and field

In 1976, Ketterle entered the University of Heidelberg, but transfered to the Technical University of Munich two years later. There, he received his master's degree in 1982. In 1986, he earned a Ph.D in experimental molecular spectroscopy at the Max Planck Institute for Quantum Optics in Garching. Since 1990, Ketterle has resided in the United States, where he works at the Massachusetts Institute of Technology. His research has focused on experiments that trap and cool atoms to temperatures close to absolute zero.

In 1995, Ketterle created a Bose-Einstein condensate, cooling a gas of sodium atoms to a temperature a billion times colder than interstellar space. Bose-Einstein condensation is a state of matter of bosons confined in an external potential and cooled to temperatures very near to absolute zero. Under such super cooled conditions, a large fraction of the atoms collapse into the lowest quantum state of the external potential, at which point quantum effects become apparent on a macroscopic scale.

Bose-Einstein condensate was first predicted by Satyendra Bose (1984-1974) in 1925. Bose submitted a paper to the Zeitschrift für Physik, but it was turned down by the peer review. He then took his work to Einstein, who recognized its merit and had it published under the names Bose and Einstein, hence the acronym.

Ketterle shared the 2001 Nobel Prize in physics with Eric Cornell (1961-) and Carl Wieman (1951-) for achieving Bose-Einstein condensation in dilute gases of alkali atoms, and for their early fundamental studies of the properties of the condensates.

After achieving Bose Einstein condensation in dilute gases in 1995, Ketterle's group was in 1997 able to demonstrate interference between two colliding condensates, as well as the first realization of an atom LAZER, the atomic analogue of an optical LASER.

1. Bose-Einstein Condensate, Wikipedia, http://en.wikipedia.org/wiki/ Bose%E2%80%93Einstein_condensate.
2. Wolfgang Ketterle, Autobiography, NobelPrize.org, http://nobelprize.org/nobel_prizes/ physics/laureates/2001/ketterle-autobio.
3. Wolfgang Ketterle, Wikipedia, http://en.wikipedia.org/wiki/Wolfgang_Ketterle.

Eric Cornell (1961-)

Achieved Bose-Einstein condensation in dilute gases of alkali atoms

Eric Allin Cornell was born in Palo Alto, California on December 19, 1961. Two years later, the family moved to Cambridge, Massachusetts. His father was a professor of civil engineering at Massachusetts Instutute of Technology (MIT), and his mother taught high school English. In high school, Eric joined the chess and math clubs, and taught himself to write programs on the school's DEC mainframe in the language, Basic.

Cornell received his B.S. in physics from Stanford University in 1985. The summer following his second year, he went to Taichung, Taiwan to teach conversational English. Six months later, he left Taiwan, first for Hong Kong and then for mainland China, where he spent another three months studying Chinese. He then went back to Stanford to finish his undergraduate work. Hel obtained his Ph.D. in physics at MIT in 1990. He then joined Carl Wieman (1951-) at the University of Colorado, and also became a physicist at the United States Department of Commerce National Institute of Standards and Technology. He and Wieman began work on the Bose-Einstein condensation.

Bose-Einstein condensation is a state of matter of bosons confined in an external potential and cooled to temperatures very near to absolute zero. Under such super cooled conditions, a large fraction of the atoms collapse into the lowest quantum state of the external potential, at which point quantum effects become apparent on a macroscopic scale.

Bose-Einstein condensation was first predicted by Satyendra Bose (1984-1974) in 1925. Bose submitted a paper to the Zeitschrift für Physik, but it was turned down by the peer review. He then took his work to Einstein, who recognized its merit and had it published under the names Bose and Einstein, hence the acronym.

Cornell and Wieman shared the 2001 Nobel Prize in physics with German Physicist Wolfgang Ketterle (1957-) for achieving Bose-Einstein condensation in dilute gases of alkali atoms, and for their early fundamental studies of the properties of the condensates.

In October 2004, Cornell's left arm and shoulder were amputated in an attempt to stop the spread of necrotizing fasciitis. He was discharged from the hospital in mid-December, and returned to work part-time in April 2005.

1. Bose-Einstein Condensate, Wikipedia, http://en.wikipedia.org/wiki/
    Bose%E2%80%93Einstein_condensate.
2. Eric Cornell, Wikipedia, http://en.wikipedia.org/wiki/Eric_Allin_Cornell.
3. Eric A. Cornell, Autobiography, NobelPrize.org, http://nobelprize.org/nobel_prizes/physics/
    laureates/2001/cornell-autobio.

Index

La'Vergne, TN USA
03 December 2009

165815LV00005B/19/P